The Human Body and Weightlessness

William Thornton • Frederick Bonato

The Human Body and Weightlessness

Operational Effects, Problems and Countermeasures

Springer

William Thornton
University of Texas Medical Branch
Galveston, Texas, USA

Frederick Bonato
Saint Peter's University
Jersey City, New Jersey, USA

ISBN 978-3-319-32828-7 ISBN 978-3-319-32829-4 (eBook)
DOI 10.1007/978-3-319-32829-4

Library of Congress Control Number: 2017942723

Printed on acid-free paper

This Springer imprint is published by Springer Nature
The registered company is Springer International Publishing AG
The registered company address is: Gewerbestrasse 11, 6330 Cham, Switzerland

This book is dedicated to Dr. Tom Moore whose unstinting and too often unrecognized work in the early Shuttle flights produced a number of the results described here.

Preface and Acknowledgements

The intent of this work is to provide a better understanding of one of the more challenging problems of current and future space operations—the reaction of the human body to a fundamental feature of the environment beyond that of Earth: weightlessness. This book was initially intended to be a small, readily accessible source of information and data on the major operational problems caused by adaptation to weightlessness. It was to be as simple as possible and comprehensible to virtually all those working in human space flight but who are not necessarily familiar with some relevant details, especially of the human body.

On beginning the chapter on space motion sickness, it became obvious that published studies and data were too diverse to be adequate for operational planning, the ultimate goal of this writing. This changed the planned format from a collation of existing studies and data to one that required the generation of applicable evidence-based information on the causes and effects of changes produced by weightlessness and a search for data on these changes. Validity and relevance were the criteria used in the selection of evidence, whether dating from before space flight or just discovered during the writing itself.

The resulting product is based on the concept that the human body can make physiological changes or temporary adaptations, which are compatible with weightlessness and a healthy body. These "somatic" adaptations are usually incompatible with normal function in 1 g and must be considered during and after return to Earth until re-adaptation has occurred. With one known exception, this is not a pathological process. Such a concept is in contrast to that of many investigators and physicians who treat these changes as pathologies to be prevented if possible.

A rational basis for the approach taken here is found in the evolution of our marine ancestors' bodies into our present form and function by innumerable species adaptations. In almost every case of somatic adaptation we took a biophysical approach, which began with the simplest functional unit in the systems involved; e.g., the hair cell's relationship to space motion sickness. Only those features directly relevant to the problem are considered. Basic science is included when deemed appropriate to provide a common basis for understanding of these adaptations.

It is written with the hope that the concepts here will be challenged where necessary and augmented and extended as additional experience allows. This is especially true in the case of the incomplete data in this book. The book could not have been done without the support of many people, a number of whom may have been inadvertently missed in the following list of acknowledgements and appreciation. The book began with the authorization and support of Michael Coats, then director of NASA JSC. Other essential contributions by people and organizations in JSC include the art and graphics skill of David Russell who painstakingly translated ideas, images, and data into the colorful graphics, which speak for themselves. Janine Bolton and Kim So of the Life Sciences library patiently supported our search for information and provided hundreds of files and sources. Mary Wilkerson in the photo archives continued her gracious tradition of skillful searches for and provision of photographic files. Some of the most valuable insights were provided by astronauts current and past and by Alan Moore and Mark Guilliam of Wyle.

This work would have been absolutely impossible without the varied and unstinting support of Sid Jones (retired) whose 38 years' experience in graphic and editorial work as a support contractor at JSC will survive indefinitely in many JSC publications.

Typing sometimes almost illegible scripts by Kathy Hernandez of a San Antonio firm has been an essential element here. My innumerable revisions were a burden to all involved.

Many basics in the work described here began with the support of General Jack Ord and by Brooks Air Force Base machine shop and were continued by the late Henry Whitmore and Whitmore enterprises. Bruce Del Mar and Del Mar Avionics contributed unique prototype hardware.

Drs. Tom Moore and John Uri at JSC provided essential support in execution, both on the ground and in-flight investigations and their documentation. Chapters on Space Motion Sickness (SMS), aspects of fluid shifts and exercise, would not have been possible without the observations and measurements of crews on STS-2 through STS-61B. Dr. Norm Thagard's clinical observations and electrophysiological recordings on STS-7 and STS-8's crew support, especially that of Cdr. Dick Truly, were crucial to the ultimate understanding of SMS. Story Musgrave's compliance measurements on later missions confirmed key portions of Chap. 4. Many other insightful crew observations are too numerous to credit.

Writing of this book was dependent on the moral, typing, and sustenance support of my wife Jennifer.

Fair Oaks Ranch, Texas, USA William E. Thornton

Contents

Introduction

This book differs from other literature on the subject in a number of ways.[1] It is devoted to a single subject, effects of weightlessness on humans, but it is sharply restricted to changes that significantly affect subjects and their operational capacity in space or on return to Earth. Conversely all of the major or significant effects are treated but in an individual and focused manner. It is not intended only for specialists in space medicine. Instead we attempted to make the material useful and comprehensible to as many as possible of those with an interest in space operations. This required several considerations.

Problems in understanding specialized language are encountered immediately. Few individuals outside medicine could locate their *levator ani* with certainty, and there are some professionals equally unfamiliar with a pilot's *pucker muscle.* Where practical, specialist terms are accompanied by a more common term. Glossaries have been added at the end of some chapters, versus the end of the book, to facilitate access to translation of unfamiliar terms. The danger in this is a mixture, which satisfies only a few.

The nomenclature used to identify the Skylab missions often causes confusion. Here, as in many publications, Skylab (SL) indicates the vehicle or the program in general. Skylab 1 refers to the unmanned launch of the Skylab vehicle, Skylab 2 the first manned mission, Skylab 3 the second manned mission, and Skylab 4 the third and final manned mission. They are identified with the abbreviations SL-2, SL-3, and SL-4 to indicate the information is mission specific.

There is always the danger of mystification, of leaving it to the experts, when complexities are encountered, especially in the human body. These complexities usually become understandable when reduced to their basics; hence sections in the basic sciences are an integral part of this work.

For those used to reviews that include as many alternative views of a phenomenon as possible, much of this may seem simplistic. Simplicity based on demonstrable fact or function has been a goal here. Wherever possible this has been augmented by color-coded diagrams.

---

[1] Unless something new or different can be offered there seems to be little point in writing.

Emphasis and interest in the various aspects of weightlessness has been shifted by operational changes over the more than half decade taken to produce this book. Initially a major portion of operation was focused on the Shuttle, which was employed primarily in its namesake role of shuttling crews, equipment, and material to and from the International Space Station (ISS). The majority of the crews involved were on short flights and short-term phenomena, such as space motion sickness (SMS) and fluid shifts, affected crew time significantly. By the end of the Shuttle program a few astronauts had begun to spend longer periods of time in flight—months and up to a year instead of days and weeks. This shifted the primary concerns to the musculo-skeletal and metabolic systems. In the course of these in-flight times, the frequently seen shifts to far vision on orbit did not resolve on return to Earth and 1 g in a few individuals and permanent changes were found in their eyes.

Chapter 1 is dedicated to weight and weightlessness and Chap. 2 to adaptation. The nine following chapters cover a particular aspect of somatic (rapid and reversible) adaptation.

Where possible chapters are organized as follows:

- A brief description of the problem in spaceflight
- Generic history of the problem with some detail of its history in flight
- A background to the causes of change is given in the most basic elemental terms
- Experience in 1 g relevant to the problem
- Results of 1-g simulation of the problem
- Flight experience with the problem including countermeasures

Chapter 5 developed as the problems with eyes and vision were disclosed late in the evolution of the book. This is the first example of a stress that exceeds the limits of reversible adaptation and produces permanent pathological changes. It is probably indirectly related to weightlessness.

While all the concepts are based on the most solid evidence available, each has been or may be verified by experiment. For example in SMS it should be a relatively simple matter for someone skilled in micromanipulation to remove the otoliths' weight and reproduce the effects seen in weightlessness.

There is a paucity or absence of much needed data in numerous areas in spite of vigorous pursuit of any valid source. More or less coincidental with the end of Shuttle, improvements in ISS onboard equipment and its use were made. Some flight data have become more accessible. There are some reports of improvement in countermeasure results, but these seem to be exceptional.

In spite of such shortcomings, the material in the book should provide a basis for the increasing knowledge and its practical application of the effects of weightlessness on the human body.

Acronyms and Abbreviations

AO	Astronaut Office
ARC	Ames Research Center
ARED	Advanced Resistive Exercise Device
ASTP	Apollo Soyuz Test Project
BDA	Bone Density Absorption
BMD	Body Mineral Density
BMMD	Body Mass Measuring Device
BR	Bed Rest
BV	Blood Volume
BW	Body Weight
BWF	Body Weight Force
Ca	Calcium
CDR	Commander
CEVIS	Cycle Ergometer with Vibration Isolation System
cm	Centimeter
cmH_2O	Centimeters of water
CNS	Central Nervous System
CO_2	Carbon Dioxide
CR	Cardiorespiratory
CSA	Cross Sectional Area
CSF	Cerebrospinal Fluid
CTZ	Chemoreceptor trigger zone or chemical trigger zone
CVS	Cardiovascular System
DEXA	Dual Energy X-ray Absorptionometry
DNA	Deoxyribonucleic Acid
DSO	Detailed Supplementary Objective
DXA	Dichromatic X-Ray Absorptiometry (radiology)
EKG	Electrocardiogram
ESA	European Space Agency
EVA	Extravehicular Activity
F	Force

FE	Flight Engineer
FGF	Foot Ground Force
ft	Foot
FWED	Fly Wheel Exercise Device
g	Gravitational constant for Earth
G	Universal gravitational constant
GI	Gastrointestinal
gm	Gram (g is the standard abbreviation for gram. In this book "gm" is used to differentiate between gram and gravitational constant.)
gx	acceleration force chest to back
gz	vertical gravitation
H_2O	Water
Hb	Hemoglobin
ICF	Intracranial Fluid
IIH	Idiopathic Intracranial Hypertension
IML-1	STS mission
IML-2	STS mission
IRED	Interim Resistive Exercise Device
ISS	International Space Station
JAXA	Japan Aerospace Exploration Agency
JSC	Johnson Space Center
kg	kilogram
L	liter
LAMMD	Linear Acceleration Mass Measurement Device
lb	pound
LBM	Lean Body or Tissue Mass
LBNP	Lower Body Negative Pressure
LOC	Loss of Consciousness
LS	Life Sciences
m	meter
m	mass
MC	Muscle Control
MECO	Main Engine Cut Off
MMD	Mass Measuring Device
Mir	Russian Space Station
mL	milliliter
mmHg	Millimeters of Mercury
MOL	Manned Orbiting Laboratory
MRI	Magnetic Resonance Imaging
MS	Motion Sickness
MS	Mission Specialist
MSK	Musculoskeletal
MSkS	Musculoskeletal System
Na+	Sodium ion
NASA	National Aeronautics and Space Administration

NMI	Neuromuscular Inhibition
O_2	Oxygen
OS	Orthostasis
PLT	Pilot
PO_4	Potassium
pQCT	peripheral Quantitative Computed Tomography
QCT	Quantitative Computed Tomography
RBC	Red Blood Cells
SAM	School of Aviation Medicine
SB	Side Bar
SCI	Spinal Cord Injury
ScopDex	Scopolamine with Dexedrine
sec	Second
SL	Skylab
SL-2	First Skylab manned mission
SL-3	Second Skylab manned mission
SL-4	Third or final Skylab manned mission
SLAMMD	Space Linear Acceleration Mass Measuring Device
SMEAT	Skylab Medical Experiments Altitude Test
SMMD	Small Mass Measuring Device
SMS	Space Motion Sickness
SPT	Science Pilot on Skylab
SRC	Short Radius Centrifuge
STS	Space Transportation System
T	Time constant
t	Time
US	United States
USAF	United States Air Force
USSR	United Soviet Socialist Republic
VO_2 max	Maximal Oxygen Consumption
WCS	Waste Collection System
WHO	World Health Organization
X axis	Vertical axis
Y axis	Horizontal axis
zero-g	Zero gravity, term frequently used for weightlessness

Chapter 1
An Introduction to Weightlessness and Its Effects on Humans

1.1 Introduction

Everyone associated with space and spaceflight has concepts of mass, gravity, weight and weightlessness. These individual concepts vary and can reach levels of sophistication that only a few can understand. In an effort to have a common set of tools with which to approach the biophysical aspects of weightlessness, a simple set of concepts, relations, and equations have been added to each chapter. These 'background' sections provide a common approach to the concepts and ideas that follow. What is included in Sect. 1.2 will often be superfluous and provide too much information for some readers. However, given the diversity of backgrounds in the intended readership, these background sections may be useful to some the authors have provided them as a useful tool. In this chapter on weightlessness and its effects, the background describing weightlessness must logically begin with clear definitions and examples of mass, weight, and weightlessness. As some readers will come to appreciate, concepts of weightlessness are often unclear or misunderstood (Fig. 1.1).

1.2 Background

The basic concepts of mass and weight are covered in many textbooks and scholarly works that address basic physics [e.g., 1, 2]. Mass, expressed in units such as kg or lb., is coherent matter that occupies a volume in space. This volume is typically expressed in units such as cubic meters (m^3) or cubic feet (ft^3). Density of mass determines the amount in a given volume. In any uniform material density will be fixed and mass will be directly proportional to volume—the more space a given substance occupies the more mass it will have. For example, 2 m^3 of iron has twice

W. Thornton, F. Bonato, *The Human Body and Weightlessness*,
DOI 10.1007/978-3-319-32829-4_1

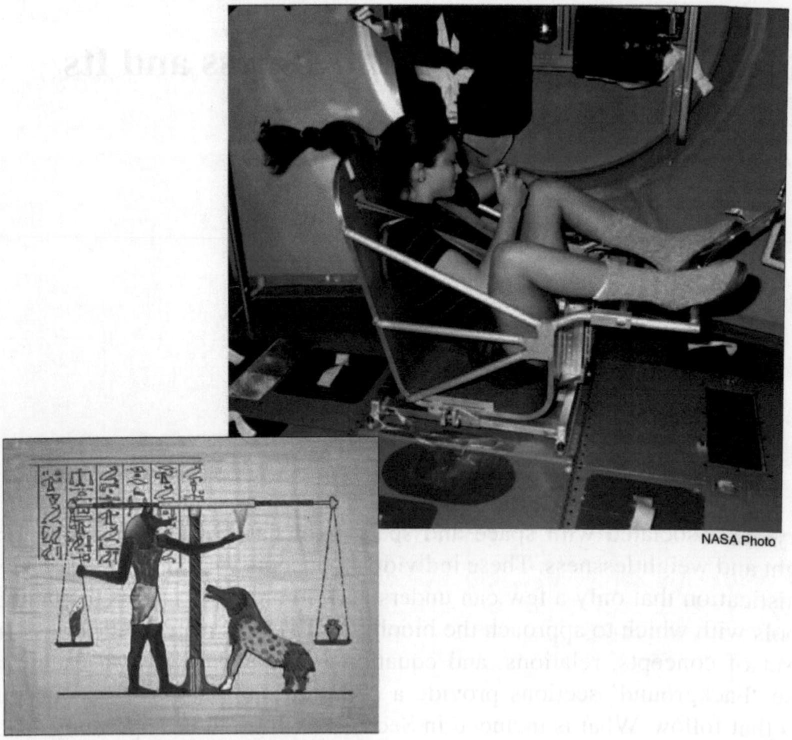

Fig. 1.1 Operational effects of weightlessness. These images illustrate the historical span of weight measurement in human technology and the recent impact of its absence in Space. Four thousand years ago, a tomb painting incorporated a commercial scale of the time into the myth that a dead Pharaoh's heart carrying the weight of his misdeeds had to be as light as a feather for him to enter an afterlife. In 1973 the first mass measurements in space were made to study the effect of weightlessness on humans. Tammy Jernigan is shown in the body mass measurement device in a Shuttle Spacelab. Over 5000 years ago, weight was an essential tool of commerce, and weightlessness was an unnoticed artifact. Neither process was understood until Sir Isaac Newton applied his genius to the apple in free fall. Actually, his genius was applied to astronomical data from orbital planets in free fall. After Newton, sustained weightlessness remained largely theoretical until aircraft became powerful enough to fly parabolas producing tens of seconds of weightlessness followed by rockets, which produced the same effect for indefinite periods

the mass of 1 m^3 and 4 m^3 of iron has twice the mass of 2 m^3. As long as the density of the material is the same, this directly proportional relationship between mass and volume will exist.

Thus a given mass occupies some fixed volume of space (Fig. 1.2).[1] The denser a given volume is, the more mass it will have. When volume is fixed, mass is directly

[1] In its normal atomic state, virtually all of mass is in atomic nuclei, and the volumes shown in Fig. 1.2 are largely empty space.

Fig. 1.2 Mass occupies volume in space. The volumes depicted are equal masses of gold (M_1) and water (M_2). In an atom volume and mass are determined by the number, type and configuration of subatomic particles (see Footnote 1). See equation below. Density: Water = 1.00 and Gold = 19.3 gm/cm³

proportional to density. For example, 1 m³ of Gold is approximately twice as dense as 1 m³ of Copper, and so 1 m³ of Gold has about twice the mass as 1 m³ of Copper.

The following formulae express the relations between mass, volume, and density.

$$\text{Mass} = \text{Density} \times \text{Volume}$$

$$\text{Density} = \text{Mass} / \text{Volume}$$

$$\text{Volume} = \text{Mass} / \text{Density}$$

Two generic properties of mass concern us here.

1. There is a force, gravity, through which each bit of mass is mutually attracted to every other bit of mass (Fig. 1.3). Bodies having more mass attract other masses more than bodies that have less mass. This is why a person can jump higher on the Moon or on Mars than on Earth. Both the Moon and Mars have less mass than Earth and so other bits of mass are less attracted to them—including humans. The human body is hence less attracted to the Moon or Mars and so a person on one of these bodies can more easily break away from it—the result: a higher jump. The spacesuits worn by Apollo astronauts on the Moon weight nearly 200 lb. on Earth, and yet because of the Moons weaker gravitation field (about 1/6 that of Earth's), astronauts moving on the lunar surface were surprisingly unaffected by the mass of these suits. This was due to the reduced mass—and gravity—of the Moon—compared to Earth.

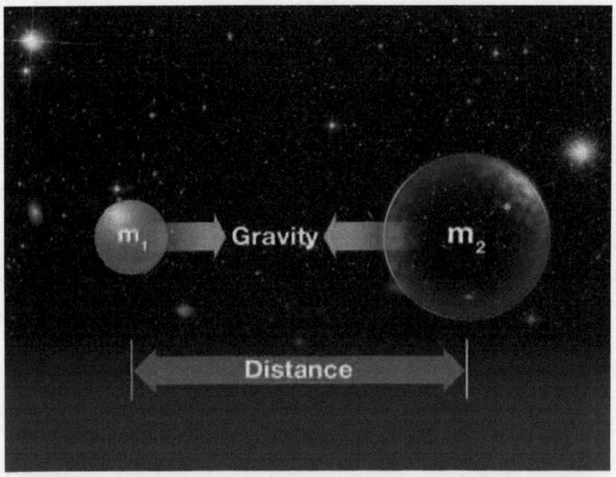

Fig. 1.3 Gravity is the mutual force of attraction between all particles of mass in the universe. See equation below. Force of this attraction between particles is given by:

$$\text{Force} = \frac{\text{Gravitational constant} \times \text{mass}_1 \times \text{mass}_2}{\text{Distance}^2} \text{ or } F = \frac{G \times m_1 \times m_2}{d^2}$$

Newton's equation describing this attraction between two bodies is:

$$\text{Force} = G \times \text{mass}^1 \times \text{mass}_2/\text{distance}^2$$

The same force is exerted on both bodies, and diminishes rapidly with increasing distance.

G, the gravitational constant, is known as "big G."

Units:

$G = 6.67 \times 10–11\ N \times kg^2/m^2$
Force = N, P(oundal)
Mass = kg, lb.
Distance = m, ft.

Although the details of big G are provided here in an effort to be thorough, these details will not be necessary to understand the material presented in the rest of this book. A general understand of mass and gravity will suffice.

2. A second property of mass for our concern is Inertia, (Fig. 1.4) a property that resists any change in a mass's velocity by generating an opposing force given by:

$$\text{Force} = \text{mass} \times \text{acceleration}.$$

In accordance with Newton's first law of motion, bodies at rest stay at rest, and bodies in motion, stay in motion at the same speed and heading, unless some other

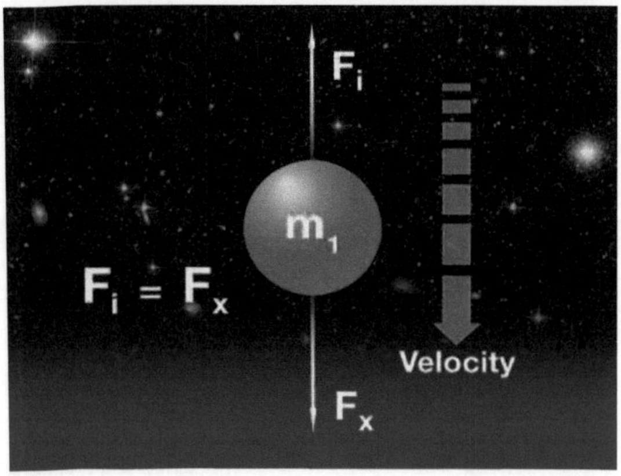

Fig. 1.4 Inertia, our second relevant property of mass, produces an equal and opposite counter force, F_i, to any other force, F_X, which changes the velocity, of the mass. It is proportional to the rate of change of velocity and, to the acceleration produced by F_X. See equation below. Force, inertia = mass × acceleration. $F_1 = m_1A$

force acts upon the body. All bodies resist change in their state of motion, but the more mass a body has, the more resistant to change it will be. This is why it is easier to push over a bag of leaves than a bag of cement. It is also why linebackers in American football tend to be the most massive members of the team. They are not easily moved. Inertia is experienced when we come to a sudden stop, or suddenly change direction at high speed, or are boosted into space. It is often measured in units of Earth's gravity, hence the term "pulling g's." Most people have experience these 'g's' such as when accelerating in an automobile or airplane—it is the feeling of being pressed back into the seat. It is also experienced in elevators and amusement rides. Albert Einstein demonstrated the theoretical equivalence of inertial and gravitational forces, but it is conceptually simpler to treat them as separate phenomena here.

1.3 Weight and Weightlessness

When small masses are on or near very much larger masses such as the Earth, we can arrange a much simpler calculation for gravitational attraction to fit the location. In spherical bodies the entire mass is treated as if it were a point at its center, so

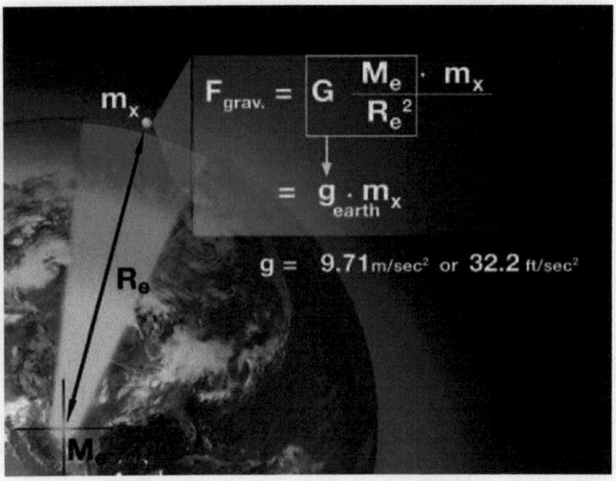

Fig. 1.5 Local gravitational forces become dominant near large congruent masses such as stars and planets, e.g. the Earth's field versus the small effects of the distant Moon and Sun. This force of attraction for small bodies on or near the larger bodies is called weight (see Footnote 2). The constant known as "little g" must not be confused with the universal gravitational constant, G. This value of g for Earth has become a reference standard for other satellites, e.g. 1/6 g for the Moon, and for acceleration forces; e.g., a 3-g turn

distance to a mass on the surface is the radius of the large body (Fig. 1.5).[2] Putting this value and the body's mass into the general equation for gravitational force gives a new constant, g, called "little g," which is unique for Earth. This g may also be calculated for other celestial bodies and is usually expressed in terms of a proportion of Earth's value (Fig. 1.6). Expressed in this way we can see that g for the Moon is approximately 1/6 of Earth's, and g for Mars is about 1/3 as strong. As noted, inertial forces are often expressed in Earth g units but must not be confused with gravitational force.

The attraction of Earth toward small objects is trivial, hence we can treat the gravitational attraction, the weight of objects, on or near its surface as a single constant.

with a magnitude of:

$$F_x = g \times mass_x$$

By measuring the force, F, with a scale calibrated in terms of mass, we have a close approximation of the measured mass[3] (Fig. 1.7).[3]

[2] Constant for most purposes, but it is modulated by the slight variation in length of the Earth's radius at different points on its surface, variation in local density and by the varying gravitational force vectors of the Sun and Moon; e.g., ocean tides.

[3] There is significant variation in g over the Earth's surface, but this is usually easily accounted for.

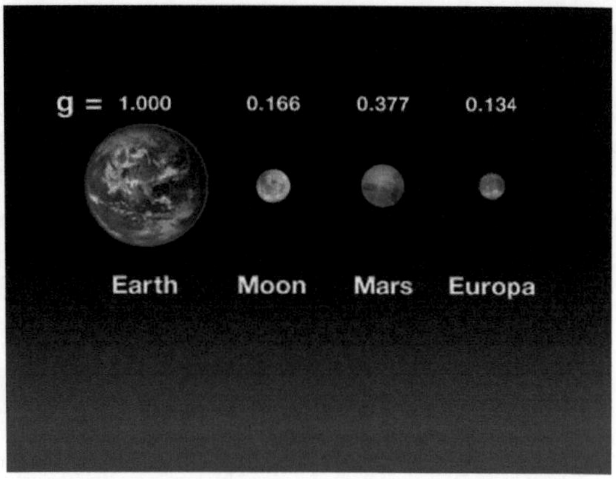

Fig. 1.6 Local g values can be calculated for other satellites. Such values relative to Earth's value are shown for three in the solar system. The Moon's gravitational field is about 16.6% that of Earths' while Mars' field is 37.7% as strong. Europa, one of Jupiter's moons has a field that is about 13.4% as strong as Earth's

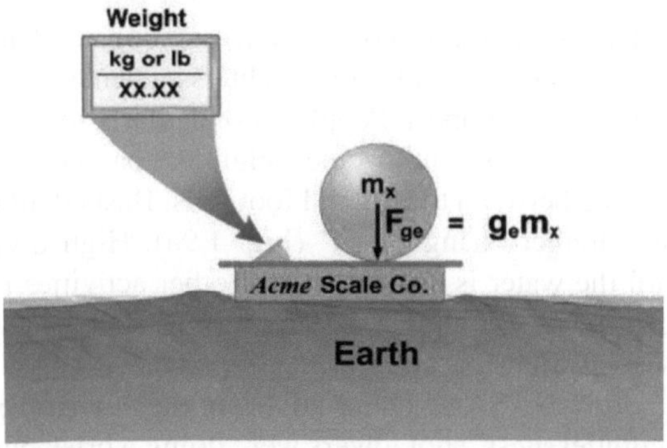

Fig. 1.7 Weight is the gravitational force on a mass when on or near a much larger mass such as the Earth. In practice it is the force measured by an instrument calibrated in terms of known mass. When properly measured, weight and mass are equivalent. For some 4000 years weight was determined by comparison to known masses using a balance. Electronic scales such as depicted here are a recent development in the history of technology

Now lift the mass, m, resting on the scale a few feet above Earth's surface and support it with a magic carpet, then drop everything (Fig. 1.8). The carpet now magically moves at the same velocity in contact with the scale and its mass, but the force of acceleration exactly offsets the weight force and m is weightless.

This is the definition of all weightlessness, on Earth and in Space: no measurable forces of attraction to other objects.

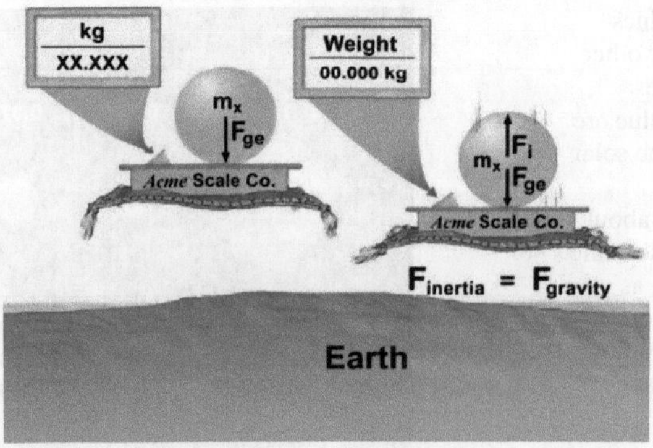

Fig. 1.8 Weightlessness is precisely illustrated by this imaginary scene. On the left, the scale of Fig. 1.7 is now being supported by a magic carpet. On the right, the whole assemblage is falling to Earth with the scale still "supported" by the magic carpet, which is now falling at exactly the same rate as the mass and scale. Since the inertial acceleration and gravitational forces are exactly equal, mX is weightless. The same condition is present when an unpowered spacecraft is falling toward Earth or other bodies, either directly or in orbit

Weightlessness has been a common, albeit, unrecognized human experience (Fig. 1.9). All non-flying land animals are weightless whenever they are not in contact with the ground or other support. People who regularly jog have typically spent many tens or hundreds of hours in miniscule weightless parabolas, albeit in intervals of a fraction of a second between toe off and foot falls. Basketball players and ballet dancers can achieve longer "hang times" (Fig. 1.9a). High diving will result in weightlessness until the water is entered as will other activities that entail leaving Earths' surface (e.g., bungee jumping, sky diving). In short, most people—almost all people—have been weightless at times.

Two centuries before NASA began to advocate the advantages of manufacturing products in weightlessness, shot towers were doing a brisk business by making spherical lead gun shot. Drops of molten lead became spherical during their weightless free fall into a water-filled catch basin (Fig. 1.9b). Drop towers with evacuated columns (Fig. 1.9c) now provide a few seconds of weightlessness for experiments on Earth, and by 1960, aircraft could provide 25–45 s periods of weightlessness (Fig. 1.9d). A "zero-g" aircraft, even in the best of hands is seldom truly weightless during parabolic flight, but freely "floating" mass in the cabin area of the aircraft can be considered "weightless." Rockets of course, and with it the emergence of human space travel, have made sustained weightlessness for humans a reality.

1.4 Weightlessness During Spaceflight

While a "magic carpet" was invoked to illustrate the physical principles of weightlessness in Fig. 1.8, Fig. 1.10 illustrates the two actual conditions that produce weightlessness during spaceflight: free fall and orbit. The physics are unchanged

Fig. 1.9 Weightlessness can be produced for brief periods on Earth. Whenever an object is not supported and falling freely it is weightless, as in the time between toe off and heel strike in jogging and running. The duration of weightlessness can be extended by jumping athletes and dancers (**a**). Shot towers (**b**) employed weightlessness to produce small spherical gunshot for almost two centuries. Drop towers, taller and with an evacuated column are now used for studies requiring a few seconds in weightlessness (**c**). Zero-g aircraft flown in a parabola can provide a close approximation of weightlessness for periods of 25–45 s. Unattached objects in its interior are truly weightless (**d**)

from Fig. 1.8. The conditions are simply more sustained leading to much longer periods of weightlessness.

Look first at an unpowered craft falling through the vacuum of space toward a significant body of mass, such as the Moon in Fig. 1.10a. Initially for simplicity, only the Moon's gravity (F_{gm}) and the vehicle mass will be considered. This is exactly the situation in Fig. 1.8, but without the carpet and scale and with a much greater distance to fall. Acceleration produces an inertial force that exactly balances the total gravitational force. In this case gravitational force and acceleration increase as the Moon is approached, increasing inertial force thus continuing to balance the forces and maintaining. In this situation, or with any other object in free fall, the balance of forces is perfect at the falling object's center of mass that is perfectly weightless. In actuality the situation is slightly more complex as there are gravitational attraction forces present from other objects such as the Sun and other planets, especially the relatively close Earth, as depicted in the insert. But these vectors summate into the single effective vector shown.

The second condition we know as 'orbit.' It is more common and more complex but again involves only the forces of inertia and gravity. Figure 1.10b depicts a

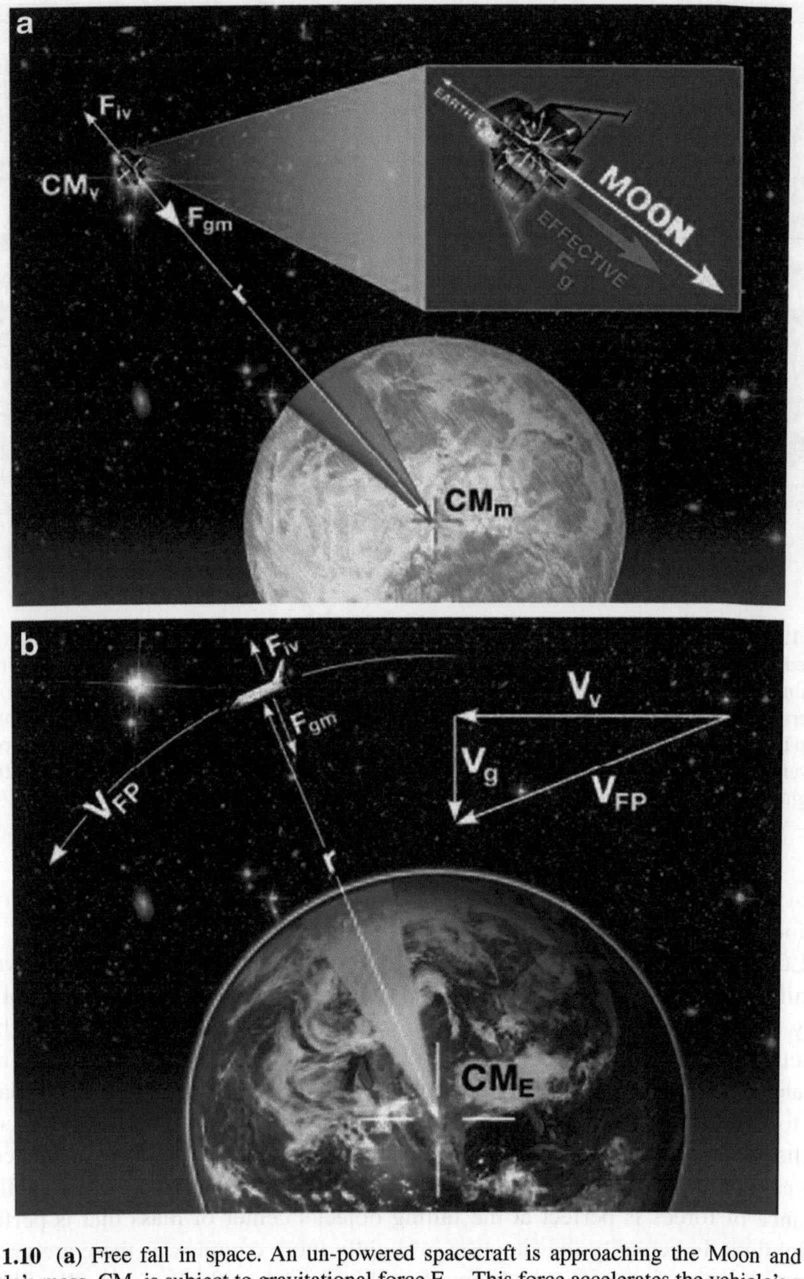

Fig. 1.10 (**a**) Free fall in space. An un-powered spacecraft is approaching the Moon and the vehicle's mass, CM, is subject to gravitational force F_{gm}. This force accelerates the vehicle's mass producing an equal and opposite inertial force F_{iv} leaving vehicle and contents weightless. Numerous other gravitational forces present are illustrated in the insert. They combine to produce a single effective vector, F_g. (**b**) Orbital weightlessness. A vehicle in a circular orbit around the Earth is constantly attracted toward the Earth's center of mass CM_e by the gravitational force F_g. This produces an equal but opposite force of acceleration of F_{iv}. These forces are exactly equal and this produces weightlessness of the orbiter

vehicle in circular orbit where its center of mass is being attracted to that of the Earth's center of mass by its gravitational force. This produces an acceleration causing the vehicle to "fall" toward Earth's center of gravity producing an equal and opposite inertial force rendering the vehicle weightless. However, to complete this picture the vehicle has a velocity that moves it to a new position as it is falling. For example, the International Space Station (ISS) orbits the Earth at a velocity that is above 17,000 mph. As ISS falls to Earth it continually moves forward to a new position that would elevate it higher in altitude were it not for the gravitational force of the Earth pulling it downward. This can be seen in the insert where the vehicle's instantaneous velocity FP is broken into two vectors, a velocity V_g produces the fall toward Earth while the vector V_v moves it away from Earth producing a combined vector VFP. The effect of these two vectors combined is a continuous circle or orbit. In short, for orbiting objects, gravitational force and inertial force are offset. When these forces are balanced a smooth orbit can be obtained. If gravitational force is stronger than inertial force, the orbiting body will eventually lose altitude and be pulled down towards Earth. If inertial force is stronger than gravitational, the body may escape Earth gravitational pull entirely.

For those interested in micro weight of objects away from the center of mass there is comment in Box 1.1. As used here, weightlessness is the absence of detectable weight.

1.5 Brief History of Major Effects of Weightlessness

The history of dreams and actuality of space travel and weightlessness is well documented and will not be repeated here. However, a brief history of the effects of weightlessness on space travelers follows.

Box 1.1 Weightless Synonyms
Weightlessness is a concise term meaning the absence of detectable weight. Some theoreticians or a rare experimentalist may need the more exacting term micro-weight to describe the situation in an orbital vehicle. There is only a single point in a vehicle in orbit where all gravitational forces are balanced exactly. Move from this point in any direction, and infinitesimal forces, local gravitation, inertial force and weight will be created. Both of these terms are awkward to use and several others are in common use as replacements. Microgravity, analogous to micro-weight, has become more popular than zero gravity, but micro- and zero gravity imply a condition of absent or negligible gravitation that cannot exist in our universe. If the implied condition were possible it would produce weightlessness, and so long as the words mean an absence of weight, it usually makes no difference. Weightlessness is the definitive term and will be used here.

Over 500 persons have flown in space. What is defined as 'spaceflight' is in accordance with the internationally recognized criterion set forth by the Fédération Aéronautique Internationale (FAI). It is any flight above 100 km (62 mi). Humans have spent a combined total time of approximately 130 years in space and approximately 100 days space walking. Four individuals have spent more than a year each in space during a single flight. In spite of this experience, there are many researchers advocating continued or expanded biomedical research before long duration missions such as to Mars are considered. Certainly there are some areas in which the lack of progress is remarkable.

The human body and many of its functions have been shaped by gravity since our aquatic ancestors left total water immersion some 375 million years ago [3, 4]. Prior to that, all life forms had developed under water for 3+ billion years. To live under gravity on earth, the evolution of aquatic animals to land animals required drastic additions and revisions such as limbs, and altered circulatory, musculoskeletal, and neurological systems. It is not surprising that many of the functions that evolved for animal bodies to cope with weight are the functions most affected by weightlessness, an environment similar to that of water. Functions originally developed under water, such as digestion, excretion, and cardiac, are least affected by weightlessness. Individual physical differences, both genetically based and acquired are common and significant, so not surprisingly there are also substantial individual differences in physiological responses to weightlessness.

Given that this is only a summary of the effects that are described in following chapters, references are not given at this point. A listing of more or less standard references for various phases of American and Russian programs is included under the reference list at the end of this chapter. A brief description of the major effects[4] follows more or less in order of occurrence. They are categorized as adaptations and accommodations. Adaptations produce physical changes in the body while accommodations are functional changes in body systems within its usual physical capacities. It is important to note that the use of the word 'adaptation' here *does not* refer to those changes that occur due to natural selection. We are not referring to evolutionary changes when discussing these *somatic adaptations*. Most of the somatic adaptations that occur during weightlessness are temporary bodily changes. A brief introduction to these adaptations and changes are described here.

1.6 Space Motion Sickness

In the early days of spaceflight, there was a tangled web of effects produced by acceleration loads, temperature, physical restraint, metabolic and psychological stresses in addition to weightlessness. However, on his second orbit during the 1965 Vostok 2 mission, Russian Cosmonaut Vladimir Titov reported symptoms of what was soon

[4]Radiation effects are not included here but may be a major factor in extended extra terrestrial flights and operations. There is a large amount of often confusing literature about this.

(mis)called Space Motion Sickness (SMS). The symptoms Titov experienced were similar to motion sickness in some ways, but in some ways different too. Both Russian and American life scientists developed elaborate but ineffective countermeasures based on 1-g theory and practice in motion sickness. NASA's first admission of the problem came during the Apollo 9 mission in 1969 when it forced a delay in an extra-vehicular activity (EVA). The incidence, symptoms and course of SMS was established on early Russian and Shuttle flights and the reciprocal process of postflight re-adaptation recognized (sometimes called entry sickiness or re-entry sickness). Subsequent extended flights have added to the experience base surrounding SMS.

In less than half of astronauts, from minutes to hours after entering weightlessness, there will be a sudden heave or two, forcefully regurgitating all stomach contents. This usually occurs without warning and often happens to those most resistant to motion sickness on Earth. It is typically followed by a general malaise, lethargy, aversion to food and hypersensitivity to movements in the pitch plane. Nausea is not common, but any stomach contents from food, drink or accumulated gastric juices and saliva will be briefly regurgitated. Motivated subjects can perform tasks trained for, but may be interrupted by another heave or two of emesis. Symptoms vary, but typically end sharply after 1–2 days. There is no recurrence in flight, but in some individuals' symptoms may recur during entry and for some time after landing. There is great individual variability in SMS effects. For previously symptomatic space travelers or first time flyers, SMS remains a confounding factor in planning operations during the initial day or so in flight. Reentry, landing, and recovery periods will lead to symptoms for a small number of those previously affected and for few were not affected during flight.

1.7 Fluid Shifts and Losses

As flights continued, a number of other signs and symptoms were regularly observed. Puffy faces and stuffy noses became routine during flight. When it became possible to remove pressure suits inflight, astronauts coined the phrase, "bird legs of space," to describe their shrunken legs—their girth significantly reduced. On landing, there was a regular, but variable weight loss, that was usually quickly regained. This loss was accompanied by a tendency to faint on standing quietly. The Gemini program (1964–1965) verified this transient postflight orthostasic intolerance.

On the final 84-day manned Skylab mission (Skylab 4, 1973–1974), inflight limb volume measurement, combined with other studies, demonstrated the shift and then loss of several liters of fluid and several hundred milliliters of blood out of the legs. This leaves the crew dehydrated and anemic on reentry and landing. Continued studies during the Shuttle program have further defined the biophysics of the problem while a number of countermeasures have been vigorously pursued by Russian and U.S. scientists including pre-entry fluid "loading", fluid shifting and anti-gravity suits. Except for individual cases, the problem is being dealt with, but at considerable inconvenience and poor efficiency. Fluid shifts have recently attracted attention as the postulated proximate cause of ocular problems in some astronauts on long duration flights. A more comprehensive consideration of these adaptations is presented in Chap. 4.

1.8 Musculoskeletal Effects

Bone and muscle loss in space are usually treated separately, but until it is recognized that bone and muscle are a functional unit, bone/calcium loss will remain an unsolved problem. Bone strength is absolutely dependent on muscle strength and usage. Based on bed rest studies dating back to World War II and subsequent research, loss of calcium from bone in weightlessness was predictable. Studies on Gemini seemed to show a large loss of bone density but these findings were later revealed to most likely be artifactual. However, a simultaneous inflight intake/output study confirmed abnormal calcium losses. On Skylab, complex studies of metabolism and bone density confirmed significant, progressive losses of bone minerals, density and presumably strength. An additional 40 years of studies have produced increased knowledge of bone and muscle losses and a variety of exercise devices on the International Space Station (ISS) that have resulted in marked decreases in loss, but efficient, effective countermeasures must be considered a work still in progress.

Loss of muscle mass and strength received virtually no attention other than a few hand grip measurements in the American program until Skylab. This may be due to mission length—compared to Skylab, previous missions were substantially shorter. It was during Skylab when pre- and postflight strength studies and crew demand for more effective exercise devices produced an ad hoc study that demonstrated the bare essentials of bone/muscle effects and countermeasures.

During the 28-day Skylab 2 (SL-2) mission in 1973, a rapid loss of strength was demonstrated in the arms and even more loss in the legs occurred without the benefits of adequate force-loading activity or exercise. Skylab 3 (SL-3), a 59-day mission, demonstrated that losses in the arms could be prevented by proper exercise. It was not possible to add adequate leg exercise then. Leg strength losses have been reduced in some cases, as on Skylab 4 (SL-4), but there are now some claims of complete preservation of muscle and bone on ISS but little reported documentation is available. This is still a major practical problem and currently 3 h per day is spent on scheduled exercise on ISS in order to counteract the effects of weightlessness on bone, muscle, and strength loss.

1.9 Cardiopulmonary Effects

This area has received great attention pre-, in-, and post-flight. While there are significant effects on the vascular system and secondary effects on the respiratory system, when the obligatory losses of blood and fluid are considered, the heart and pulmonary systems' essential functions of pumping blood and air are little affected by weightlessness. Moreover, it was reported on Skylab that regular application of Earth-equivalent metabolic loads would maintain their Earth equivalent pumping capacity. For example, the cycle ergometer, while inadequate for application of locomotor force loads, is capable of providing adequate metabolic (aerobic) loads. Subsequent repeated analysis of these data have called the initially reported SL results into question. Reported ISS studies indicate that cardiorespiratory function can be maintained by adequate aerobic exercise in flight.

1.10 Metabolic Effects

Prior to Skylab, the amounts of food consumed in space plus the confusion of post-flight weight losses from fluids left the metabolic costs of living and working in space unknown. The largest, most complex life science study on Skylab included metabolic studies. A carefully composed and meticulously measured basal diet, based on bed rest studies was imposed, beginning weeks before flight, during flight and weeks after, with a core diet of 2000 calories with 70 g of protein plus vitamins and minerals. Any caloric deficits were to be covered by lemon drops and sugar cookies. All feces and urine were collected and analyzed, and weights or masses were measured daily.

This diet protocol was part of the 56-day Skylab Medical Experiments Altitude Test (SMEAT) that was conducted at NASA's Manned Spacecraft Center in Houston in 1972. One crew member was large and physically active and lost 17 lb. while on the diet. All crewmembers on SL-2 showed the same monotonic pattern of weight loss caused by inadequate intake. On SL-3 the diet was increased with reduced loss of weight. On SL-4 the diet was increased again and this proved to be adequate for one crewman who was the first to not lose weight in spaceflight. At least one crewman rebelled against the food provided on Mir and demonstrated a significant metabolic loss. A study on ISS reported similar but smaller losses. A recent ISS report is more or less consistent with Skylab experience, but the low resolution of procedures and instruments used allow only estimates. Two body mass measurement devices (BMMDs) have been used on ISS but no in-flight calibration studies have been reported.

In summary, metabolism in weightlessness is surprisingly high but is readily met by providing adequate food acceptable to crewmembers. Monitoring of body mass should be routinely used.

1.11 Neuromuscular Adaptation

One of the memorable postflight effects of weightlessness is one's first attempts to stand after landing. As Ken Mattingly (Apollo 16, STS-4 and STS-51-C) described his experience, "You might as well have left the (restraint) straps on." But one rapidly readapts and is usually able to walk off after short flights, albeit with a transient sensation of being in 2 g. This is a universal phenomenon with a well-established neurological basis that has not been reported in other studies to date. It may be coupled to effects seen in some vestibular studies. This is of less concern now that Shuttle is no longer flying but may be important for future space vehicles and in some automated landing situations.

1.12 Effects of Accommodation

The previous effects of weightlessness all required more or less rearrangement of the body to optimally accommodate the absence of weight. These range from altered neuronal synapses, connections between nerve cells, accomplished in hours, to changes in bone composition and structure over months. There are other striking changes in response to weightlessness that occur within, that are accommodated by, the body's usual configuration. No significant time is required for these changes to occur on transition from 1 g to weightlessness and vice versa.

1.13 Anthropometric Accommodations

Inflight there are significant reductions in girth and increases in standing height that are discussed in detail in Chap. 10. A change in height of two-plus inches is not insignificant to a closely fitted, unyielding space suit fitted in 1 g. A bodily accommodation perhaps related to increases in height is the loss of spinal curvature. The familiar S-shaped curve of the spine is 'straightened out' during spaceflight.

1.14 Postural Change

After numerous indications that 1-g furniture and many positions were poorly tolerated in weightlessness, the basic posture of weightlessness was finally documented on the third Skylab (SL-4) manned mission. On Earth the body at rest seeks positions with the least expenditure of effort, be it lying, squatting, sitting or standing. It does the same in weightlessness where muscles are relaxed to reduce forces to a minimum. Needless to say, knowledge of this posture is critical to well designed accommodations in space.

1.15 Other Changes

Other memorable postflight effects include the stagger when walking, understeering on turns with door post collisions, and the half-conscious sensation of floating in bed the first night back. Effects increase with time in weightlessness and have

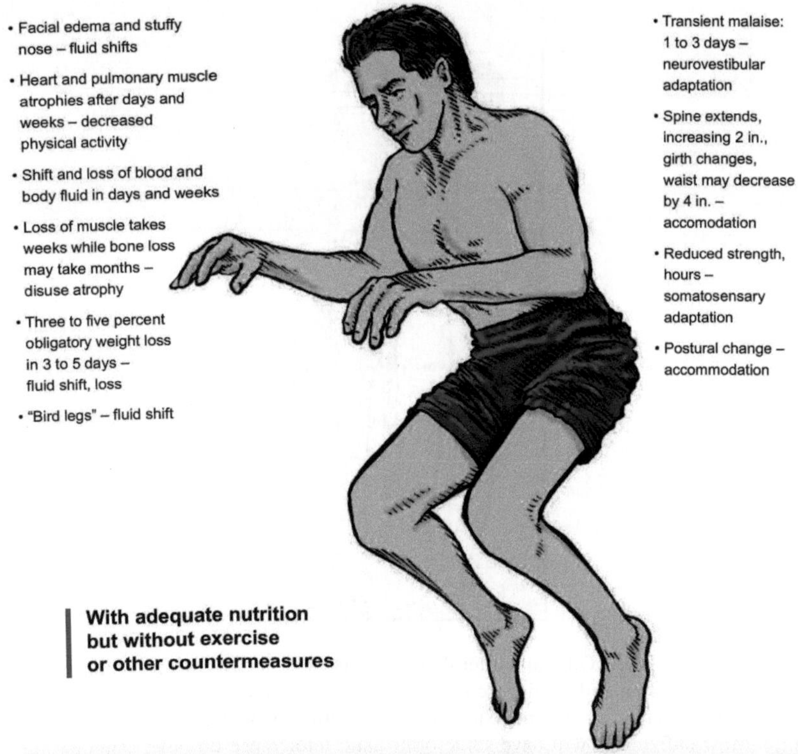

Fig. 1.11 Inflight. Most of the inflight effects are perceived in the first days or weeks of a flight. Less than half of astronauts will experience SMS, but everyone notices the rush of blood to the head with stuffy nose and change in the sense of taste. Sleep may be difficult in this early period. Almost everyone recovers from SMS, usually within 2 days, and there is no documented evidence of recurrence. Facial edema with congested head veins continues, although the sensation of fullness decreases. Unless there is some problem, such as with an individual's food intake, there are few physical problems with extended stays in weightlessness; however, adaptation continues in many areas, and the effects will become significant on return to 1 g

reached disabling levels postflight in some individuals. Such effects on motion and orientation have been studied in- and postflight, but the complexity of the central nervous center and our lack of understanding still make this an unpredictable area. They are not covered in this work.

1.16 Summary of Effects

A summary, including these effects, is shown in Figs. 1.11, 1.12, and 1.13. A number of astronauts with and without corrective lens, have suffered visual disturbances from changes in the eyes' global geometry inflight. This is now a major concern and effort in the NASA life sciences community that will be covered in Chap. 5.

Fig. 1.12 After short flight. On return after flights as short as 3–4 days, effects of the rapid inflight adaptations in the nervous and cardiovascular systems will be felt. This includes a rapidly clearing sense of increased weight of self and external objects with loss of strength, and some locomotor instability. Most individuals will have some orthostatic intolerance on quiet standing and especially on first leaving their seats post landing. These symptoms will be somewhat more prevalent and severe with flights of up to 2 weeks. Postflight symptoms rapidly improve, usually in hours with adequate rehydration. Even after a week there may be reduced exercise tolerance that will increase with flight duration unless inflight exercise is used

Additional Background

The following are more or less standard sources on changes seen in- and postflight.

- Biomedical results of Apollo, NASA SP-368. Washington, DC: U.S. Government Printing Office; 1975.
- Biomedical results from Skylab, NASA SP–377. Washington, DC: National Aeronautics and Space Administration; 1977.
- Buckey JC. Space physiology. USA: Oxford University Press; 2006.
- Evidence report, risk of spaceflight-induced intracranial hypertension and vision alterations. In: Human research program health countermeasures element. National Aeronautics and Space Administration; 2012.
- Extended duration orbital medical project final report 1989–1995, NASA SP–1999-534. Houston: NASA Johnson Space Center; 1999.

Months to years

- Transient weakness, sensation of increased weight – somatosensory adaptation
- Orthostasis, faintenss with increased heart rate, decreased blood pressure dehydration, reduced blood volumd from fluid shift, loss
- Locomotor instability – neurovestibular and somatosensory adaptation
- Reduced strength and muscle mass, especially in legs; disuse atrophy
- Reduced exercise capacity caused by reduced capacity of heart, pulmonary system – secondary to disuse; dehydration and anemia from fluid shift, loss
- Reduced bone mass especially in feet, legs, pelvis, and spine – reduced muscle loads
- Anemia; approximately 10% loss of red cells – fluid shift, loss
- Some individuals had motion sickness for more than one to two days on return – neurovestibular adaptation
- Frequent development of in-flight visual changes in week to months, which resolve postflight. A small number of males have developed permanent visual and ocular changes after flights of six months or more.

Fig. 1.13 After long flight. Transient weakness and perceived heaviness has now merged with weakness from loss of muscle, especially in the legs and back. Muscle loss is very dependent on the individual amount, level and type of exercise countermeasures. Without proper locomotor exercise few individuals would be able to walk after 60–70 days inflight. Endurance exercise capacity will also be a function of inflight exercise and mission duration. As flight duration increases a small but increased number will experience vertigo and postural instability for up to a week post landing. Laboratory studies regularly reveal anemia, that reaches a stable maximum of approximately 10% of blood volume after 30–60 days inflight. This is rapidly reversed after return to Earth. More or less loss of bone mass is also regularly seen in leg, pelvis and lower spine with smaller losses in the arms and shoulders. The degree of loss depends on the individual, inflight exercise, and flight duration. Most or all of the loss is replaced in months. Eleven individuals who were in orbit greater than 6 months developed changes in the optic nerve and related regions. This is under intense study

- Gemini mid-program conference, NASA SP-121. Washington, DC: National Aeronautics and Space Administration; 1966.
- Skylab medical experiments altitude test (SMEAT), NASA TMX-58115. NASA Lyndon B. Houston: Johnson Space Center; 1973.
- Effects of microgravity. In: Space biology and medicine Vol III: Humans in spaceflight, Book I, AIAA Inc., Reston VA; 1993.

References

1. Halliday D, Resnick R. Fundamentals of physics. New York: John Wiley & Sons; 1981.
2. Serway RA, Moses CJ, Moyer CA. Modern physics. Cengage learning; 2004.
3. Shubin N. Your inner fish: a journey into the 3.5-billion-year history of the human body. New York: Vintage; 2008.
4. Shubin N. The universe within: the deep history of the human body. New York: Vintage Books; 2013.

Chapter 2
Basic Mechanisms

2.1 Introduction

Even before sending humans into space much thought went into how the human body is affected by weightlessness. Research on how weightlessness affected living bodies was first conducted with animals. Additional lessons have been learned during the first 50+ years of human spaceflight. However, a considerable amount of uncertainty still exists about the effects of spaceflight (and weightlessness) on the human body. These effects range from a transient complex of gastrointestinal and central nervous system symptoms through anemia to loss of weight, bone and muscle. More recently, the effects of weightlessness on the eyes and vision have become an area of intense study. Even more uncertainty exists regarding the origin of some of these effects.

If the effects of spaceflight on the human body are approached in the context of human development, a simple, comprehensive pattern becomes apparent (Fig. 2.1). Weightlessness effects should be studied in the context of the basic mechanism of *adaptation*, changes in an animal's body to better fit it for survival in its environment [1]. There are two types of adaptation that pertain to this approach, *species* (or permanent) and *somatic* (or temporary).[1] Species adaptation is the process that was described by Darwin in his theory of evolution [2]. Our ancestral fish were species adapted to the absence of body weight that was caused by buoyancy of the seawater in which they lived. When their descendants evolved and left the seas for dry land, species adaptations allowed them to accommodate weight, and over the next 375 million years, the unique form and function of our bodies was shaped by constant interaction with the forces of weight [3].

[1] *Permanent* only in the sense that untold numbers of generations are required to change them. *Temporary* in that they are limited to the individual and not encoded and passed on to the next generation.

W. Thornton, F. Bonato, *The Human Body and Weightlessness*,
DOI 10.1007/978-3-319-32829-4_2

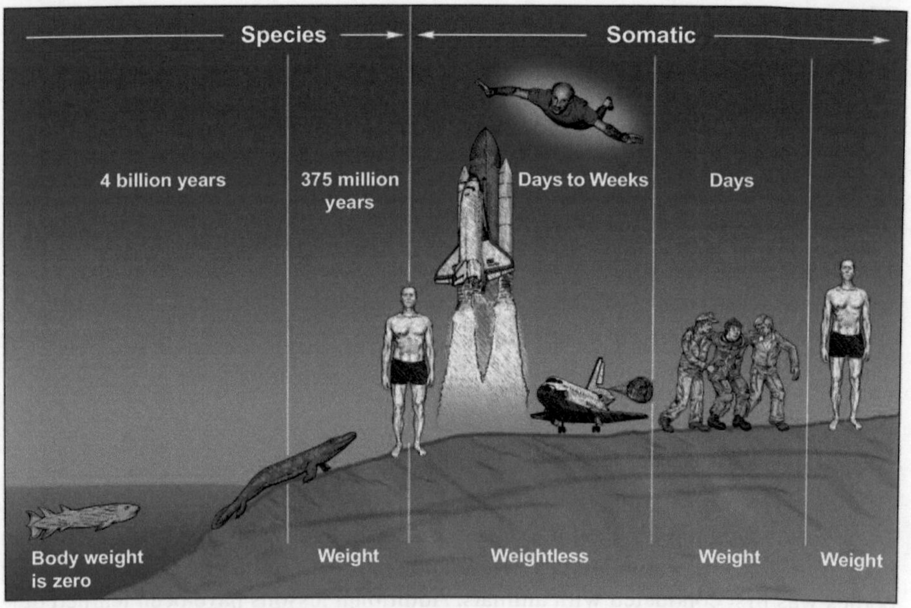

Fig. 2.1 Adaptation. This is a pictorial representation of the significant phases in the long history that precedes, and determines, an astronaut's responses to weightlessness. A surprising amount of every human's DNA is inherited from the ancient fish that left the buoyancy of seawater for the weight of dry land [1]. Hundreds of millions of years adapting to the body weight produced a great deal of additional DNA that is now part of the human genetic makeup. This body inherited the capacity to make temporary changes called "somatic" adaptations to cope with changing environmental stressors. Weightlessness produces a number of such changes that compromise our fixed adaptations to weight. These compromises become obvious on return to Earth. Depending on flight duration, they are erased in days to months after return to the familiar stresses of weight—until the next flight. Some changes can be permanent. It was arguable whether a subject as arcane as species adaptation belonged in a book dedicated to operations in space. However, this process determined the human body's response to weightlessness by producing another mode of adaptation to improve survival. Identified as somatic adaptation, this process allows rapid changes in the body to meet changing stresses. Species adaptation provides a rationale for changes occurring primarily in some systems, the systems that are species adapted to weight. When our bodies are returned to an environment with effects similar to the ancient seas, it is the anatomical and physiological features that were added to cope with weight, especially the legs, that suffer most. "Bird legs of space" were noted by astronauts as soon as they could remove their space suits and noticed their legs' reduced girth. It is no accident that underwater simulations have become an essential component in the space program. For these reasons and more, species adaptation is briefly described as an introduction to the changes that have to be understood and dealt with

When weight forces are removed in spaceflight, leaving many of our adaptations to weight almost useless, the body adopts a series of somatic or temporary adaptations to weightlessness that compromise our species adaptations to weight. These compromises are immediately felt on attempting to stand after landing back on Earth and become more obvious as a return to daily life and its activities with the normal gravity that is experienced on our planet. Over the next minutes to days post landing—or sometimes more, the stresses of weight produce re-adaptations to weight.

The first step in understanding this sequence of events is an appreciation of the adaptation processes in general that are introduced in this chapter. Innumerable species adaptations over 375 million years transformed our ancestral bodies into what they are today. This chapter defines species adaptations to weight and describes the process of somatic adaptation in general. Rules of somatic adaptation, as we understand them, are given and examples of both are provided.

2.2 Species Adaptation

Species, or permanent adaptation, has resulted in all the species of life on Earth. The characteristics of these species encompass a vast array of sizes, shapes and capabilities. At its simplest species adaptation is a complex process and is still poorly understood. It has entailed untold millions of generations. Small inheritable changes of benefit to the organism are incorporated into the DNA of successive generations by natural selection. These benefits increase survivability of the organism in its environment. Our bodies have untold numbers of such changes to better suit them to life in 1-g and the forces of weight. Figure 2.2 is an example of this process. For purposes here, we do not need to know details of this process. However, it should be noted that benefits in one environment may be, and often are, liabilities in other environments. For our purposes, benefits that serve us well in Earth's 1-g environment often will not serve humans well in weightlessness.

All parts of our bodies have been affected by weight and hence can be affected by weightlessness, but the following systems are the ones primarily affected

Skeletal
Muscular
Nervous
Cardiovascular
Metabolic (indirectly)

While we do not need to know details of these species adaptations we need to understand one important result. *All our systems have an inherited limited range of performance and maximum performance within this range is determined by the usual loads the system is subjected to; e.g.,* bone strength depends on its usual maximum force loading. This is *somatic adaptation* in response to tons of strength adapted muscle forces.

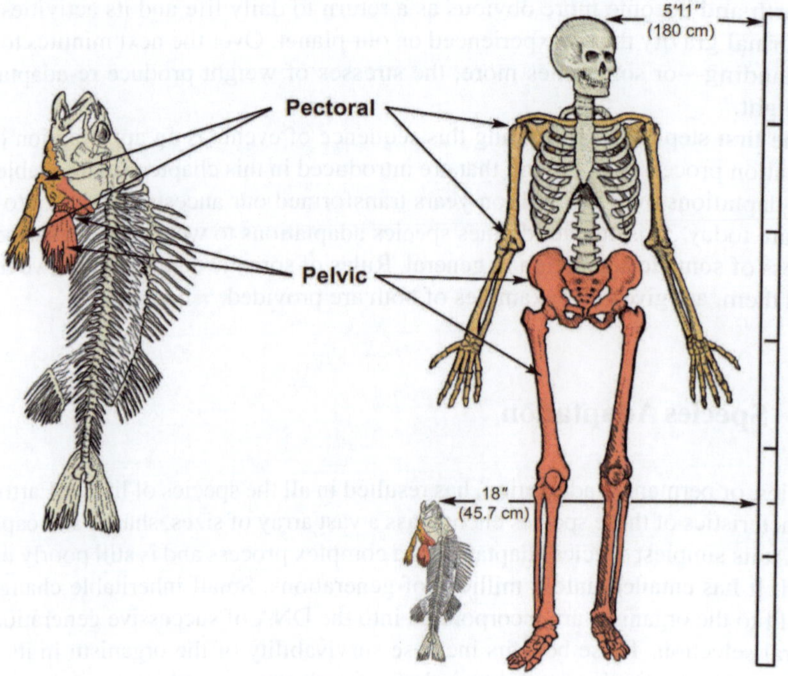

Fig. 2.2 Species Adaptation. An example of conversion of small body parts into major components. Small bones adequate for propulsion in water were transformed by adaptation into the major bones of the body necessary for locomotion under 1-g. N.B. Each of us has undergone a remarkably similar process during our early embryonic life, in days vs. hundreds of millions of years [3]

2.3 Somatic Adaptation

Somatic adaptation[2] [3], is the process by which an individual's body and physiology change for its benefit, usually in response to a changed environment (Fig. 2.3). This concept is key to understanding life in weightlessness. All the major bodily changes we see in and after spaceflight are somatic adaptations to weightlessness and re-adaptation to weight upon returning to Earth.[3] While such changes are very specific and will be discussed individually, there are some common features of somatic adaptations that will now be examined as a general process.

"Feedback" control is essential to somatic adaptation and is a principle that will be encountered repeatedly in the following chapters. An imaginary mechanical example is shown to illustrate feedback control and it is followed by a more realistic

[2] Unless noted, when used here adaptation is somatic.

[3] In space, the body adapts to meet its needs in weightlessness without concern for returning to 1-g where it will readapt.

Fig. 2.3 Somatic adaptation. This large Sumo wrestler has spent much of his life in training. Adaptation to the stresses of this training has produced a body mass approaching a quarter ton, and muscles that can produce the thousands of pounds of internal force that can move his body with remarkable agility. His skeleton has bones that safely support the several tons of leveraged forces frequently imposed on them (see Chap. 7)

example from life. These principles are employed in countless ways throughout the body. Simple feedback control loops are everywhere in living, mechanical and electrical processes. James Watts' flyball governor illustrates the principles of such a loop, and the need for adaptive feedback (Fig. 2.4) [4].

A steam engine's power and speed depends on its steam input, controlled by a valve as shown in Fig. 2.4a. A "governor" consisting of rotating weights is spun by connection to the engines' output. It is connected to the engine's inlet control valve such that as the engine speed increases or decreases, the balls' centrifugal force opens or closes the valve to produce whatever power the load requires to maintain a constant speed. There are untold numbers of such basic control loops in our bodies, from the cellular to system level, as we will see when systems are examined in more detail. This is *not* adaptation and will be called accommodation here (see Chap. 10).

Now, take the same engine and add loads until it runs almost constantly at maximum power. It will wear out or fail prematurely. Conversely, if loads are usually small, efficiency will be low. At this time for purposes of illustration, we do something mechanically impossible (Fig. 2.4b). Take some point of stress in the engine, such as inlet steam pressure, and input this signal to a process that evaluates both pressure and pressure duration and then increases or decreases size and power of the engine to meet its load requirement efficiently. This is adaptation—not currently practical in our technology—but a process that is very much a part of biology. Somatic adaptation has been essential in animals since life began and it is possible in life many cells are constantly renewed, usually by replacement of their proteins.

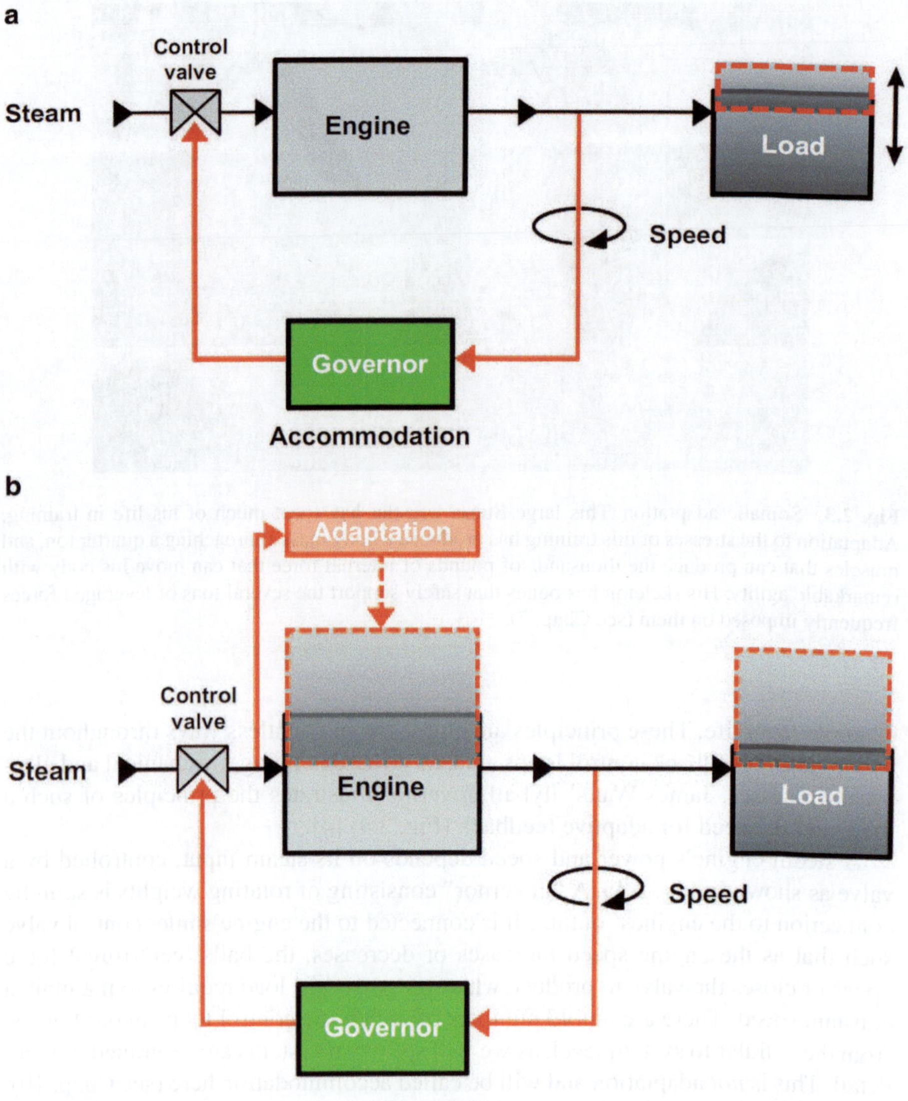

Fig. 2.4 (**a**) Illustration of a mechanical feedback control loop. A steam engine is driving a load that varies (*dotted lines*), and will vary engine speed if steam input is constant. A centrifugal governor, driven by the engine and connected to the steam inlet valve, increases steam and engine power when load is increased and speed slows. When speed increases, steam is reduced resulting in decreasing engine power and speed. Increasing or decreasing steam input and power output through a speed control feedback loop thus maintains adequate speed and power. (**b**) Adaptive feedback. The same engine and control in (**a**) now has a second more complex feedback loop shown for illustration of somatic adaptation. This loop can theoretically increase or decrease the size and power of the engine. Its input is steam pressure and time. If loads frequently cause near maximum pressures, then size and power of the engine are increased optimizing life and efficiency

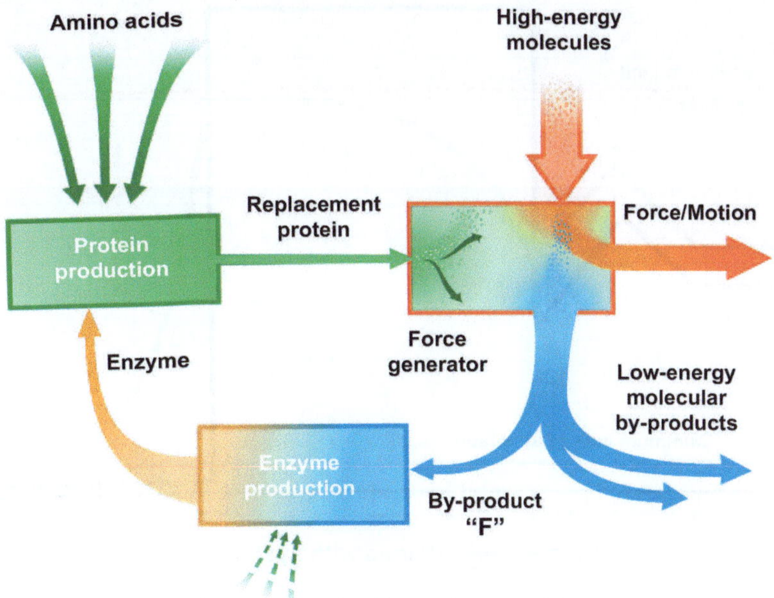

Fig. 2.5 Illustration of adaptation in a muscle cell. Renewal rate of the replacement proteins determine size and strength of the force generator. Replacement rate is in turn determined by the usual maximum force loads on this muscle cell through the force byproduct/enzyme production feedback loop

Our muscle proteins are replaced in weeks, all our red blood cells in 120 days, and our bones in years.

Figure 2.5 is an example of one such adaptation illustrating strength adaptation in a muscle cell. Chemical energy is released from carrier molecules by reaction with other molecules in the organic 'machinery' of the muscle that transforms this energy into force and motion when required. This machinery is continually renewed by removal of its old protein components and replacement of them with new protein from amino acids produced by sites in the muscle cell. The rate at which new proteins are produced and installed in the muscle determine its size and strength. This rate of production depends on trace amounts of enzymes that facilitate the amino acid reactions to produce proteins.

Mixes of chemical byproducts are produced in the generation of force. The amount of some of these byproducts depends on the level of force generated. One or more of such byproducts then participate in enzyme production, increasing it with increasing force levels. This in turn increases muscle protein production and capacity of a muscle to produce force, it becomes stronger, and coincidentally larger. Conversely, if force demands and production is decreased, the process is reversed.

Demands for muscular force can vary from those required of an accountant to those of a piano mover. Also, an individual may change his force needs, the accountant may begin weight training. Such adaptations can, within limits, accommodate

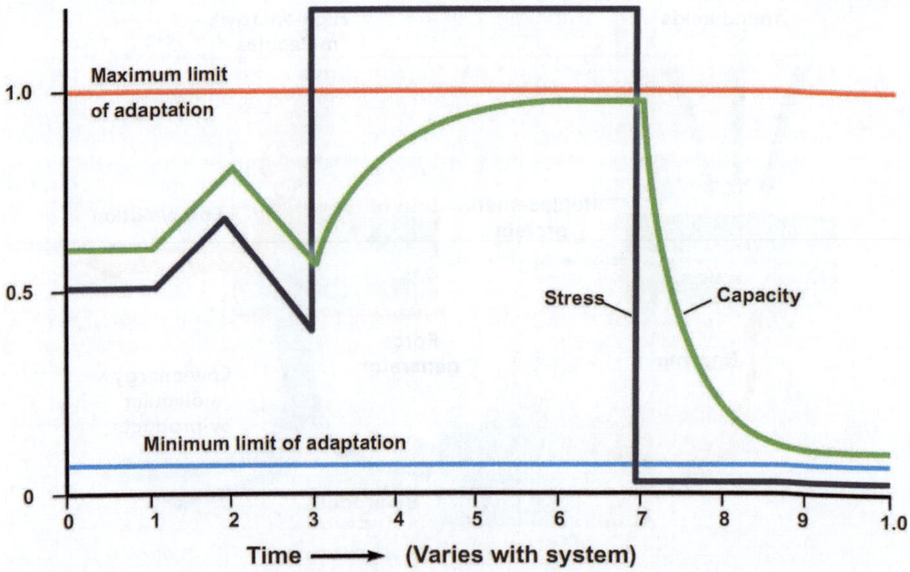

Fig. 2.6 Graphic depiction of characteristics and limits of somatic adaptation. There are absolute maximum and minimum adaptation limits on every organ and system for every individual: heart, lungs, muscle, glands, bones, etc. (Individual maximum and minimum limits are first determined by genetics and altered by age and health.) Examples of maximum limits can be seen in Olympic athletes. Minimum limits may be found in a person in an extended coma. Capacity, *green*, which is determined by stress, *black*, normally varies over a limited range. There is a time lag between the change in stress and system response, the response time. It appears to be longer with increases than in losses of capacity. Although poorly quantitated, we know there are large differences between systems (Fig. 2.7). The nervous system has components that can adapt in minutes, while changes in bone can take years or more. Time profiles of adaptations are generally exponential and of the form $\mathbf{e}^{-t/T}$. The stress profile, in *black*, is illustrative only to show a systems change with time

such changed demands. Analogous processes are present in all the cells of the body. The significance of this general process is that the body has the means to adjust (within limits) the capacity of the body to meet a wide variety and range of stresses.

There are absolute limits con and rules of operation in this system (Fig. 2.6) including:

1. **Maximum capacity limits** are inherent in all adaptive systems. These limits are genetic but may be modified by age and health.
2. **Minimum capacity limits** are also genetic. Adaptation is biphasic and can decrease capacity, but not to zero. So long as tissue is present, some capacity remains even if all stress has been removed.
3. **The level of adaptation** is proportional to the usual maximum stress level. Experience gained during the Skylab missions demonstrated a loss of leg muscle with the inadequate force loads of cycle ergometry. Conversely, maintenance of arm strength was achieved with added exercise [5].

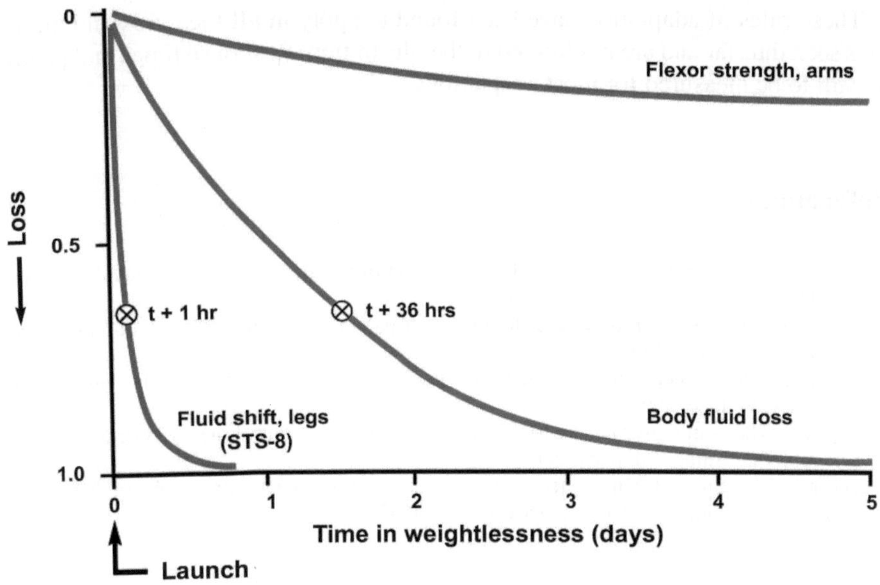

Fig. 2.7 Profiles and response times of three adaptations. The most rapid time shown here is shifting of fluid from legs into the body [5]. This process has a time constant, T, the time for 63% of the change to occur, of 1 h. These measurements, made by one of the authors (WT) during a Space Shuttle flight (STS-8), match the proposed exponential response within a few percent. Loss of this fluid from the body typically requires 3–5 days. In this case, the time constant is approximately 36 h. Adaptation of muscle is much slower and has never been properly measured. The portion shown here is extrapolated from Skylab measurements with an estimated time constant of weeks [6]. Individual response times vary within large limits. This illustrates change of two capacities with time, which have been normalized to the fraction of total change, 1.0, with time

4. **Specificity of response** is inherent in the process. In the example shown, force-capacity strength is based on a unique byproduct of force, so force is the unique stimulus of strength. If some other stimulus such as velocity of motion or number of repetitions is applied to the muscle, unless there is also an increase in force loading, there will be no response in force capacity; i.e., in the strength of the muscle involved. A common mistake made in counter measures is use of an inappropriate stimulus, especially in exercise.
5. **Response time** refers to the time it takes a system to adapt. Response time varies from system to system. Response time can be hours in some areas of the nervous system or years in some areas of the skeleton. It also depends on whether capacity is being increased or decreased.
6. **Profile of response time** appears to be exponential. Until adequate data to the contrary is obtained, it is assumed to be $\mathbf{e}^{-t/T}$ where t is time and **T** is a system constant in time units. Engineering convention is to define response time as the point at which **t = k, e**$^{-1}$. At this point there has been a gain or loss of 67% of a system's capacity through adaptation. Figure 2.7 lists estimates of response times for various systems.

These rules of adaptation have been found to apply in all the cases and all systems seen thus far and are not limited to the illustration. Response times and profiles are still to be measured for most adaptations.

References

1. Kirschner MW, Gerhart JC, Norton J. The plausibility of life: resolving Darwin's dilemma. New Haven: Yale University Press; 2006.
2. Darwin C. On the origin of species by means of natural selection. 1859 ed. Joseph Carroll. Toronto: Broadview; 2003.
3. Shubin N. Your inner fish: a journey into the 3.5-billion-year history of the human body. New York: Vintage; 2008.
4. Lenon G. The Watt Governor. In: Mathematical problems in control theory. Singapore: World Scientific Publishing Company; 2001.
5. Thornton W. Rummel J. Muscle deconditioning and its prevention in spaceflight. In: Biomedical results from Skylab, pp. 191–7, NASA SP-377; 1977.

Chapter 3
Space Motion Sickness and Vestibular Adaptation to Weightlessness

3.1 Introduction

Emesis, or vomiting is a debasing and debilitating act, especially for those with claims to the "right stuff." It is bad enough to have to reach for the airsick bag in the rear seat of a gyrating fighter jet, but to fill a spacecraft's motionless[1] flight- or mid-deck with floating globules of half digested food and noxious odors is worse. Yet this often happened, usually in the first few hours of spaceflight, even to commanders, and often to those who were resistant to sickness in aircraft and other vehicles. Vomiting during spaceflight usually happens without warning and often too suddenly to get a bag up to one's mouth in time.[2] Emesis is only the beginning of what is known as *space motion sickness* (SMS). It is followed by increasing malaise, loss of appetite, lethargy, loss of interest and drive, with somnolence and variable headache. Body and head motions, especially pitching motions, are unpleasant and, if persisted in, can produce nausea, prolonged vomiting and other symptoms of motion sickness (MS). Motivated and trained crewmembers can work effectively during this period, albeit with varying amounts of discomfort. All of these symptoms typically dissipate within 2 days, and do not recur in flight. It is not surprising that early astronauts were not eager to report SMS, especially since nothing could be done about it except make it a part of their personal records. It was referred to only half jokingly as the "dreaded space motion sickness." Somewhat less than half of first time space flyers will experience SMS until vestibular adaptation removes the cause.

Everyone exposed to weightlessness undergoes adaptation of the vestibular system, whether they perceive it or not. This adaptation will sometimes result in symptoms upon returning to Earth. On return to g loads, some who may or may not have

[1] There is no local motion, such as in ships and aircraft on earth, to produce motion sickness.

[2] This is projectile vomiting—sudden, without nausea or retching, and emptying the stomach in one or two heaves.

W. Thornton, F. Bonato, *The Human Body and Weightlessness*,
DOI 10.1007/978-3-319-32829-4_3

experienced SMS, will experience vestibular symptoms because of their adaptation to weightlessness. Many of these symptoms are similar or the same as those experienced during SMS and can include disorientation, ataxia, nausea and vomiting. Symptoms may continue until readaptation to 1 g is complete.

After half a century of human spaceflight SMS remains an enigma and is often misunderstood, without reliable prevention, predictability, or cure, it remains a frustration to astronauts, mission planners and controllers. This is true in spite of SMS being one of the most talked about and most studied problems associated with weightlessness. With few exceptions, all work on SMS to date is based on the premise that it is a form of MS and that motion is a necessary cause of SMS [1]. Pre- and inflight adaptive efforts, some of them mandatory and most of them onerous, have been for the most part operational failures. Two medications have been shown to be somewhat effective. One medication that is used for in motion sickness, can delay SMS onset for those who are susceptible. The other medication has been moderately successful in relieving symptoms when administered via intramuscular injection.

If the SMS problem is approached as the product of evolutionary and adaptive processes, and the sensory functions understood in quantitative biophysical terms some understanding of why SMS results may be gained [2]. The following text pursues this course and the authors propose a hypothesis as to what causes SMS.

3.2 Brief History

The first reported symptoms consistent with SMS occurred in 1961 on the second orbit of the fourth manned space flight. Cosmonaut Gherman Titov aboard the Vostok 2 reported disorientation and other symptoms that some experts in MS had predicted. This set of symptoms was referred to as Space Motion Sickness (SMS). In 1968 during the Apollo 8 mission Commander Frank Borman reportedly experienced nausea and vomiting, perhaps the result of SMS. During the Apollo 9 mission in 1969 Astronaut Rusty Schweickart terminated an extra vehicular activity (EVA) early presumably because of SMS symptoms. It is possible that SMS was experienced even earlier on Apollo 7 but there are no easily obtained official reports of SMS symptoms at that time. Investigators have reported six mild, two moderate and three severe cases of SMS in 33 Apollo crewmen. Inflight adaption procedures were counterproductive and several anti-motion sickness drugs produced varying results [3]. A large study of SMS on Skylab used a rotating chair in three experiments to replicate 1 g MS studies, including provocative head motions [4]. These experiments demonstrated that by the fifth day of the mission all crewmen were remarkably resistant to MS. Extensive 1 g studies on Earth and simulations of SMS were also produced by inflight reports. These studies generated numbers of papers, conventions, and new theories, plus numerous plans and scheduled complex studies with large experimental hardware for the Shuttle program.

During the Shuttle era, early unverified reports of SMS allowed media and investigators to paint pictures of a Shuttle being flown into the ground or worse by an astronaut still suffering effects of SMS. With the support of SL-4 Commander Ken

Mattingly, one author (WT) was able to obtain electrooculograms (EOGs) of individuals with and without SMS using existing hardware on Shuttle without waiting 2 years for planned studies and complex hardware. It was rapidly shown that trained astronauts could perform demanding tasks in the midst of SMS. Over the next 2 years it was shown that SMS and MS are different processes. Over the life of the Shuttle program there were many and varied studies, a few more unproven hypotheses and continued pursuit of MS as the problem. Inter-muscularly injected Phenergan proved to be efficacious for some individuals.

3.3 Background

To understand SMS we must first accept that in order to move effectively humans (and other animals) must be oriented to the surrounding world and especially to the gravity vector. Large and complex portions of our nervous system are devoted to this orientation. Inputs from sight, hearing, sense of touch, pressure and force are coordinated with signals from gravitoinertial sensors in each inner ear. These inner ear sense organs provide primary information to the *vestibular system* on body orientation and self-motion. They are both sensitive and fragile, and in spite of being in the body's most protected location, are sometime subject to injury. They are also subject to organic disturbance from infections (especially viral), minor anatomic failures, chemicals (especially alcohol, drugs, illicit and therapeutic) and fluid problems. Alan Shepard's and John Glenn's careers were disrupted by vestibular disorders. Many other astronauts have been temporarily grounded for one of the more minor vestibular organ problems.

Relatively minor disturbances in these sensors can produce error signals large enough to cause disorientation and incoordination possibly resulting in falls and serious injury should the affected person continue his or her usual activity. Fortunately we inherited a safety net from a species adaptation, along with our vestibular organs, from our ancient ancestral fish. Vestibular organs of humans so closely match those of ancient fish that species of ancient fish are flown for study. An ancient fish with a vestibular disturbance that affected its motion had a short life expectancy. Species survival was improved when the vestibular system was connected to the gastrointestinal tract, which already had an unlikely but effective failsafe system, the *emesis system*, which controls nausea and vomiting and their side effects.

However, when humans became sailors, boat motions caused sensory conflicts, which also triggered the emesis system to produce nausea and vomiting. Seasickness was probably the first in a series of motion sicknesses perhaps first coming about more than 4500 years ago with the invention of the first watercraft. As other modes of transport were developed, carsickness and airsickness appeared. More recently *simulator sickness* and *cybersickness*, associated with visually simulated self-motion was added. These phenomena have been extensively investigated in theory and practice, especially during wartime. When spaceflight and weightlessness became possible, experts in the field of MS made many predictions about the effects of space flight.

Titov's reported vomiting on the second orbit of his 1961 spaceflight [5] triggered a rush of activity in this area. Russian and U.S investigators concluded that symptoms reported were caused by MS, named it space motion sickness, and applied all that had been learned in more than a half century of work on MS to the problem. Preflight MS adaptation "trainers" and in flight adaptation "exercises" were developed and made part of crew training and operations. The Russian trainer produced violent motions and MS, and remains a part of their program [6]. The use of Scop-Dex, a combined drug (Scopolamine and Dexedrine) effective for some MS, was once encouraged for U.S. space crews. NASA's first admission of SMS occurred on Apollo 9 when Schweickart's EVA was shortened because of adverse symptoms [7].

On Skylab a large number of psychophysiological vestibular studies were conducted during each mission. These studies included the rotating chair and head motions to induce MS [8]. Significant findings included the sudden development of immunity to MS after 2–3 days in weightlessness with rapid recovery from SMS at the same time. There was also anecdotal evidence of significant differences in the symptoms of SMS versus MS. It was at about this time that neuroscientists produced definitive 1 g laboratory studies of the vestibular sensors and apparatus adequate to understand the response to weight and weightlessness [9].

The Shuttle program initially continued the Apollo/Skylab approach to SMS. After several significant cases of SMS and some off nominal entry/landing incidents a rapid response objective inflight project was initiated and directed by the astronaut office with the support of the Johnson Space Center and NASA Headquarters life sciences organization. An initial goal of the study was to determine whether SMS was a functional disorder induced by weightlessness or a malfunction produced by increased intracranial pressure from shifts of fluid [10]. A variety of physical and electrophysiologic studies found no evidence of organic malfunction. A number of functional anomalies were found, including loss of gastrointestinal motility, which explained the characteristic vomiting of SMS. Hypersensitivity to pitching motions was also found. This would explain the more usual signs of MS such as persistent nausea and vomiting sometimes seen in SMS. It was also demonstrated that motivated, trained individuals could adequately perform planned tasks during SMS.

Spacelab became operational in 1985 allowing for large, elaborate studies with equipment long planned for by established investigators. These studies included those conducted during the Neurolab mission which was dedicated to the study of animal and human neurology in weightlessness. Results of this work are published, but they are usually complex [11]. Clément and Reschke have provided a good summary of them [12]. None of several new preflight adaptation protocols, simulation systems, or inflight counter measures were effective. MS remains the most widely accepted theoretical basis for SMS today [1], despite its inability to account for a good portion of observed SMS phenomena.

However, applying an evolutionary and biophysical approach to the available knowledge and data concerning SMS allows one to reasonably conclude that:

- Weightlessness induces anomalous static and dynamic signals from the gravitoinertial vestibular organs.
- These anomalous signals produce somatic adaptations in the vestibular system.

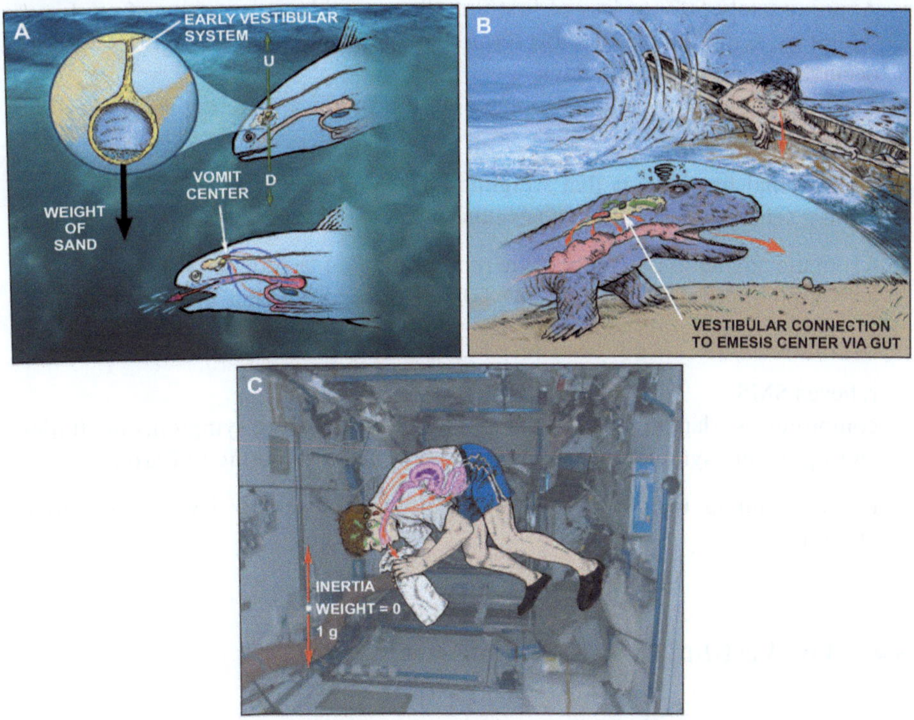

Fig. 3.1 Origin of MS and SMS. This figure illustrates the pre-historic origins of the primary component the GI (gastorintestinal) component of both MS and SMS, and its human effects. (**a**) Motion sickness and SMS begin with two of the oldest and most essential systems in any mobile animal, orientation and digestion. In the upper fish there is an antecedent of the otolithic vestibular system. Grains of sand can move freely in a fluid-filled sphere that has inner walls covered with pressure sensitive neurons. The weight and location of the sand are detected by these neurons to provide orientation to the gravito-inertial vector (U–D). Digestion, shown in the lower fish (**a**), is normally a progression of food down the digestive tract shown in *red*. An added safety feature evolved whereby neurons in a vomit center detect any noxious or dangerous substances that when ingested cause distension or other disturbances (*blue arrows*) and then cause rapid ejection of the stomach contents by retrograde gastrointestinal action by motor neurons (*red arrows*). With continued species adaptation the orientation system became the more effective vestibular system but is subject to damage that causes disorientation. At some time a connection was made to the GI system such that when vestibular signals were dangerously scrambled by injury or disease, the vomiting center was activated causing the animal to remain inactive until the vestibular system recovered (animal in lower **b**). This connection increased species survival and remains a species adaptation connecting our much more complex human vestibular and vomiting systems. Human development of transport with artificial environments also scrambled vestibular signals triggering the vomiting systems as depicted in the upper portion of (**b**). This produces the family of MSes including sea, train, car, air and simulator motion sickness. On removal of weight, (**c**), a portion of the vestibular sensors generate a completely scrambled signal that indirectly triggers the primitive vomiting first seen in the fish in (**a**), plus other symptoms we know as SMS. It also produces hypersensitivity to motion that can produce symptoms of MS

- Approximately 50% or less of first time flyers will have a variety of unpleasant inhibitory symptoms until short term somatic adaptation to weightlessness has occurred, usually within 2 days.
- Static and dynamic vestibular organ conflicts produce different responses from the emesis system.

The rational for SMS is that weightlessness produces two major errors in the otolithic g sensors that initiate two events[3]:

1. a somatic adaptation by neurologic reprogramming to relieve the errors;
2. activation of the emesis system in <50% of fliers, producing SMS symptoms.

Reprogramming

1. relieves SMS;
2. compromises adaptation to gravity and g loads, producing symptoms in all fliers and significant symptoms in a few individuals upon returning to Earth.

Reprogramming that occurs with continued exposure to 1-g reverses these compromises.

3.4 The Vestibular System and Orientation

In addition to spaces in the inner ear, the human vestibular system physically occupies several areas of the brain. It provides essential coordination and control of numerous functions including eye motion and muscle function and coordination. It is the primary source of information for head and body orientation and motion.

A fundamental head-tail axis develops after a few divisions of the fertilized human egg cell. Virtually all features and functions of the body are subsequently built on and organized around this axis. Any effective interface with the external world is dependent on orientation of this body axis with the world, which in turn, is organized by and around the gravity vector (up and down) (Fig. 3.2). Several inputs to the vestibular system provide indirect information about orientation to the external axes including vision and body senses of weight, force and pressure. Dominating these inputs is that of the vestibular organ sensors in the inner ear that directly sense orientation of the body's axis relative to the gravity vector.

This vestibular information is so crucial that when it is distorted as in injury or illness, a type of "fail safe" system triggers various pathways to the gastrointestinal (GI) and other systems to inhibit the affected individual's activity until recovery or reprogramming occurs (Fig. 3.1b). In weightlessness these gravity dependent sensors become non-functional, generate erroneous and conflicting signals, and trigger

[3] The otolithic sensor is the portion of the bilateral inner ear vestibular organs that detect gravity and linear and angular accelerations.

Fig. 3.2 An extreme example of orientation demands on the vestibular system. In spite of a rapid divergence between body axis and the g vector, multiple adjustments are being made to maintain a complex interface with the gravity dominated environment. Muscles are extending the neck to provide a stable platform for the head's sensors. Ocular muscles have rolled and turned the eyes to precisely track the thrown ball's target. Corrections have been made to the throw for body angle and motion at release. The right thigh is being flexed for the impending ground impact. All of this is dependent on the vestibular system and its motion and gravity sensors. Our concern is with a small but crucial portion of this system when it is not moving

the failsafe response producing SMS. This chain of events will now be followed through the systems primarily involved, beginning at its source and continuing to its resolution [2].

3.5 Vestibular Sensor Function and Malfunction

SMS is triggered by the nature and arrangement of the sensory cells that detect orientation of our head to Earth's gravity vector, or more simply, to tell the body which way is up or down. Tens of thousands of these microscopic hair cells are present in the vestibular organs of each inner ear, and it is necessary to understand their input/output characteristics to understand SMS (Fig. 3.3).

A series of mechanically connected elastic microtubules or cilia project from the cell body that contains an electrochemical pulse generator whose frequency depends on position of the cilia, which are "spring loaded" to maintain the neutral position shown. Note that a significant signal is present at this position. The cell responds to changes in cilia position only in the axis shown.

Figure 3.4 illustrates the arrangement that makes these cells sensitive to their orientation to a gravity vector. This arrangement is also sensitive to acceleration, but that is not a consideration here. A layer of gel covers the cilia and supports a layer

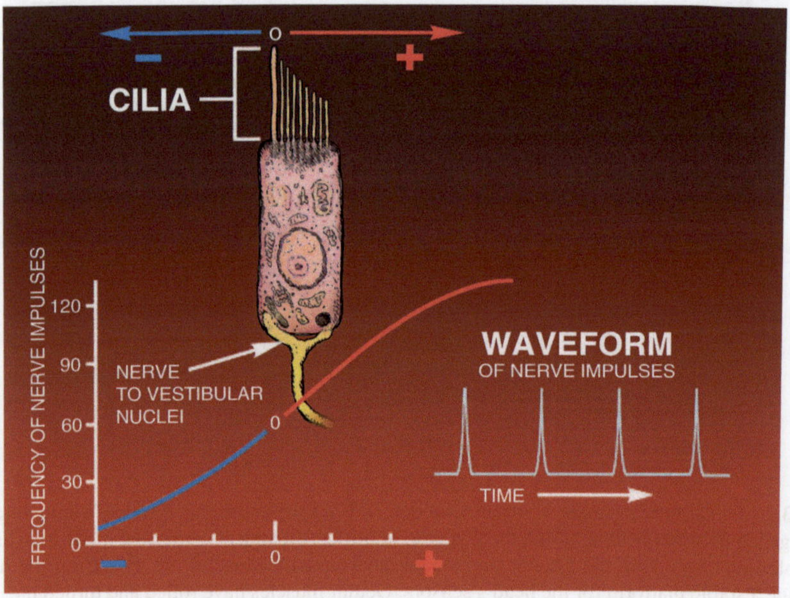

Fig. 3.3 Hair Cells are extraordinary transducers, microscopic in size and sensitive to displacements as small as the diameter of an atom. They are the basis of all the vestibular organs, seen in Fig. 3.5, and are the source of SMS. Hair cells have a variety of characteristics, but this illustration is generic. They send a continuous stream of pulses via their axons to the vestibular system. Frequency of these pulses is determined by position of the interconnected bundle of hairs (cilia). Sensitivity is limited to the axis shown. A significant point in our investigation is that the cilia are elastic and "spring loaded" to maintain the position shown, an important signal position

of *otoconia*, in Greek: ear stones. While this position does not deflect the cilia and the cell body is not sensitive to weight, the cell produces a major signal indicating that it is aligned with the axis of the stones' weight vector. When this arrangement is tilted, weight of the stones shifts the layer of gel and associated cilia. Output frequency of the hair cell is now a function of cell angle to gravity (weight) vector.

Vestibular otolith organs, (Fig. 3.5), located in each inner ear, consist of two tiny curved sheets of hair cells whose cilia support a layer of otoconia. Curvature of the sheets and cell orientations insure that no matter what head position takes place, its tilt angle to the gravity vector will be sensed and sent to the vestibular system regions in the brain. On Earth the thousands of frequency modulated signals from otolith hair cells will be processed by the brain according to their orientation, and all will show a congruent deflection of cilia by weight of their associated otoconia. This provides the necessary orientation of head and body to the gravity vector on Earth, which is typically shown in Cartesian coordinates (Box 3.1).

The trigger for SMS can be easily seen by taking two hair cells from such an array, one oriented horizontally and the other vertically, and placing them first in 1 g and then in weightlessness, (Fig. 3.6). In 1 g the signals are congruent with a single orientation to the gravity vector, which occurs with the head upright. Now place the hair cells and their otoconia in weightlessness. Elasticity of the cilia move them

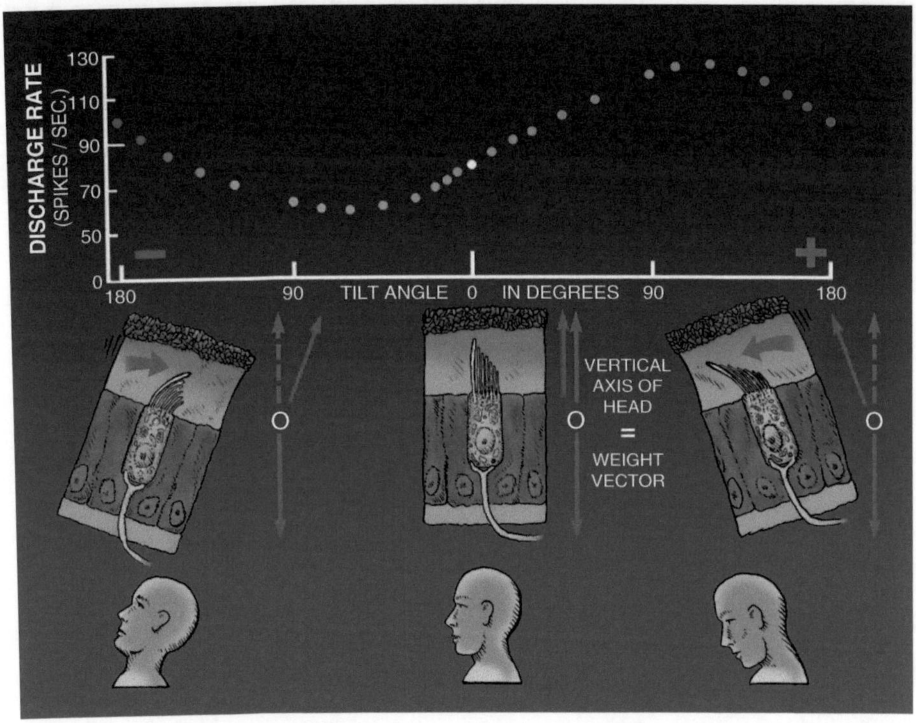

Fig. 3.4 Coupling the cilia of a hair cell to a weight makes it sensitive to gravity as shown. In this arrangement the elastic cilia are embedded in a layer of gel that supports a layer of dense microscopic crystals. When tilted, the weight of the stones displaces the gel and cilia, producing a change in the signal as shown on the graph. Also shown on the graph is a measured plot of a cell's response to rotation in 1 g. It is the expected sinusoid of the vertical component of a weight on a staff rotated around a horizontal axis

Box 3.1 Vestibular Sensor Function and Malfunction

For this to be a practical book instead of a 500 page text, it is necessary to concentrate on only the essentials necessary for understanding the problem. There are two types of hair cells with an amazing range of properties, but a prototypical cell with basic properties is used to illustrate the problem.

and their stones into the neutral position. But now both cells indicate that they are normal to the gravity vector, an impossible situation on Earth with its 1 g environment. In reality all sensors are each indicating a vector according to its own physical orientation, an even more impossible situation. In the absence of weight, all "gravity" transducers have moved to their zero displacement, midrange output positions and will stay there until returned to weight. *This could reasonably be assumed to be the primary cause of SMS and associated MS.*

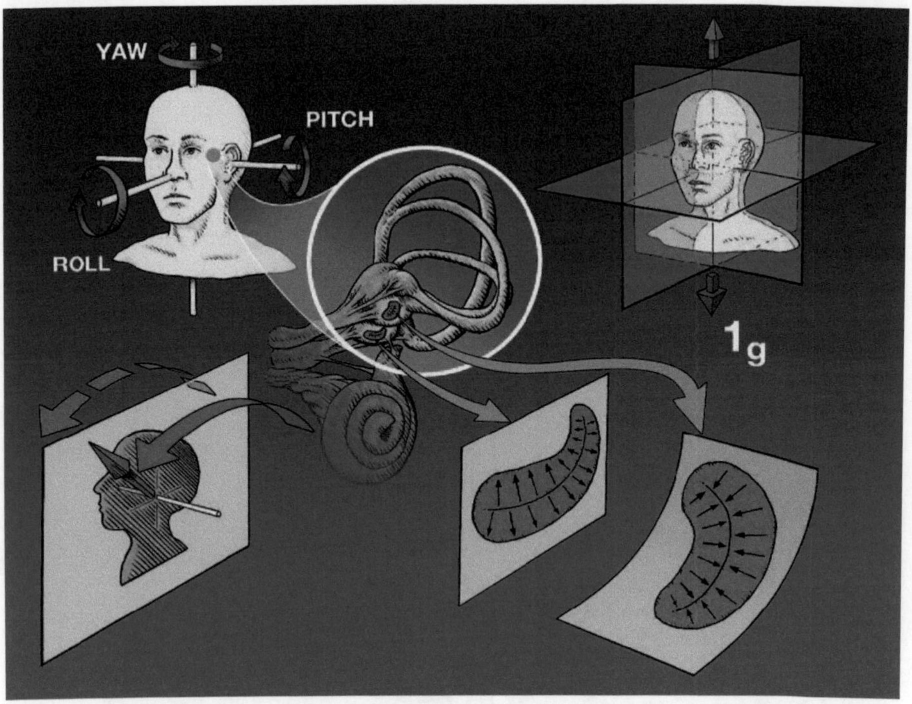

Fig. 3.5 Bilateral vestibular organs in each inner ear consist of the cochlea for hearing (*shaded*), semicircular canals for sensing rapid angular motion (*blue*) and otolith organs for sensing orientation to the 1-g vector and changes in this orientation (*red*). Orientation is sensed by some 20,000 hair cells in two monolayer sheets covered with layers of gel and "ear stones" and attached to the curved walls of two chambers such that every orientation in a 360° sphere is sensed. This is usually arranged in the three-plane coordinate system for discussion. The otolith sensors also detect linear motion, but that is not addressed here. A second vestibular sensor organ, the semicircular canals (*blue*), detects angular motion by means of hair cells and inertia of fluid in the canals during head rotation. Otolith sensors also detect angular motion during such changes (sketch in lower left) and they are combined in the vestibular system. Otolith sensors are weight dependent and disabled by weightlessness. The mechanisms and consequences of this are discussed in the following pages (colour figure online)

3.6 Sensory Conflict

Neurological signals throughout the body are being constantly crosschecked for validity. Crosschecking of the Shuttle's computer signals was a relatively crude process by comparison but analogous. When signals from human sensors disagree, they are said to be in conflict. Such conflicts can generate a new signal to produce protective and corrective actions. While details of sensory conflict are unknown it is a concept that fits many situations, including this one.

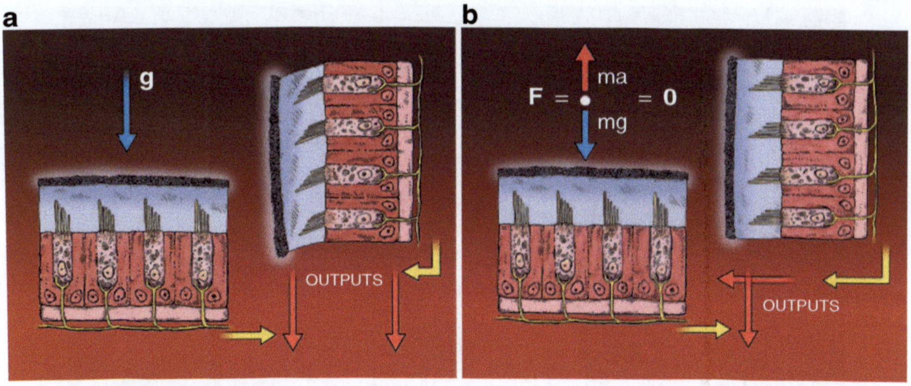

Fig. 3.6 The trigger for SMS can be illustrated with two hair cells in an otolith gravity sensor, one in the horizontal plane and the second in the vertical plane. Outputs from the hair cells go to vestibular nuclei, not shown, which compare their gravimetric signals to their respective mounting axes and produce the resulting vectors, shown in *red*. Part (**a**) shows their response with the head erect in 1 g. Both indicate a vertical downward vector. When the they are in weightlessness, part (**b**), there is no displacement of the cilia that are now in their neutral positions, normal to their respective mounted orientations. Signals now produced by the vestibular nuclei indicate two head positions, an obvious impossibility on Earth and a major conflict

Most of the conflicts postulated to cause MS on Earth in 1 g are between different sensory modalities; e.g., between visual and vestibular sensing. When sensory conflict is referred to in the context of MS, what is typically meant is that an inter-sensory conflict has led to MS. In contrast the conflict just examined is within a single sensory modality, within the otolith gravity sensors. This produces the unique symptoms of primary SMS and what can be referred to as an intra-sensory conflict. We will also examine a conflict between two modalities of sensing angular motions within the vestibular organs, between the semicircular canals and otolith sensors—another intra-sensory conflict. This bimodal conflict produces the symptoms of MS. These documented differences are explained by differences in neurological pathways.

Conflicting signals from the array of gravity sensors trigger a safety shutdown of the subject's normal activities and initiate processes to correct the problem. GI and other protective effects of this sensory conflict will now be examined.

3.7 The Emesis (Vomiting) Center in SMS

The alimentary canal, now our GI system, is one of the oldest systems in the history of animal development. The ability to quickly remove material from this system in retrograde fashion, by vomit, improved the survivability of many species, and still does. As the complexity of the GI system increased through evolution, a neurological center developed to control vomiting: the emesis system. A survival advantage

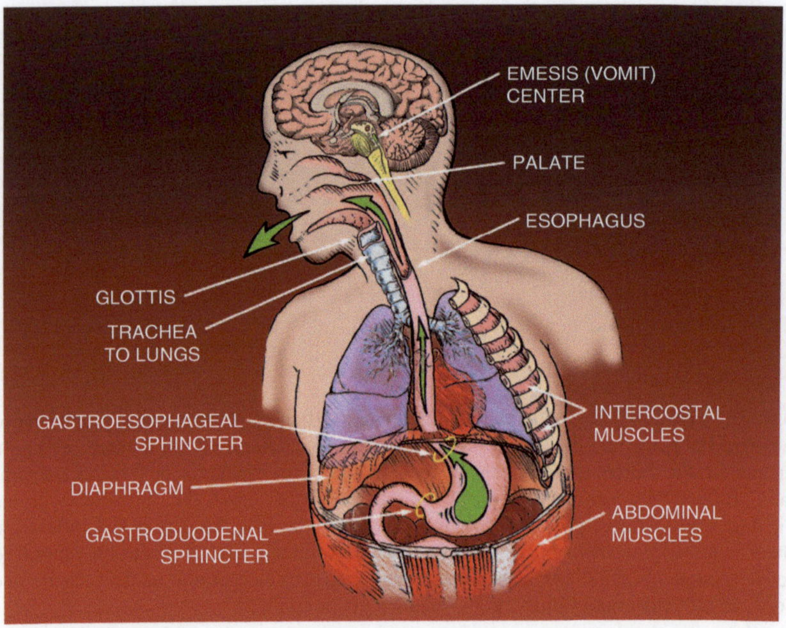

Fig. 3.7 Vomiting requires a surprising number of rapid, powerful, and coordinated-actions to force a viscous mass of vomitus out of the stomach and upper intestine through the esophagus, throat and mouth without contaminating the airways and lungs. It begins with a large inspiration, closure of valves to bronchus and nose, followed by powerful contraction of diaphragm, neck, and trunk muscles with opening of the stomach's valves. This involuntary action is under the control of the vomit center

of also using this system to inhibit a subject's activity during a period of vestibular malfunction was briefly explained in the background section.

Vomiting is a complex, coordinated activity (Fig. 3.7) involving several actions to eject the vomitus while keeping it out of the lungs. These actions are controlled by the vomiting center, a bilateral collection of neurons in the brain stem called nuclei, located very near the vestibular nuclei in the brain stem (Fig. 3.8). This center receives sensory signals from much of the GI tract, especially the stomach [13]. If for any reason contents do not leave the stomach, digestive juices collect and distend it. Signals sent directly to the vomiting center can cause "projectile" vomiting without nausea, the vomiting that can occur as part of SMS.

In those affected by SMS, all bowel sounds cease within tens of minutes to an hour after becoming weightless indicating absence of bowel activity [14]. Normal gastrointestinal activity that propels gastric contents down the GI tract is controlled by signals from the autonomic nervous system. Conflicted orientation signals from the otolith organs cause the vestibular nuclei to trigger sympathetic nervous activity through this path, which stops movement in the GI system. The resulting distension from immobile food and gastric secretions soon triggers the vomiting center to pro-

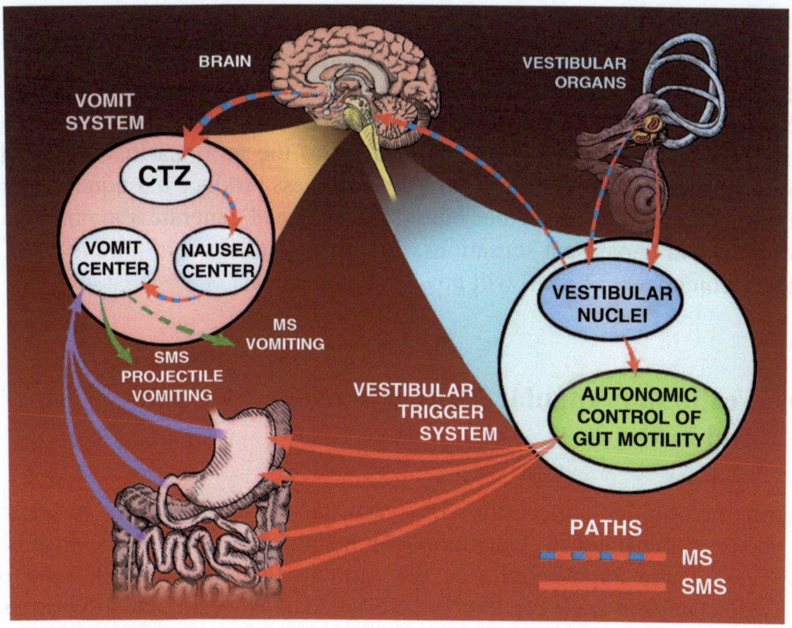

Fig. 3.8 The human emesis (vomit) system has added other inputs and functions to the vomit center. A nausea center can produce sensations of nausea, which, if severe, can trigger vomiting. It has a Chemical Trigger Zone (CTZ), that detects harmful agents in the blood and also produce nausea and vomiting. This is the pathway in MS, shown as *broken lines*. The intrinsic otolith pathway (*red*) for SMS vomiting first stops normal GI motility through slow autonomic nerves (*long red*) and the resulting buildup of pressure episodically and briefly triggers (*purple*) the vomit center

duce the projectile vomiting. The vomiting will only recur if more digestive juices collect, usually in hours, unless food or drink is put into the stomach, which will quickly produce distension and vomiting. This process produces stomach awareness and loss of appetite, but usually stops short of nausea; however, other inhibitions are added to GI symptoms.

3.8 Other Inhibitions in SMS

Increasing somnolence, oftentimes variable headache, and resistance to all mental and physical activity, especially activities that require head motions follows the initial bout of vomiting in SMS. There may be feelings of warmth and flushing, but this is variable. Inhibitory signals produced by the orientation conflict are transmitted from the vestibular nuclei to the autonomic nervous system and to conscious centers of the brain to produce the described sensations. The sum of these effects is to inhibit all subject activity as a safety measure until the conflict is resolved.

3.9 MS in SMS

There are many published reports of symptoms of MS in space equivalent to those of MS on Earth. Except for contrived motions such as the rotating chair, there are no motions inside a normal spacecraft equivalent to the motions that cause MS on Earth. However, the otolith failures in weightlessness produce a unique situation in which normal, even trivial angular motions of the head generate a major intra-sensory conflict and unpleasant sensations, that if continued would produce MS. This secondary component of SMS will now be examined.

3.10 Vestibular Angular Motion Sensors, on Earth and in Orbit

On Earth, the otolithic organs (Fig. 3.5—red) sense head orientation references to the g vector and changes in orientation.[4] By nature of the sensor, response is relatively slow. Three angular sensors, the semicircular canals, with rapid direct response to change in angle to rotation were added to the vestibular array by species adaptation (Fig. 3.5—blue). The resulting signals from otolith and canal sensors have overlapping response times and are congruent in 1 g and complement each other. In weightlessness the otolith sensor is fixed, indicating no motion is taking place during angular head motions while the canals are unaffected and produce normal signals. Thus, during SMS, any angular head motion produces a second major sensory conflict, and the symptoms of MS it produces are added to those of existing SMS symptoms.

3.11 The Emesis (Vomiting) System in MS and SMS

In addition to direct inputs from the GI tract in humans, the vomiting center receives inputs from other components of the emesis center, a chemoreceptor trigger zone (CTZ) and nausea center (Fig. 3.8) [15]. A variety of substances in the blood, morphine for example, can activate the CTZ that can in turn activate the vomiting and nausea centers.[5] Unlike the all or none projectile vomiting seen with the intra-otolith conflict, this pathway is slower and more variable, depending on the nature and strength of the stimuli. This is also the pathway used by the vestibular system to produce symptoms of MS.

---

[4] Changes in the orientation signals generate primary signals of angular head motion.

[5] Stimuli from tastes, odors and visual scenes can also stimulate the nausea center, even to the point that it stimulates the vomiting center.

In addition, MS stimulates the autonomic nervous system, producing salivation, pallor, cold sweating, GI motions and bowel sounds. Many areas of the brain are stimulated by MS sensory conflicts [16]. This may also produce something called the "sopite syndrome" characterized by somnolence and withdrawal. Adaptation to the continued otolithic conflict appears to be achieved by neurological reprogramming in the vestibular system. Within 3 days (maximum) this conflict is resolved and the safety inhibitions removed. This adaptation also inhibits MS. As noted, some of this reprogramming may produce conflicts on return to g forces during entry and landing.

3.12 MS and SMS in Spaceflight

The following is a summary of the effects of weightlessness in those affected by SMS before adaptation. Table 3.1 lists characteristics and symptoms of SMS and MS. This summary is complicated by current terminology.

All symptoms of SMS are produced by the functional failure of the gravity dependent otolith organs in weightlessness, which produces the impossible indications of the body orientation. This intrinsic primary error is fixed and cannot be changed by position or motion. The resulting intrasensory conflict is continuous and functionally paralyzes the gut, producing occasional brief bouts of vomiting, usually without significant nausea.[6] During this time there is anorexia and stomach awareness usually short of nausea. A variety of other feelings that also inhibit mental and physical activity are present.

The fixed orientation signal also produces a second major conflict but only with angular motion signals from semicircular canals, which occur with normal and even small head movements. Unpleasant sensations are produced by this conflict causing avoidance of all such motions wherever possible.[7] Continuation of angular motions can produce the crescendo of symptoms characteristic of MS on Earth—excess salivation, pallor, cold sweating, increasing nausea, continued vomiting, often ending in dry heaves. Abnormal GI activity may be induced and bowel sounds increased. Cessation of motion removes this conflict, and the symptoms recede more or less slowly in decrescendo fashion. MS is usually accompanied by feelings of malaise.

Such a difference in symptoms is the result of orientation conflict signals from the otoliths going more or less directly to the gut and motion conflict signals going from vestibular system through the emesis center. These secondary, transient symptoms of MS may be superimposed on the constant primary symptoms produced by the fixed orientation errors (Fig. 3.9).

---

[6] Such vomiting is seen only in serious disease on Earth.

[7] An Apollo crewman, instructed to make repeated sets of head motions to "expedite adaptation" reported, "I did it once, and it made me so sick to even think about it that I wasn't about to do it again."

Table 3.1 Characteristics of motion sickness and space motion sickness

	Motion sickness	Space motion sickness
Onset	Ramps up, adding symptoms in sequence: salivation, pallor, cold sweating, stomach awareness nausea, vomiting. Rate depends on stimulus intensity	Typically, sudden projectile vomiting without warning, minutes to hours after weightlessness begins
Duration	Depends on duration and on intensity of stimulus. Most adapt with continued stimulus in days to weeks. Some never do	8–72 h, typically 24–36. Always resolves
Time profile	Trapezoidal; height, slope depend on intensity of stimulus	Similar to step function; constant unless provoked by motion stimuli
Gastrointestinal (GI) symptoms	Stomach awareness, anorexia, nausea and vomiting, which may become continued, severe with retching	Stomach awareness, anorexia. Nausea, usually mild if present. Vomiting brief with one or more heaves, periodic (hours)
Autonomic nervous system activity	Skin pallor, cold sweating. GI system has abnormal activity	No change or flushing, warmth. GI system constant ileus
Central nervous system	Variable. Some develop "Sopite syndrome" with somnolence, lethargy. Often physically and mentally disabled	Somnolence, lethargy, variable headache, averse to all physical and mental activity. Can perform trained tasks
Resolution	Decreasing symptoms over many hours. Symptoms to nausea and vomiting may recur after cessation of very long stimulus. Mal de embarqment (MS on leaving ship after long voyage). Sp X	Once begun, recovery is rapid (1–3 h). Can recur in a few individuals on return to g loads
Incidence	Highly variable with intensity	<50% (depends on criteria)

3.13 Vestibular Adaptation and Resolution of SMS

While usually considered fixed, the nervous system is in fact the most adaptive system in the body. Neurological adaptive processes are still poorly understood, but the significance here is that adaptation, apparently by a process equivalent to modification of a computer program, occurs in a short time and is remarkably effective. For example, final resolution of the primary phase of SMS occurs within 3 days with restoration of normal GI activity and hunger instead of aversion to food. Inhibitory feelings are replaced by anticipation, and one becomes remarkably resistant to MS. Moreover, the fix is permanent for so long as the subject remains weightless, the problem does not recur. There is a continued silent adaptation of the vestibular system for an undetermined but long time in weightlessness. This adaptation is stored in the system's memory. Now the question becomes how does this reprogrammed system function inflight and during and after return to Earth?

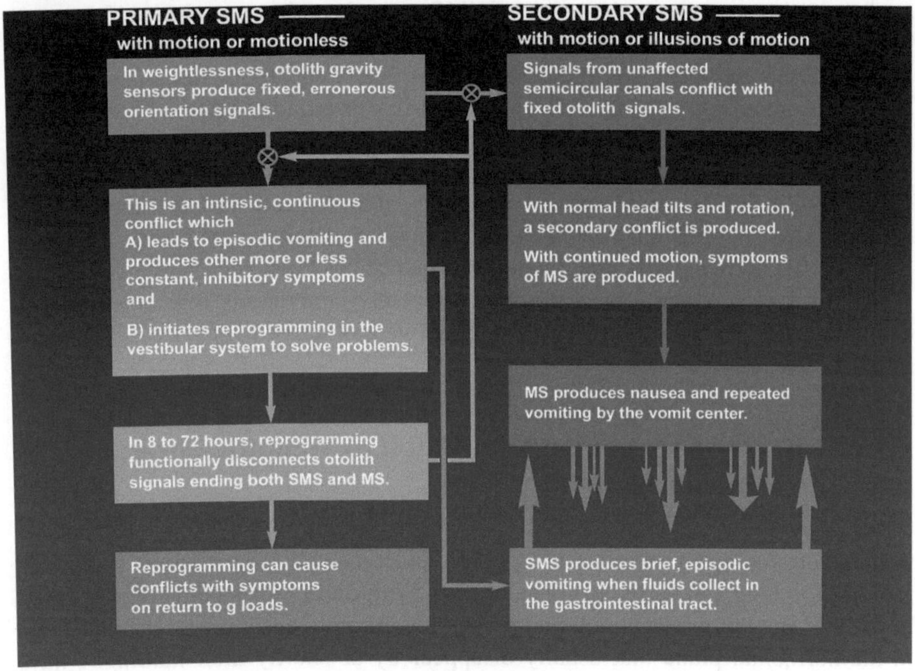

Fig. 3.9 Major events in primary and secondary SMS

3.14 Other Vestibular Effects and Adaptations Inflight

Soon after weightlessness in space is experienced a variety of sensations, usually involving the XY plane, have been reported including inversion, tumbling, and a variety of tilts, lasting from minutes to hours. Apparently this is resolved by the vestibular system rapidly making the individual's head/foot body axis the primary orientation reference. Two pieces of evidence for this are the ability of one in weightlessness to orient external objects as desired and the ability to work comfortably in any desired orientation relative to other objects and people.

For example, the Shuttle was frequently inverted above the earth with its occupants upside down. If one chose, this could be perceived as the earth above and the Shuttle beneath with one's feet "down," and head "up." This is possible only if the visual field is restricted as by a window. A full view, as in EVA in low earth orbit, provides a visual signal strong enough to restore geocentric orientation. Inside the craft one can work in any position with one's head always "up." If one rotates to a new orientation, it takes a second for the exterior to rotate and "lock into" the new orientation. One becomes physically egocentric in weightlessness. Note: By choosing to ignore the otolithic inertial, visual, and somatosensory inputs, many tertiary vestibular effects considered operationally insignificant are deliberately omitted here.

A recurring concern is the effect of SMS and adaptation on crew performance. There is no question that during SMS motivation is sharply reduced. In a small number, SMS has been disabling, but it has been repeatedly demonstrated that astronauts are able to capably perform trained-for tasks and to respond competently to unexpected anomalies. This includes tasks requiring complex and sensorimotor performance from manipulation of the Remote Manipulator System to blood draws.

In depth investigation of isolated modalities of motor and sensory function have demonstrated changes that have been widely published, but compensation is a major ability of the human body. Integrated performance has been such that investigators searching for decrements in performance recently concluded "few unambiguous examples exist." None of the "few examples" were described [2].

3.15 Effects of Vestibular Adaptation to Weightlessness on Return to Earth

A complete listing of effects during entry, and especially the theories to explain them, is too long and varied to consider here. With onset of g loading during entry, horizontal head turns (yaw motions) may produce some sensations of displacement in the horizontal plane that usually disappear by or shortly after landing. I have observed that if one is naive enough to persist in head turns, MS is likely to ensue. As usual there are a very few who will develop MS during entry, plus a few more on attempts to resume their usual activities after landing. The probability of this occurrence and the severity of disorientation and MS appear to be positively correlated to mission length. These effects do not seem to be related to susceptibility to SMS inflight. Incidence of such postflight events is rare in my experience. Objective studies of postflight effects have been difficult to find.

The most significant concern is occurrence and severity of any vestibular effect in a pilot flying the entry. This remains a well-kept secret, but there have been enough credible in-house rumors of such events to make it likely that at least a few incidents have occurred. Apparently there have been none that jeopardized the flight.

Immediately after landing, effects of a number of adaptations are more or less present in everyone. These include neuromuscular, cardiovascular and after longer flights, musculoskeletal adaptations. In many cases, such as the usual lateral instability in walking, these effects cannot be easily separated from vestibular adaptation.

The "door frame" effect, turning into an open door at right angles and over or undershooting the door opening and colliding with its frame, is likely a remnant of the canal/otolith conflict. It is more difficult to imagine the mechanism of a fairly common postflight event. This common phenomenon is making a sudden change of position while asleep and sometimes dreaming of floating in flight. The pilot of SL-3, an unusually powerful individual, heaved himself clear of his bunk and landed on the steel deck of the recovery ship.

Some effects persist for a considerable time. Perception of aircraft motions, especially pitching motions, is blunted for weeks. There is great resistance to airsickness during this time but not to seasickness. This is consistent with the MS adaptation that is for angular conflicts not for the motions present in a capsule on an ocean. If one had SMS on an initial flight, there are some data that suggest symptoms are likely to be reduced on subsequent flights.

3.16 Variability

Variability in all aspects of this problem is a major hindrance to its study. For example, no reliable indications of SMS susceptibility have been found. In spite of several notable cases of individuals who were remarkably resistant to MS having severe SMS, no statistical correlation was found between this, or any other marker. Absence of SMS on a prior flight does not insure its absence on subsequent flights nor does its presence on prior flights guarantee its recurrence. If anything, postflight variability is worse. This has made planning for studies difficult.

But it gets worse. Onset, duration, severity and number of SMS symptoms all cover a wide range. Manifestations of MS and SMS are virtually all psychophysiological with its inherent variability. The absence of bowel sounds, first noted by Norm Thagard on STS 7, was proven to be the most reliable indicator of the presence or absence of SMS as demonstrated by pre-, in-, and postflight recordings from a population of 17 subjects with and without SMS [2, 14].

Variability of the investigators and their investigations is large. There are still no consistently used criteria or standards. When all this variation is considered, the lack of consistency in reported studies and data is not surprising. For example the reported incidence of SMS varies from 36 to 80% depending on the study. This variability in reporting accounts for the absence of demographics in this chapter.

3.17 Countermeasures, Simulations and Medication

Attempted countermeasures have usually been based on subjecting individuals to the stimulus producing SMS before they flew, for "preflight adaptation." Based on the hypothesis that SMS is MS in space, a variety of devices known to produce MS were applied to cosmonauts and astronauts. Investigators typically applied simultaneous conflicting motions or visual illusions. The favored NASA device was the rotating chair in which tilting one's head could produce more or less controlled MS to all stages including nausea and vomiting. Preflight training sessions in this device were initially strongly encouraged in the Shuttle program. The Russian program still uses an especially brutal preflight "trainer." A complete list of such devices is too long to include, especially since they all failed for an elemental reason. An adaptive stimulus must be specific. Simply provoking the vestibular system to trigger MS adapts it to only the stimuli provoking the response, not to the sensory conflicts produced by weightlessness.[8]

Analogous to the formal training above was that conducted somewhat informally by astronauts. A tradition of wringing out a jet the day before a mission was expanded into wild maneuvers whenever possible weeks to months before a flight, even on cross country flights. Those who did not get sick on their mission probably propagated the myth that violent aircraft maneuvers preflight prevented SMS while those who did conveniently forgot as much of the experience as possible. Until some way is found to remove weight from the otoconia in 1 g there appears little chance of preflight adaptation for SMS.

What about "zero-g" flights? Even if the tens of seconds of weightlessness in each parabola were effectively summated, the shortest time known for SMS adaptation is 8 h, and the usual is more than 24, or almost a thousand parabolas minimum (each offering 25–45 s of weightlessness). Weightlessness during parabolic flight is also different from weightlessness in space in that periods of weightlessness are separated by periods of hypergravity of about 2 gs toward the bottom of each parabola. This cyclical type of weightlessness interspersed with intervals of hypergravity is very different than the sustained weightlessness that is experienced during space travel.

Medications for MS could fill a small pharmacopeia. None is completely effective, and all have side effects. Scopolamine (Hyoscine hydrobromide) has been most widely used by NASA. It can be found in Solanaceae (nightshade) family of plants and can be classified as an anticholinergic drug. Scopolomine has been used in combination with Dexedrine, an "upper" to counteract its "downer" effects. Reported effectiveness of "ScopDex" in SMS varies but is significant. Its use was strongly encouraged on Apollo and Skylab but was forbidden to pilots, commanders, and flight engineers on Shuttle. Its mode of action is largely unknown, but it apparently blocks vestibular signals to conscious areas of the brain and signals from brain to vomiting center. When effective it delays SMS, which follows its usual course when this drug is stopped.

Promethazine (Phenergan) is an old drug with antiemetic and sedative properties used in surgery and MS. The most experienced flight surgeon on NASA's zero-g aircraft used it in severe cases of airsickness and recommended it for relief of symptoms of established SMS. It was first given as a buttocks intramuscular injection in 1991 and then used in severe SMS either as a suppository or arm injection. Reports of its effectiveness are mixed but worthy of note. It appears to be the current favorite and noninvasive modes of administration are in work.

[8] A favorite story (WT) of adaptive specificity is that of an astronaut who spent countless miserable hours in the rotating chair tucking his head for the Coriolis effect to produce MS. When he finally became "immune" to the effects, he reversed the direction of rotation and found head tucks made him as sick as ever.

3.18 Re-adaptation to 1 g

On return to g forces and weight, the adaptations to weightlessness must be suppressed or removed and an adaptation to 1 g restored. These processes remain largely unstudied during atmospheric reentry and exit from the spacecraft after a short flight. Moreover, it is difficult to distinguish vestibular effects from effects of adaptations in other systems, especially the neuromuscular. It appears that the adaptations to weight are recalled from memory and again become operational more or less rapidly, at least for most, who remain well oriented during atmospheric reentry and spacecraft exit. This is not so for everyone. A few remain hypersensitive to normal head motions that in rare cases caused MS during entry and afterward. In this subpopulation, such effects usually clear in hours, but in rare cases may take days. Complete reversion to 1-g adaptation takes some time. Unless one concentrates on it, walking is unsteady after egress. Hours after landing from a 7 day mission, one astronaut asked another several tables away for a can of Coke. The second astronaut responded as he had in the past week on orbit and with the usual force. The tossed can hit the floor halfway between, with a fountain of foam. Objective studies show more or less vestibular effects on orientation for individually variable lengths of time. For example, the vestibulo-spinal reflex, which coordinates landing after a jump or fall, may be blunted for a week after a mission [9]. Apparently memory of the adaptations are suppressed not destroyed and fade slowly with time. As previously mentioned, it is a common experience that SMS is less severe on subsequent flights, even with years of delay.

3.19 Operational Considerations

The termination of the Shuttle reduces some of the importance of SMS in current operations. Also, unpredictable performance in approximately half the crewmembers for up to 3 days is of less concern with longer times in orbit such as missions aboard ISS. However, some lessons learned on Shuttle should not be forgotten. If for any reason it is necessary for new crewmembers to immediately engage in some significant activity there are two cardinal rules—train, train, train for the task and don't expect more than the task trained for in subjects with sickness. In rare cases a subject will be incapacitated, and this leads to the next rule. Fully train back-ups for a significant task—a good rule at any time. The evidence-based etiology described here opens an opportunity for quantitative human and animal investigation of a major neurological system in an unconvoluted fashion. Current miniature instrumentation can allow great savings in the process.

Other specific considerations include:

- Variability of SMS is such that the previous presence or absence of SMS is not an absolute guarantee predictor of what may happen in the future.

- The rule that prohibits EVAs for the first 3 days of a spaceflight is a good one.
- The only certainty in this process is that with rare exception it will be over in 2–3 days. Effects of nausea and vomiting on fluid and electrolyte balance are discussed in the next chapter.
- Effects of vestibular adaptation to weightlessness during reentry, spacecraft exit and postflight become more significant with longer flight duration, especially for anyone with piloting or other operational duties. Again variability is such that back-up is essential here.
- Return to duty with driving and flying after 3 days and a flight physical exam has worked well for 1–2 week flights, but longer flights will require caution and possible longer periods of restricted duty. Automated evaluations of various neurological and motor functions have been used to aid such exams, but should not become a substitute for demonstrable performance.
- NASA was fortunate never to have had an unaided emergency egress from Shuttle. Vestibular effects could have had serious consequences in rare individuals. In-flight adaptation of other systems can lead to significant postflight effects in most if not all individuals.
- Finally, be cautious to generalize results from 1-g studies of SMS or accept any claims for efficacy of drugs or other countermeasures unless demonstrated in extended spaceflight.

3.20 Summary, Present and Future

Termination of the Shuttle program produced a major loss of operational and investigator interest and support in SMS. In addition to the large reduction in number of subjects available, the fraction of time lost to SMS on extended missions is now small. Evidence of this was the absence of SMS in a review of the first decade of ISS [17]. Long duration flights are now the primary interest and vestibular readaptation to g loading is the major concern [18]. Orientation studies have increased accordingly. We look forward to SMS experiences and studies from the Chinese space program. One recent paper by Chen and colleagues supported the theory of subjective vertical conflict and applied it to SMS [19].

As technology in digital computing advanced, it became a potentially valuable and extensively used tool in modeling and simulation of MS and SMS. Although mathematics is the rigorous language of science, ideas expressed in it still require validation by measurement of the actual phenomena, There are more published engineering and scientific “Just So Stories” of SMS than Kipling ever created.

The future of human spaceflight currently appears to be very dependent on commercial developments, especially suborbital or short tourist flights. Thoughts of a space bus load of flight naive individuals with expensive tickets always recall the sight and sounds of a veteran pilot and astronaut with thousands of hours experience vomiting just after main engine cut off (MECO) on his first launch and then progressing to 2 days of SMS. One of the authors (FB) is currently a co-investigator on a project that entails the measurement of physiological and psychological responses to simulated 'tourist' spaceflights. The g profile of future flights is accurately replicated using a long arm centrifuge. However, what cannot be replicated is the 4–6 min interval of weightless that space tourists are expected to experience. We cannot know at this time how much of a problem SMS will be for space tourists but anecdotal accounts of SMS sometimes rapidly affecting professional astronauts should be considered.

Much work remains to be done in all these areas. An area essential to this work has seen remarkable progress but MS/SMS investigators have shown no interest in it. The micromanipulation and neural circuit monitoring of hair cells offers a means of demonstrating the otolithic sensory conflict of SMS in 1 g [20]. This could also provide the first scientific validation of the sensory conflict hypothesis, a lasting benefit to neurology—and justification of the untold millions spent on this problem.

Glossary

Atmospheric reentry Return to Earth in a spacecraft.

Autonomic nervous system Predominant portion of nervous system that functions automatically and unconsciously.

Emesis (syn.: vomit) Ejection of material in stomach/bowel through mouth.

Emesis Center Collection of nerve cells in brain stem that coordinates, controls vomiting.

Emesis System Three collections of interconnected cells (emesis center, nausea center and chemical trigger zone, or CTZ) in brain stem that receive a variety of inputs whose outputs control gastrointestinal (GI) sensations, nausea and vomiting.

Fail safe A system mode that protects a system from further damage in event of component or subsystem failure.

Glottis Valve in throat normally open to air for lungs that closes against liquids or solids going into or out of the esophagus.

g vector (as used here) Magnitude and direction of gravitational force of attraction of the Earth's mass on or near its surface.

Motility Spontaneous motion, in this case of the GI tract that normally propels food through the tract.It is under neurological and hormonal control.

Motion sickness (MS) Also known as kinetosis, a more or less common set of GI autonomic and unpleasant sensations triggered by the vestibular system in response to motions or simulated motions outside the subject's usual experience.It is used here in that context, but MS in weightlessness is produced by abnormal sensitivity to motion from failure of the otolith sensors that continues until reprogramming occurs.

Nucleus (pl., –ei) In this context a collection of nerve cells, often in a nodule, usually interconnected and with a system function. Multiple inputs from sensors and other cells may be input, processed, integrated and used to produce outputs.

Orientation As used here, it is spatial orientation, usually the relation of body axis to g vector.

Otolithic Related to the vestibular sensors that detect gravitational orientation by weight of dense crystals once called otoliths (Gk.: ear stones).

Spacecraft exit Exit from spacecraft on Earth.

Space Motion Sickness (SMS) GI autonomic signs and symptoms produced by a intrinsic conflict in the vestibular otolith sensors produced by weightlessness. It is resolved by reprogramming in the vestibular system in 1–3 days. The conflict is independent of motion, and a better term is needed; however, SMS is reluctantly used here since it is in universal use.

Vestibular Name derived from the convoluted spaces in bone housing the vestibular organs that comprise the inner ears, but it now applies to the ventricular system.

Vestibular organs Consist of three sensory systems in each inner ear, which are cochlear for hearing, otolith for orientation to the g vector and for sensing linear and angular motions, and semicircular canals for sensing angular motion.

Vestibular system Consists of the vestibular organs, bilateral vestibular nuclei in the brain stem and numerous areas in the upper brain.

References

1. Lackner JR, DiZio P. Space motion sickness. Exp Brain Res. 2006;175(3):377–99.
2. Thornton WE, Bonato F. Space motion sickness and motion sickness: symptoms and etiology. Aviat Space Environ Med. 2013;84(7):716–21.
3. Homick J, Miller E. Apollo flight crew vestibular assessment in biomedical results of Apollo, NASA SP-368, National Aeronautics and Space Administration; 1975.
4. Graybiel A, Miller E, Homick J. Experiment M131: human vestibular function in biomedical results of Skylab, NASA SP 377; 1977.
5. Clément G, Reschke MF. Neuroscience in space. New York: Springer; 2010.
6. Linenger J. Training Russian style. In: Off the planet. New York: McGraw Hill; 2000.
7. French F, Burgess C. In the shadow of the moon: a challenging journey to tranquility, 1965–1969. Lincoln: University of Nebraska Press; 2007.
8. Graybiel A, et al. Susceptibility to motion sickness. In: Johnson RS, Dietlein LF, editors. Biomedical results from Skylab (NASA SP–377), vol. 36. Washington, U.S. Government Printing Office; 1977. p. 74–91.
9. Fernandez C, Goldberg JM. Physiology of peripheral neurons innervating otolith organs of the squirrel monkey. I. Response to static tilts and to long-duration centrifugal force. J Neurophysiol. 1976;39(5):970–84.
10. Clément G, Reschke MF. Neuroscience in space. New York: Springer; 2008. p. 79.
11. Buckey JC Jr, Homick JL. "The Neurolab Spacelab Mission: Neurospace Research in Space" (NASA SP-2003-535). Houston: NASA JSC; 2003.
12. Clément G, Reschke MF. Neuroscience in space. New York: Springer; 2008. p. 85.
13. Purves D, Augustine GJ, Fitzpatrick D, Hall WC, LaMantia AS, McNamara JO, Willams SM. The vestibular system. Neuroscience. Sunderland: Sinauer Associates; 2004.
14. Thornton WE, Linder BJ, Moore TP, Pool SL. Gastrointestinal motility in space motion sickness. Aviat Space Environ Med. 1987;58(9 Pt 2):A16–21.

15. Guyton AC, Hall JE. Physiology of gastrointestinal disorders. Textbook of medical physiology. Philadelphia: Saunders Elsevier; 2000. p. 764–70.
16. Napadow V, Sheehan JD, Kim J, LaCount LT, Park K, Kaptchuk TJ, Rosen BR, Kuo B. The brain circuitry underlying the temporal evolution of nausea in humans. Cereb Cortex. 2013;23(4):806–13.
17. Hayes J. The first decade of ISS exercise: lessons learned on expeditions 1–25. Aerosp Med Hum Perform. 2015;86(12):A1–6.
18. Clément G, Wood SJ. Rocking or rolling–perception of ambiguous motion after returning from space. PLoS One. 2014;9(10):e111107.
19. Chen W, Chao JG, Wang JK, Chen XW, Tan C. Subjective vertical conflict theory and space motion sickness. Aerosp Med Hum Perform. 2016;87(2):128–36.
20. Salvi JD, Maoiléidigh DÓ, Fabella BA, Tobin M, Hudspeth AJ. Control of a hair bundle's mechanosensory function by its mechanical load. Proc Natl Acad Sci. 2015;112(9):E1000–9.

Chapter 4
Fluid Shifts and Loss

4.1 Introduction

Volume and distribution of intra and extra vascular fluid volume is the first and primary consideration of maintenance of human ability to function in an upright posture under gravitational and inertial loads, especially with static vertical posture in a gravitational field such as on Earth. It makes no difference whether a person working in a field has reduced total body fluid because of sweating all morning or an astronaut who has just returned to Earth from a week in orbit suddenly stands. If blood volume is too small to maintain an adequate cardiac filling pressure, both the worker and the astronaut are subject to the same laws of biophysics and may even experience *syncope*, a transient, self-limited loss of consciousness that is commonly referred to as 'fainting'. Allowing individuals who have fainted to lay supine for a few minutes typically causes consciousness to return. Give them water over a brief time and the field worker can stand to eat his sandwich and drink his soda for lunch while the astronaut can walk out and wave to his admirers.

Postflight weight loss and *orthostatic intolerance* (OSI) were the second major effects of weightlessness encountered in the early manned space programs [1]. A typical astronaut's postflight reaction might have been, "What the…? Can't stand still without getting dizzy. Just took eight[1] gs and coulda' pulled the wings off that 102 (aircraft), five days ago."[2] By the next day these symptoms disappeared for most and there was no problem. This was a common postflight experience for astronauts and cosmonauts on recovery after early spaceflights. Such symptoms

[1] On reentry, g vectors were chest to back (gx) and well tolerated.

[2] On the day before launch astronauts warmed up by taking an aircraft to its limits in acrobatics.

W. Thornton, F. Bonato, *The Human Body and Weightlessness*,
DOI 10.1007/978-3-319-32829-4_4

usually resolved in hours. Inflight they had stuffy noses, puffy faces and when pressure suits could be removed, had noticeably reduced leg volume, which was promptly dubbed "the bird legs of space". Postflight there was a significant loss of weight, most of which was recovered in a matter of hours. Cosmonauts were also susceptible.

Astronauts and cosmonauts were familiar enough with another aspect of the same phenomenon, blackouts and loss of consciousness (LOC) while "pulling gs" in high performance aircraft or centrifuges used for training. In both situations the final cause is the same, inadequate blood flow through the brain. However, the situations are very different. When under increased g-loading, LOC is the result of overloading a normal cardiovascular system. Blood is drawn away from the head (and brain) and down towards the feet. The phenomenon of g-induced LOC, (G-LOC) was reasonably well understood and led to methods for keeping the blood up into the head, such anti-g straining maneuvers that entailed tensing muscles in the legs and buttocks to force more blood up into the head. However, potential LOC following four to five days of spaceflight was new and not well understood.

While it was soon assumed that body fluids were involved, this was not confirmed until SL-4 (1973–1974) when inflight measurements were made allowing the current theories of the shifts and loss of body fluids to be produced. The resulting decreased g tolerance was initially of concern for piloted Shuttle entry and landing. With the use of "g-suits" developed in World War II this did not become a significant problem. These suits are comprised of tightly fitting trousers that can squeeze the legs and can at least partially prevent blood from pooling in the lower body. A g-suit can help maintain adequate blood in the head, reducing the chance of fainting. In addition, both U.S. and Russian programs soon began to fluid load individuals by having them ingest salt and water just prior to entry into Earth's atmosphere at the conclusion of a flight.

There are persistent rumors of orthostatic difficulties on reentry in very few commanders, but if true, they are well suppressed. In a few crewmembers, especially after longer flights, complete recovery from these effects has taken days instead of hours. However, the problem of postflight OSI, or actual loss of consciousness on standing immediately after landing, is still significant. While the g-suit and other countermeasures were usually adequate to support Shuttle crews on reentry, it is fortunate that emergency egress was never required. It is likely that some of the crew could not have safely performed the egress because of hypotension plus cardiorespiratory effects (discussed in Chap. 8). Although reentries are now less demanding on pilots, both pilots and crew face problems of hypotension on egress from the vehicle, especially after long flights for today's countermeasures remain incomplete. Based on attempts to remedy this problem to date, an understanding of the fundamentals also remains incomplete. Another part of the problem may over reliance on research conducted using head-down bed rest and small tail-suspended animals as simulations for studying weightless effects.

4.2 Brief History

Fainting has been a more or less familiar experience throughout human history. Dehydration, anemia, blood loss, severe emotional stress, and illness can reduce cardiac output when standing, reducing blood flow to and through the brain sufficiently enough to cause dizziness and LOC. When aircraft developed enough power and became sturdy enough to bear g-loads in World War I, blackouts or transient LOC from high-g maneuvers were encountered. In World War II and its more advanced (and faster) aircraft such blackouts became a serious problem. Centrifuges were developed to study the effects of g-loads on humans, and from these studies a number of countermeasures were developed. Best known and most effective of these is the anti-g suit, a garment that applies transient pressure to a pilot's legs and lower trunk during g-loads, forcing blood upwards towards the brain reducing the probability of LOC.

When manned spaceflight first began in the early 1960s, the launch vehicles used had been designed to carry bombs, not humans.[3] They produced force loads of up to 12 g. Ablative heat shields allowed the capsules to return to Earth rapidly but under even larger g-forces. Astronauts and cosmonauts were placed in form fitting couches transverse to the g-forces, and they sustained such loads without undue difficulty. Such chest-to-back g-forces (gx) are tolerated better by humans than the head-to-toe forces (gz) that typically result from aircraft maneuvers.

Inflight sensations of head fullness, (called "a rush of blood to the head" by cosmonauts), stuffy noses, and "bird legs" became routine during weightlessness. Earth's gravity no longer pulled blood and other bodily fluids down towards the feet. The result during weightlessness is a sudden shift of fluids upwards towards the head, a redistribution of fluid that remains as long as weightlessness conditions remain. After flights of several days, returning crewmen felt the sensations produced by decreased blood pressure and increased heart rate. There was at least one postflight faint in a hot shower. These symptoms typically decreased in a matter of hours. Weight losses were variable and significant, yet weight was quickly regained once back on Earth. There were also indications of blood volume loss. In short, astronauts returned with transient orthostatic hypotension, dehydration and anemia, which were an unanticipated effects of spaceflight and weightlessness.

By the time of Gordon Cooper's 34-hour 1963 Mercury-Atlas 9 flight, there was enough concern to institute pre- and postflight tilt table tests to study these effects, which were found to be individual and variable but characteristic of orthostatic hypotension. Tilt table tests were continued through the Gemini project with similar findings [1]. By 1969 a simple stand test replaced the tilt table procedures on Apollo 9, 10, and 11 [2]. No testing was conducted for Apollo 13 and 14. More complex, less comparable lower body negative pressure tests were initiated late in the Apollo program, and an apparently successful anti-g suit demonstration was conducted or Apollo 17 in 1972. Postflight leg volume measurements showed no change. It was concluded that loss and gain of body fluids were involved, but how? About this time the Soviets began to "fluid load" their cosmonauts with salty bouillon prior to reentry in an attempt to compensate for fluids lost during spaceflight.

Skylab had an array of cardiovascular instruments on board including a lower body negative pressure (LBNP) device. But until ad hoc studies on SL-4 (1973–1974), there was no explanation of the effects observed. During SL-4, one author (WT) obtained a series of leg and arm volume measurements, pre-, in-, and postflight, that showed a rapid inflight loss and postflight gain of measured leg volume [3]. Concurrent in-flight body mass losses were consistent with this loss of mass that was recovered in 1 or 2 days postflight [4]. Other studies confirmed a loss of blood volume, but no change in cardiac function was found. This provided a basis for the now accepted concept of a fluid shift followed by a loss of fluids in weightlessness, which produces postflight OSI.

There have been a great number and variety of subsequent studies of OSI and other aspects of the cardiovascular system, which arguably is the most studied human biological system in spaceflight. As already noted, numerous countermeasures were proposed, and a number of them were flown. For every inflight study there were ten or more bed-rest flight-simulation studies [5], most with the subjects tilted head downwards at an angle of 6°. Such simulation studies cannot truly 'replicate' weightlessness on Earth. However, bed-rest studies do redistribute body fluids using gravity to draw blood and other fluids away from the feet and legs.

During the Shuttle program, G-suits and fluid loading plus innate tolerance of g-loads were typically enough to protect the majority of fliers during reentry and postflight prior to loss of Challenger in 1986. However, without an inflated g-suit, LOC on standing following reentry was not uncommon. The problem of postflight OSI was exacerbated by the addition of a pressure suit for reentry after the loss of Challenger. Inadequate cooling and increased body temperatures with sweating became frequent additional stressors. A few fliers have experienced significant postflight OSI symptoms for a number of days, and there have been rare cases of disabling orthostatic intolerance as the diversity of background and experience of fliers increased. Studies suggest that women are more susceptible to orthostatic intolerance effects. It is often difficult to separate these effects from vestibular and other effects seen postflight. As the Shuttle program progressed, fluid loading volume, variety and added salt was increased and ingested just prior to reentry. The only details of this, and problems associated with it, that could be found, are in a personal account [6]. OSI plus other adaptations almost certainly would have disabled some Shuttle crewmembers had an emergency egress ever been necessary.

[3] The Atlas rockets used for the Mercury project and the Titan rockets for the Gemini project were both originally designed as intercontinental ballistic missiles (ICBM). There intended purpose was to carry warheads long distances—not humans.

Longer duration flights on Shuttle/Mir flights (1995–1998) and now on the International Space Station (ISS) produced a large increase in post flight frequency, severity and duration of OSI. A complex comparison of OSI after 12-day Shuttle flights and ISS flights of up to 177 days was recently published [7]. Based on the data presented from that study some conclusions can be drawn. There was a sharp increase in the number of astronauts who failed to complete the tilt table test after the long flights and their recovery was delayed. Finding changes in some crewmembers' vision after long duration flight has brought new attention to fluid shifts, to the extent that head down long-term bed rest studies were suspended briefly until new safety monitoring could be put in place. Recently, studies of mice and rats with their hindquarters suspended by their tails appear to be the favored simulation of fluid shifts studies in 1 g.

4.3 Background

The problem of postflight OSI may be seen in perspective by examining the evolutionary species adaptations of our ancestral fish as they evolved to live on dry land and eventually become *Homo sapiens* [8]. Eons later a few of us adapted somatically to weightlessness in spaceflight (Fig. 4.1c). What happened when these ancestral fish left the sea for dry land? Under water, reduction in blood pressure from weight of the blood (hydrostatic pressure—Box 4.3) is exactly offset by the weight of the surrounding water so no effective reduction in blood pressure occurs with changes in orientation. On land, when the brain was elevated above the heart, its blood pressure was reduced by unbalanced hydrostatic pressure. This required a variety of circulatory adaptations to maintain consciousness. Snakes that climb trees adapted to this problem by moving their hearts nearer to their brain. Other animals with long vertical necks increased their blood pressure.

Humans had to overcome this problem plus others; including the addition of two large legs to our ancestral fish body, and these required changes in our circulatory and neuroendocrine systems to support them on Earth. The two large legs form more than a third of normal body mass and contain a significant portion of a person's total blood volume when standing erect. Adaptation to accommodate this was accomplished through several anatomical changes and a complex network of reflexes to maintain volume and pressure.

Inflight, cardiovascular problems are minimal, but in addition to shift and subsequent loss of fluids including blood, gain of some of the unused control systems is reduced by adaptation. On return to Earth's 1 g environment, the maintenance of an adequate blood pressure to sustain flow through the brain depends on provision of an adequate input pressure to the heart, but the inflight loss of large amounts of blood and fluid from the legs often simply overwhelms the capacity of compromised controls to compensate. Postflight orthostatic stresses are seen more or less

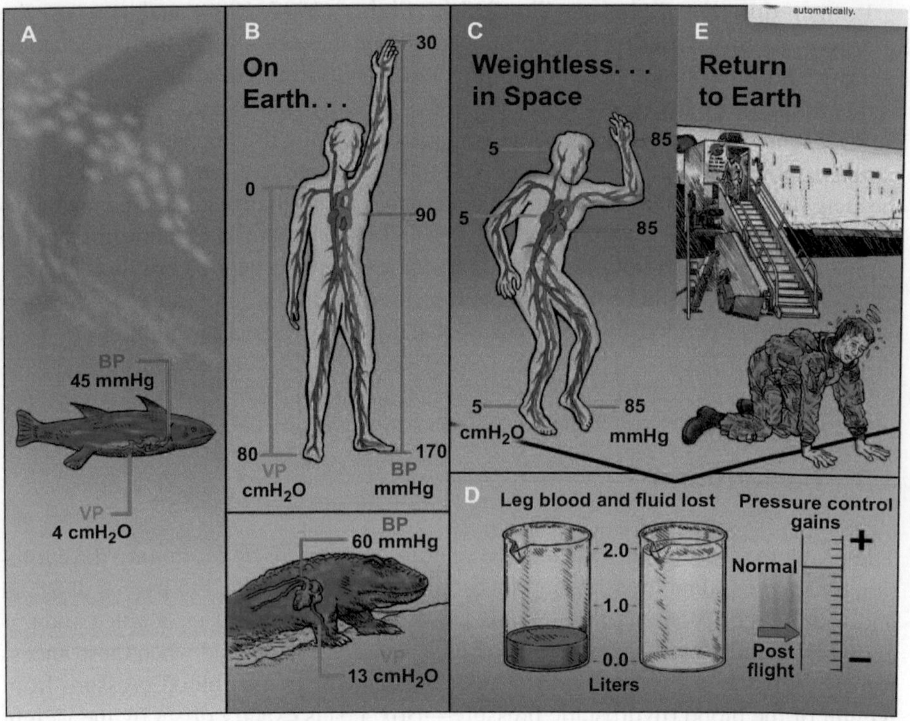

Fig. 4.1 Origins of fluid shift and loss, and sequelae. Changes (**a**) to (**b**) are species adaptation and from (**b**) to (**e**) are somatic adaptation. BP indicates arterial blood pressure and VP indicates venous blood pressure

regularly and can result in loss of consciousness. The initial and primary problem is biophysical, replacement of lost fluid, while functional physiological changes in the control systems are secondary.

4.3.1 Circulatory Essentials

Maintenance of blood flow to the brain is our major concern here. A brief explanation of the cardiovascular mechanisms used to maintain adequate brain circulation on Earth follows. Sidebars are included to insure we share common terms and concepts.

The functional chain of events leading to postflight OSI begins with the basic arrangement of our circulatory system, which was well established in our ancestral fish long before they set fins on solid ground. The function of fish and human circulatory systems is to provide each of the billions of tightly packed cells in a body with their necessary nutrients and oxygen, while removing wastes and other cell

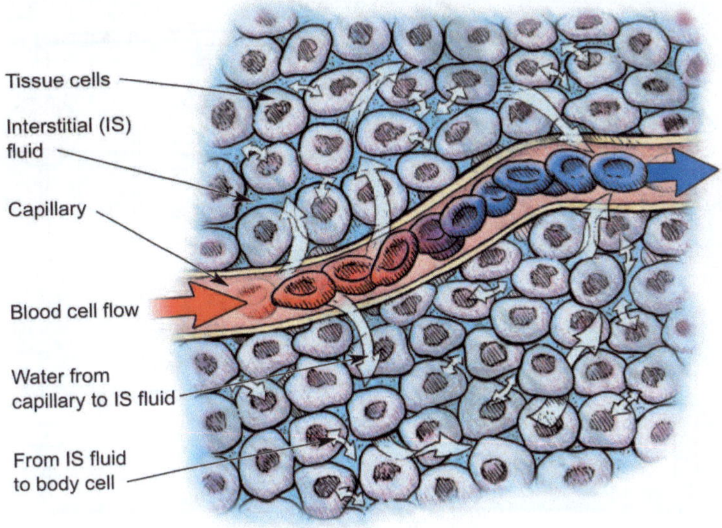

Fig. 4.2 Diagrammatic view of fluid between blood cells (*red/blue*), interstitial fluid (*light blue*) and intracellular fluid (*pink*) exchanges at the cellular level. There is a progressive pressure drop in capillaries. As the capillary enters the tissue, water leaves capillary blood (*large arrow*) under pressure into the interstitial fluid. As pressure falls in the capillary, interstitial water reenters it. Changes in blood and extra-vascular volume depend upon the net exchange of this water. Water is also exchanged from interstitial to intracellular fluid across cell walls (*small arrows*). Although the volumes exchanged are microscopic, so are distances and exchange times are microseconds. This plus the large surface areas, tens of meters, can result in changes of large total volumes in minutes

products. Microscopic transport to do this for all tissues is illustrated in Fig. 4.2 [9]. There are three fluidcompartments in the microcirculation separated by membranes; *intracellular*, *interstitial* or between cells, and *intravascular*.

Water, the major component of blood and tissue fluid, passes freely in either direction across both cell and capillary walls and membranes. A vast array of substances, from blood cells to potassium ions, are selectively blocked or passed through the cell walls and capillary membranes. Rate of passage of these substances depends on their concentration and on selective resistance of the walls separating the compartments. In the illustration, color of the red blood cells changes as they give up oxygen that readily passes into cells, while carbon dioxide passes from tissue cells to red (now blue tinted) blood cells. Water, shown by white arrows, freely passes out of the higher pressure arterial side and into the lower pressure venous side of the capillary generating a flow in the interstitial fluid. Sodium is a major component of blood and interstitial fluid, but it is excluded and pumped out of cells as an essential part of their life processes.

Circulation of substances over microscopic distances is rapid and done in milli- or microseconds. The total area of cell and capillary walls is large, estimated to be 800–1000 m^2 in humans. This allows large volumes, especially of water, to be transferred quickly from one compartment to another. If capillary flow is stopped,

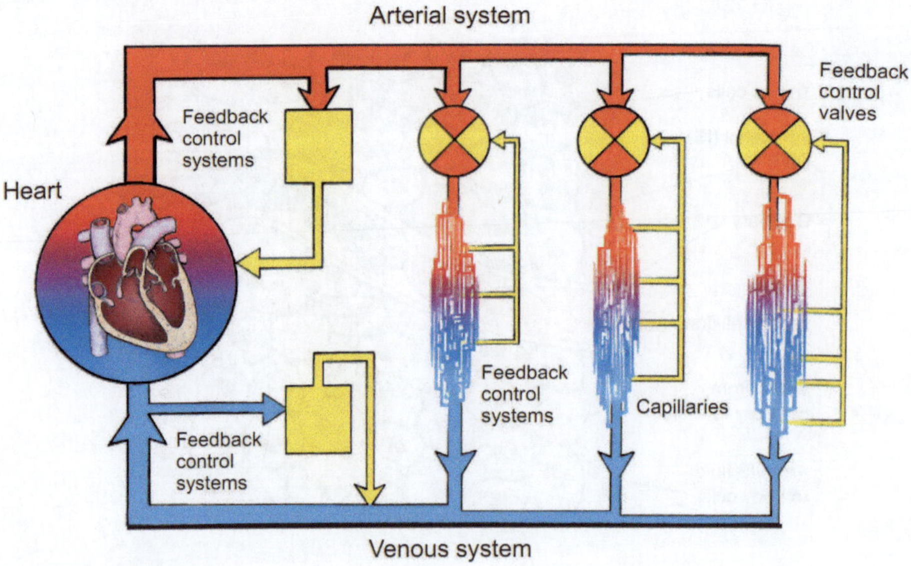

Fig. 4.3 Basic Animal Circulatory System. A pump, the heart, provides blood to the arterial system under sufficient pressure to provide capillary flow through all the tissues. Varying needs of the tissues are met by the equivalent of valves between the arterial supply and capillaries. These valves control flow by way of feedback loops. Blood is returned to the heart by the low-pressure venous system also under control of reflex systems, which maintain adequate input to insure sufficient flow to meet the needs of the body. A second loop controls blood pressure. This is a schematic and single units represent multiple reflex loops. Lungs and pulmonary circulation are not shown

disrupting oxygen transport, some cells in the brain will cease to function in seconds causing LOC. If flow stops for minutes, some cells begin to die.

Figure 4.3 is a simplified diagram of the cardiovascular system that maintains flow through the capillaries [10]. We also inherited this system from fish [7]. The heart is a pump that sends blood under arterial pressure to the supply lines, the arteries. Tissue capillaries are connected in parallel to this line through valves that control the flow through them according to the needs of the tissue they are supplying and neurological signals from the brain. A low pressure venous system then collects and returns blood to the heart. Reflex loops (shown in yellow) control arterial pressure by altering heart rate and force, vessel size and flow distribution. Other loops indirectly control blood pressure by control of blood flow and fluid volumes. It should be noted that without an understanding of pressure, volumes and their relation, there is no chance of truly understanding the phenomena we will examine [11]. Equally important is an understanding of hydrostatic pressure. For this reason, pressure units for the circulatory system are described in Box 4.1. Arterial pressure is addressed in Box 4.2. Hydrostatic pressure as it related to gravitational fields or acceleration are described in Box 4.3. Compliance, the change in fluid volume in an elastic container that occurs with changes in pressure are shown in Box 4.4.

Box 4.1 Pressure Units for the Circulatory System

Pressure is defined as force/area and expressed in a variety of units. Forces producing pressure come in many forms, mechanical, fluid, liquid, etc., but the concern here is with liquids. In the physical sciences conventional units are used for hydrostatic pressures, but in medical science the pressure is usually given as the height of a column of a specified liquid whose pressure is equal to that being measured.

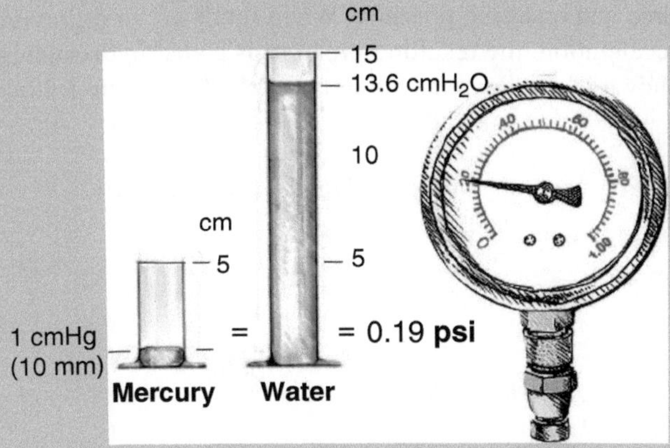

Box 4.2 Arterial Blood Pressure

Arterial blood pressure is produced by periodic ejection of blood into the elastic vessels producing a pressure pulse superimposed on a more or less steady component. The mean or effective blood pressure is given by the equation shown and is the pressure used here.

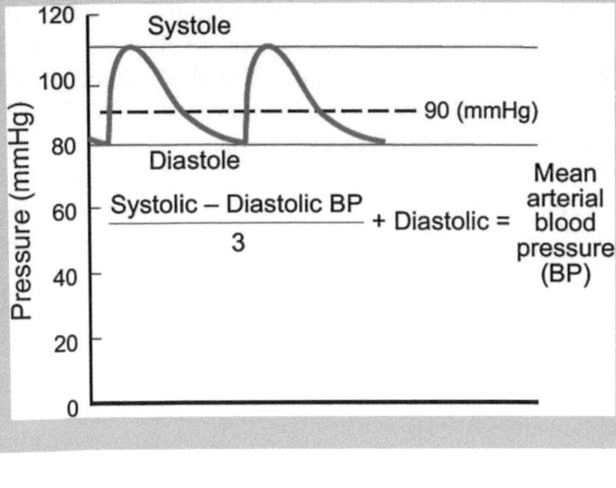

Box 4.3 Hydrostatic Pressure

When a force is evenly distributed over an area, the force per unit area is defined as pressure.

$$Pressure = \frac{Force}{Area}$$

Any number of means—electrical, mechanical or weight of fluid—can produce force and resulting pressure. When fluids are in a gravitational field or under acceleration, the resulting pressure is called hydrostatic and is produced as shown in the figure as column 1 according to the following simple equations:

$$Force = Density \times Volume \times Local\ gravity$$
$$= \rho \times \mathrm{a} \times \mathrm{h} \times \mathrm{g}$$

and

$$Hydrostatic\ pressure = \frac{\rho ahg}{\mathrm{a}} = \rho hg$$

For convenience, fluid pressures are often expressed in terms of the height of a specified column of liquid, measured in units of cm H_2O, mmHg, etc. To illustrate the contribution of hydrostatic pressures to all liquid pressures in a g field or under acceleration such as in humans on Earth, a single closed vertical column of blood or water under a fixed pressure from a pump is separated into two hypothetical columns each containing only a single pressure source to show how they combine into the pressure that we can measure.

Column 2 in weightlessness has only the pressure generated by the pump, 90 mmHg, throughout the column directed as shown.

In the hypothetical hydrostatic pressure, column 3, all pressures above and below the reference level are directed downward. Thus pump pressures will be opposed by hydrostatic pressure and reduced according to height of column above the pump and to the g load. This produces the effective force that can be measured as shown between the columns.

Below the reference level the pressures' directions are congruent and add to produce the effective pressures shown. This is analogous to standing blood pressures on Earth (Fig. 4.9), and to the pressures on the Moon and Mars shown in Fig. 4.16

(continued)

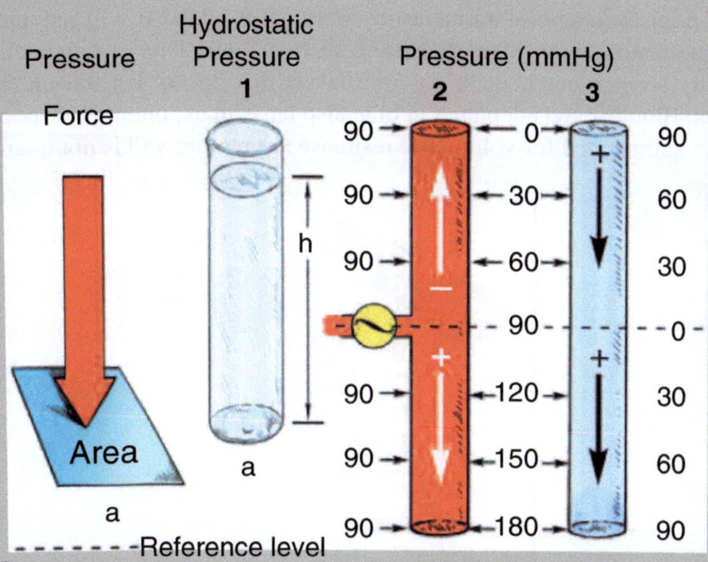

While hydrostatic pressure is simple in principle, combining it with other pressures is sometimes confusing, yet understanding this combination under various conditions is essential to understanding fluid shifts associated with spaceflight. It is easy to combine force vectors, but sometimes visualization leads to better understanding (see figure above).

Box 4.4 Compliance (Volume to Pressure Relation)

Compliance is a specialty term not familiar to many, but it is a relation necessary for understanding the fluid shifts in the human body on Earth and in space. Hence this definition: compliance is a change in fluid volume in an elastic container occurring with a change in pressure,

$$Compliance = \frac{\Delta volume}{\Delta pressure}$$

The figure on the right illustrates compliance by filling a red balloon in weightlessness with a fluid. Initially there is a rapid change in volume with small pressure change as the flaccid balloon fills, but with increasing volume it exerts increasing back pressure until its elastic limit is reached. Increasing pressure will make the balloon burst.

(continued)

If prior to this point an inelastic safety net is added it will prevent further volume change with increased pressure (red line). This arrangement is frequently seen in our bodies; e.g., the fascia that covers leg muscle compartments. Blood vessels depicted in blue also have an exponential response. This curve is illustrated for volumetric response to pressure and is not quantitative.

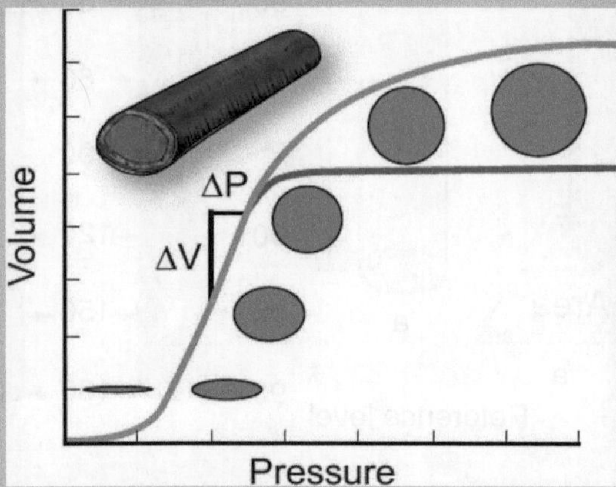

Figure 4.4 shows the critical relation between consciousness and blood flow through the brain. This flow is maintained constant over a wide range of arterial blood pressures [12]. In this subject, regulation began to fail at a pressure of approximately 60 mmHg. At this point the subject would have noted symptoms. If pressure is further reduced, LOC follows rapidly. Dependence of the brain on a critical minimum arterial pressure is obvious. Arterial pressure is in turn dependent on venous pressure, which determines the filling of the heart at every beat or stroke. If venous pressure is reduced, the stroke volume of blood pumped by each beat of the heart[4] is reduced, and the only way cardiac output and arterial pressure can be maintained is by an increase in heart rate. An increased heart rate is the first indication of OS, but venous pressure is in turn a function of blood volume, and blood volume is absolutely dependent on the body's total water. A number of control systems have evolved to maintain adequate pressure and flow to the brain under all conditions. Some are complex and a few are still incompletely understood.

4.3.2 *Circulatory Control Systems*

Circulatory control systems are described here in the following three categories: (1) control of water volume and the concentration of Na+ in it by control of intake and output; (2) control of compliance, (Box 4.4), of blood vessels and distribution of

[4] Cardiac Output = Stroke Volume × Heart Rate.

cardiac output; and (3) control of cardiac pumping. Human circulatory controls are numerous, overlapping, and integrated. Control signals may be chemical substances transported by the blood stream or neurological impulses. Because of their numbers and complexity only representative samples are shown.

4.3.3 *Examples of the Systems that Control Blood Volume and Pressures*

Fluid Volume Control Systems: Total body water, approximately 40 L, is the sum of the water volumes in blood, interstitial fluid and intracellular fluid. Ultimately this volume controls blood pressure to the brain and its level of consciousness (Fig. 4.4).

Total volume of body water is determined by the difference between oral intake, urinary output, and other losses. Thus to maintain a relatively constant fluid volume

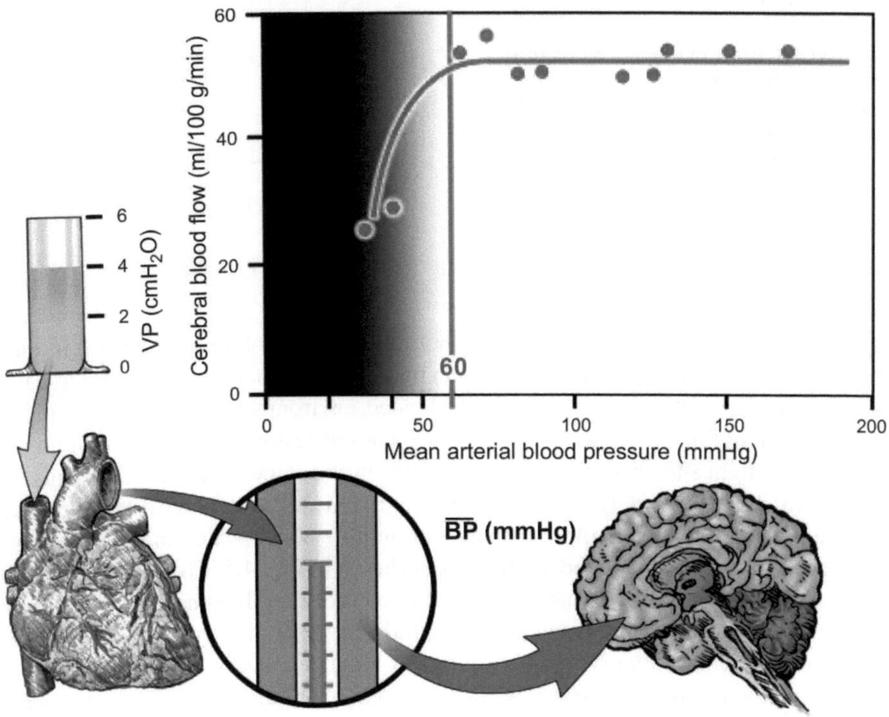

Fig. 4.4 The brain, consciousness and key circulatory pressures. Note the sharp transition at a critical level of blood pressure only slightly less than the usual pressure at brain level when standing. The illustration shows the dependence of blood pressure on a much lower venous filling pressure that in turn depends on the volume of water in the body. In the graph consciousness is indicated by white versus unconsciousness indicated by *black* (color figure online)

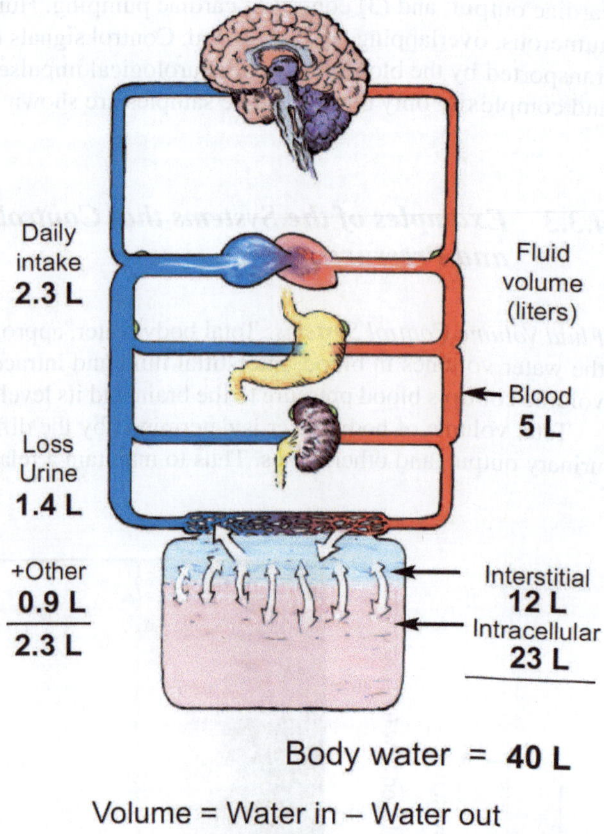

Fig. 4.5 Body water, volumes and intake-output. The circulating fluids of interest to us are virtually all water. Even blood with its cells is treated as water here. Their volumes are functionally divided into the three divisions or compartments shown: blood or vascular; interstitial or between the cells; and intracellular or within the cells. Since water is freely exchanged between all compartments blood volume and the heart's pumping capacity is dependent on total body water volume. This volume is determined by differences between intake of water and water in food and losses through urine, feces, sweat, and expired respiratory moisture

and blood pressure, water intake and output must be controlled, even in fish in their more or less constant environment. Human bodies are far more complex and live in a more variable, complex, dry, and weight loaded environment. While the details of this network of fluid volumes and pressures are not necessary to the understanding of their role inflight and on return to Earth, understanding some of the basics is required.

Figure 4.5 shows a typical volume and the flows that control the volume. Figure 4.6 is a schema of the circulatory system in which water and Na+ volume control mechanisms are added to the system shown in Fig. 4.5. The oldest and most basic of these controls are water intake under control of a thirst center[5] in the more primitive area of the brain, balanced by the kidneys' excretion or retention of water as urine. The thirst center receives a number of inputs including fluid volume. There is a typical daily loss of water to sweat, feces, and respiration of 0.9 L. Urine clears the body of much of its soluble waste using another 1.4 L of water per day. A reduced total volume of water triggers a signal to the brain resulting in a sensation of thirst in the brain until a sufficient volume of water is ingested to restore the volume of body water, normally a more or less regular event throughout the day.

[5]The thirst center in the brain is triggered by blood pressure and blood volume sensors, that decrease thirst and increase urine production when pressure and volume are increased.

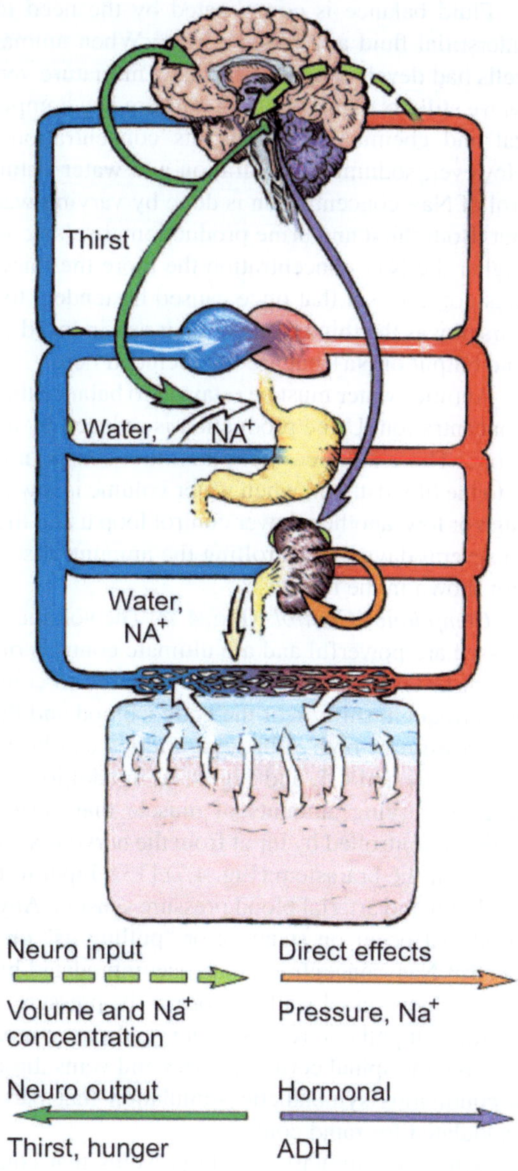

Fig. 4.6 Body water volume control loops. Ultimate control of blood pressure is exerted by water volume whose primary controls are thirst and urine production, a matter of life or death. Thirst control is by way of a center in the brain that senses blood volume and electrolyte concentration. Control of urine production begins in the kidney and is sensitive to blood pressures and composition. The anti-diuretic hormonal loop for restriction of urine production is shown. Additional urine control loops are not shown here. Osmotic pressure of Na+ plays a key role in volume control

However, it is a fact of life that much more than the necessary replacement volume is often consumed. Water must often be lost to maintain the balance. To accommodate excess fluid intake, kidneys can increase their urine production eight times or more through the action of a pressure sensitive flow control mechanism within the kidney. Rate of water excretion by urine is exponentially proportional to blood pressure; e.g., an increase in systolic blood pressure from 120 to 140 mmHg doubles urine output. Operational significance of this will be seen in attempts to water load astronauts prior to reentry.

Fluid balance is complicated by the need to control sodium concentration in interstitial fluid and blood. (Note: When animals left the salty sea in which their cells had developed, they carried a miniature version of its environment with them, as we still do.) Since sodium ions are key components in many of the body's physical and chemical reactions, its concentration must be closely controlled [13]. However, sodium concentration and water volume are interdependent. Initial control of Na+ concentration is done by varying water volume through intake and output. Both thirst and urine production are also controlled by Na+ concentration. The higher the Na+ concentration the more the thirst center will be activated. (Note: It was not altruism that once caused bartenders to put free pretzels and nuts on their bars, it was the thirst for beer their salt induced.) Slower mechanisms control intake and output of Na+ but are not included here.

At times water must be retained to balance the volume equation or to control Na+ concentration. Urine production is reduced by a hormone labeled anti-diuretic hormone. There are areas near the thirst center in the brain that release this hormone into the bloodstream when water volume is low and must be retained. If total Na+ is high or low, another slower control loop using the hormone aldosterone corrects this in several days by controlling the amount of sodium lost in the urine. This loop is not shown in the figure.

Compliance Controls (Fig. 4.7): The volume control mechanisms previously discussed are powerful and are ultimate controls of blood pressure, but they are slow. Much faster response times are required to control pressure during activity in 1 g with frequent shifting of the body's blood and fluids into and out of legs.

Pressure in both veins and arteries can be modified by tension in their walls, (Boxes 4.4 and 4.5, and Fig. 4.7). In addition to elastic components, these vessels contain varying amounts of muscle that is also elastic, but muscle elasticity is actively controlled by input from the nervous system under control of the vasomotor center in the brainstem (Fig. 4.7) [13]. Input to this center is from several areas of the brain and arterial blood pressure sensors. Any decreases in blood pressure in the head that occur on standing or "pulling gs" or from changes in blood volume to control Na+ concentration are accommodated by vasomotor stimulation of muscle in the blood vessel walls to correct the pressure change. Vasomotor signals are sent by the sympathetic (constricting) and parasympathetic (relaxing) nervous systems through the spinal cord to arteries and veins throughout the body. Normally there is a continuous sympathetic stimulation that partially constricts the vessels and is modulated for rapid control.

Selective stimulation of large veins that contain significant volumes of blood will make either more or less blood available to the heart for pumping, which is a critical factor in blood pressure. Another factor in blood pressure is the flow of blood through the various tissues of the body (Fig. 4.3). This flow is also controlled by the vasomotor center through constriction of the vessels into and out of selected tissues such as in the kidneys and GI tract. Finally, the pumping of the heart can be directly changed by a special set of nerve fibers from the cardiac portion of the vasomotor center.

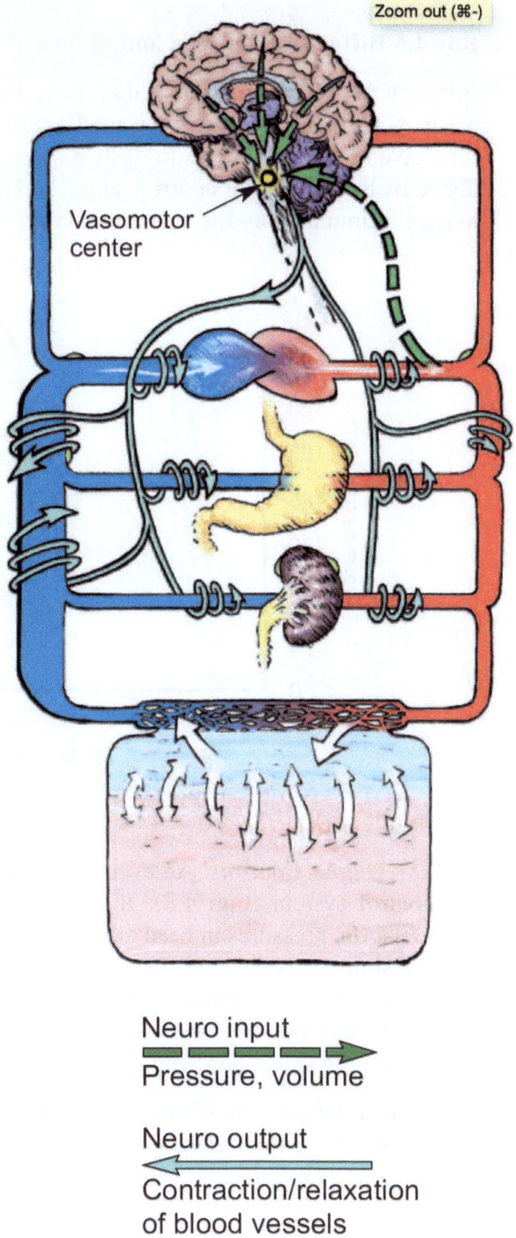

Fig. 4.7 Vascular pressure control by compliance. The next level of control is through compliance (Box 4.6) of the blood vessels through both hormonal (not shown) and neurological control loops. Vessels can relax or contract to provide the necessary volumes and pressures to maintain circulation of blood. A complex vasomotor center in the brainstem is the primary center for these and other functions

Control signals for these more rapid mechanisms are combinations of rapidly acting hormones such as norepinephrine and even faster neurological loops, that act in seconds to control [14] blood vessel diameter. Neurological control lines are shown in the Fig. 4.7 as green vasomotor loops around the blood vessels.

Box 4.5 Differences in the Compliance of Large Arteries and Veins

Both arteries (red) and veins (blue) have layers of muscle and elastic tissue, but they are much thicker in the smaller arteries and accept a smaller volume for a given pressure. The short dashed lines are with muscle relaxed, not stimulated by the vasomotor control center, and the long dashed lines are when the muscles stimulated by the vasomotor center.

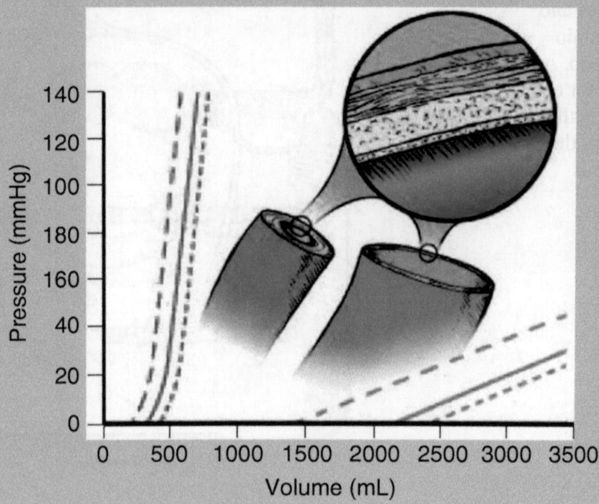

Cardiac Output Controls: An even more rapid response is found in the baroreceptor control system (Fig. 4.8). It is part of the vasomotor center that senses changes in aortic pressures in head and neck regions, especially pressure changes in the carotid sinus at the level of the angle of the jaw, and responds in a fraction of a second. Any increase or decrease in blood pressure causes an immediate appropriate change in the heart's force and rate of pumping. These signals are sent from the brain stem via the spinal cord and vagus nerve to the vasomotor center and heart.

Control System Adaptation: Gain of the basic total body water control loops remains constant, but that of the other loops adapt. A sustained change in blood pressure will produce an adaptation to the change after one to two days but remains sensitive to rapid changes. Such adaptation is significant to our interests here and will be discussed in the postflight segment.

Summary of Cardiovascular Control: Blood pressure is ultimately determined by the total volume of body fluid, primarily water. A complex series of overlapping integrated control systems maintains this fluid volume and also compensates for transient changes in pressure produced by activity. In the sections that follow, effects of activity and response of these systems on Earth, in spaceflight, and on return to Earth are examined.

Fig. 4.8 The final level of control is by rapid changes of the heart's pumping, especially its filling and rate. This is implemented through neurological and hormonal loops in the baroreceptor center in the brainstem and closely integrated with the vasomotor center. Gains of the fluid intake/output loops are very large under all circumstances. The other gains are adaptive to the environments they encounter, albeit at differing rates. In weightlessness they are generally reduced, often leaving individuals subject to postflight orthostatic hypotension

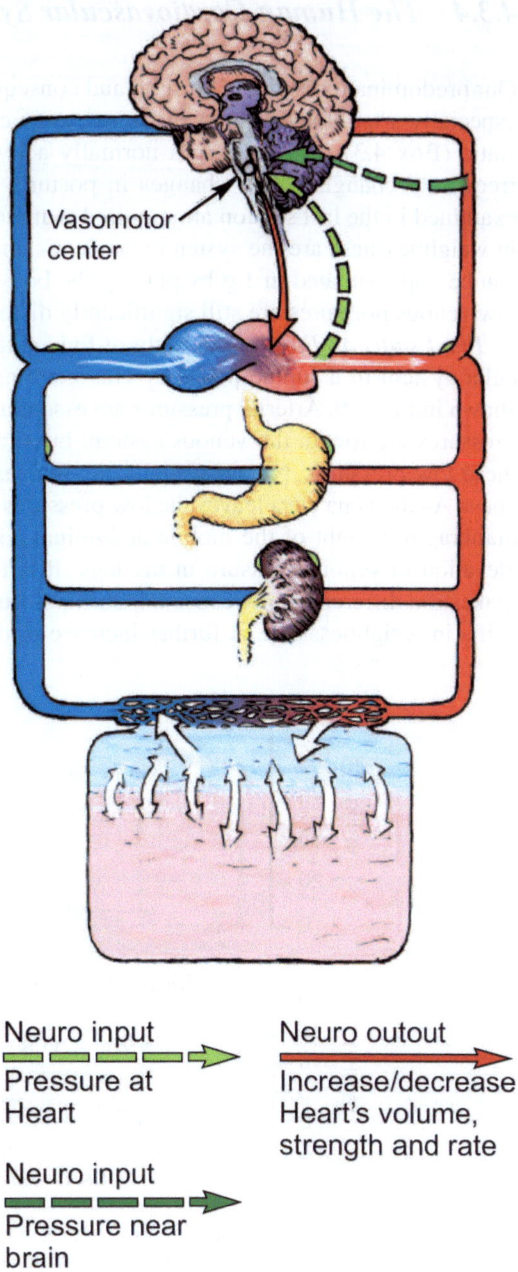

4.3.4 *The Human Cardiovascular System on Earth*

Our predominant posture is upright, and consequently our cardiovascular pressures, especially arterial and venous pressures, are a combination of systemic and hydrostatic (Box 4.3) pressures. In a normally active person hydrostatic pressures are frequently changing with changes in posture, and action by the reflex loops we examined in the last section are required to maintain adequate blood pressure. Only in weightlessness are the systemic pressures unaffected by hydrostatic forces. This can be approximated in 1 g by placing the body in a supine position. However, the low venous pressures are still significantly distorted by hydrostatic forces.

Fluid shifts on Earth: Our study of fluid shifts begins by placing the cardiovascular system in a supine position. The resulting arterial and venous pressures are shown in Fig. 4.9. Arterial pressures are essentially the same throughout the system. Pressures are low in the venous system, but there are two significant variations in these low pressures. Note especially the increased venous pressure just below the chest. As the vena cava leaves the low pressures of the chest and passes through the diaphragm, weight of the mobile abdominal contents is imposed on it causing an elevation in venous pressure in the legs. It is this slight increase in pressure that makes the difference between small shifts of fluid seen in bed rest versus the large shifts in weightlessness. A further increase in this pressure is seen in veins of the

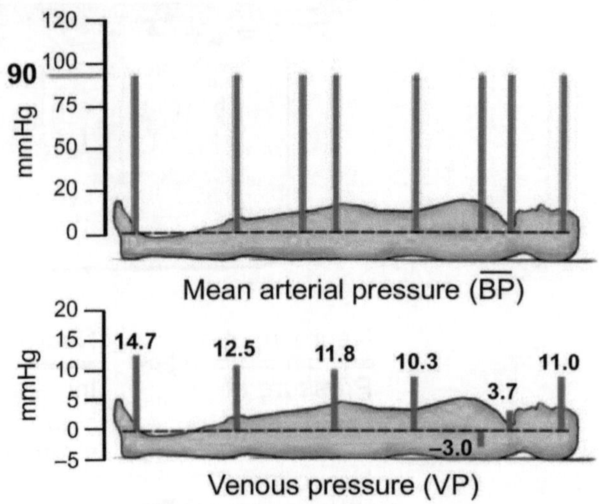

Fig. 4.9 Blood pressures when supine in 1 g. Except in very small arteries, mean arterial pressures (*red*) are essentially the same everywhere. Not visible at this scale are pressure losses of 2 and 1 mmHg at foot and head levels from flow resistance in smaller arteries. Venous pressures (*blue*) are much lower but significantly increased by hydrostatic effects. Note especially the sharp increase in pressure as the vena cava passes the diaphragm; the negative intrathoracic pressures becomes positive under the weight of abdominal contents. This presents a dam to venous return from the legs elevating their pressure in 1 g. There is also a slight elevation of pressure in the head veins. These are directly measured pressures (color figure online)

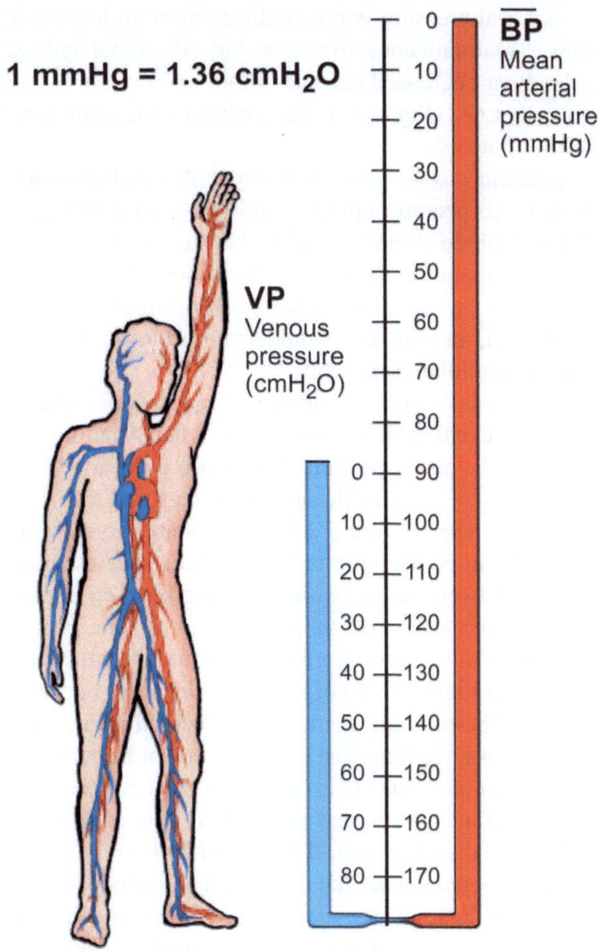

Fig. 4.10 Blood pressures when standing erect in 1 g. This subject's hydrostatic effects (See Box 4.3) reach the level of arterial pressure, thus doubling total arterial pressure at foot level and reducing it to zero only 40 cm above his raised hand. Venous pressures are relatively more disturbed. Although pressure in superficial head and neck veins is zero, intracranial venous pressure while problematic does not appear to be negative. The pressures shown are a combination of measured and extrapolated values and are illustrative not quantitative

foot, probably from constricting foot tissue. An increased pressure is also present at head level. This may be an artifact of measurement site; e.g., a small vessel with high resistance, or from an as yet unrecognized cause. In this posture, tissue fluids and blood volumes are at their maximum, and both heart rates and blood pressures reach their minimum, especially during sleep.

Figure 4.10 shows the pressures that are present when a person stands. Venous and arterial pressures increase linearly below the heart until at foot level venous pressures are equal to supine arterial pressures and arterial pressure is doubled. This is the result of hydrostatic pressures adding to arterial and venous pressures below the heart while reducing them above the heart (Box 4.3). A few centimeters above the heart, venous pressure is zero and veins collapse allowing venous pressure to drain from the head without back pressure.

Arterial pressure is now reduced at brain level to a point near that necessary for flow to maintain consciousness. Any additional hydrostatic pressures from increased g-loads or decreased cardiac output and pressures from reduced blood volume can produce LOC (Fig. 4.4) unless pressures are maintained by the previously described reflex controls.

Adaptations to vertical posture: A number of adaptations were necessary for humans to become upright with our long and large legs for bipedal locomotion. Arterial blood pressure had to be increased to override the opposing hydrostatic forces to maintain the brain's blood flow. If the standing pressures in the legs were applied to "normal" vessels like those in the upper body, both veins and arteries would fail, and the legs would swell grotesquely from fluids forced into interstitial and intracellular compartments.

Three major and several minor adaptations, when aided by the control systems, largely prevent such adverse effects when standing. Evolution in large animals such as the horse and especially the giraffe overcame the same problem by keeping as much muscle as possible out of the legs and near heart level and then binding the legs with inelastic fascia. Human legs are more complex (Fig. 4.11). Blood is prevented from pooling in the legs by three species adaptations. Major leg vessels are protected from the increased pressure by their enclosure in powerful leg muscles (Fig. 4.11c)—the first major adaptation. The second adaptation is that smaller superficial veins have increasing amounts of muscle in their walls as they progress from heart to foot level (Fig. 4.11a). Leg veins become more like arteries,[6] and arteries have increasingly thick walls. These thicker, stronger walls can withstand the higher pressures that exist below the heart. Leg muscles are also located in inelastic fascial compartments (Fig. 4.11b) to limit collection of blood and fluid in them.

A third adaptation and an effective line of defense is a pump system in the leg veins and muscles (Fig. 4.11c). Veins, both deep and superficial, have one way valves directed toward the heart such that venous blood can only flow out of the legs toward the heart. The major veins are surrounded by the most powerful muscles in the body that on contraction shorten and bulge against the veins they surround with large constricting forces that move blood up and out of the legs into central circulation. A few contractions of these muscles, only one or two steps, can reduce venous pressure from 90 to a few mmHg at foot level. In addition these capacious central leg veins have frequent connections to the superficial veins through one way valves such that blood outside the compartments can flow freely into them until removed by the next contraction of muscles.

In spite of the species adaptations to minimize the increase of leg fluid volumes from hydrostatic pressures, there is an almost immediate increase of 400 mL blood into the legs on standing, that is followed by hundreds of mL of fluid over the next several hours[7] [15]. This is a functional loss of almost 10% of total blood volume. Blood volumes are adapted to maintain a volume to accommodate this. At this point blood pressure is maintained by the previously described control systems.

---

[6] The choice of lower leg veins to replace coronary arteries is not a matter of convenience but a necessity.

[7] Other investigators quote larger volumes, but the volumes used here are based on careful and calibrated measurements on Earth and in flight.

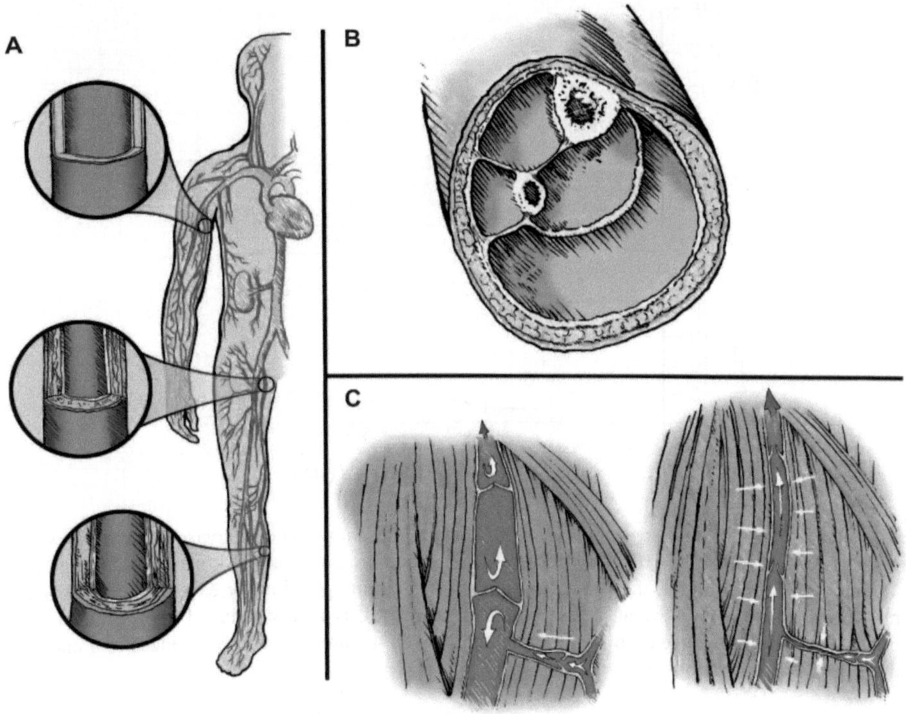

Fig. 4.11 Human legs have three natural countermeasures against hydrostatic forces in 1 g. (**a**) Veins near the surface have walls of increasing thickness as standing pressures increase from thigh to foot. All the major vessels are surrounded by leg muscle (**c**). (**b**) Muscles are enclosed by inelastic fascia that limit fluid absorption and compliance. (See Fig. 4.14.) (**c**) A combination of contracting muscle pressure and one-way valves form an effective pump that in a few steps on Earth will empty the legs of venous blood; first the central veins are emptied then connections to the surface veins allow them to empty into the central veins and also pump this blood back to the heart

In addition to blood, some hundreds of milliliters of fluid are shifted into the interstitial and intracellular volumes in the legs under the increased blood pressure. This effectively reduces total fluid volume and triggers the body's volume reflexes to increase volume by thirst and drinking and by reducing urine production.[8] Throughout a normal day these reflexes are exercised.

4.3.5 Fluid Shifts in Weightlessness

During Shuttle launches significant fluid shifting began when the crewmembers were strapped into their seats, supine with their legs elevated. Launch g loads were modest and eyeballs in (+gx) and had little effect on fluid shifting during the eight

[8] Breakfast usually has beverages and often salty items.

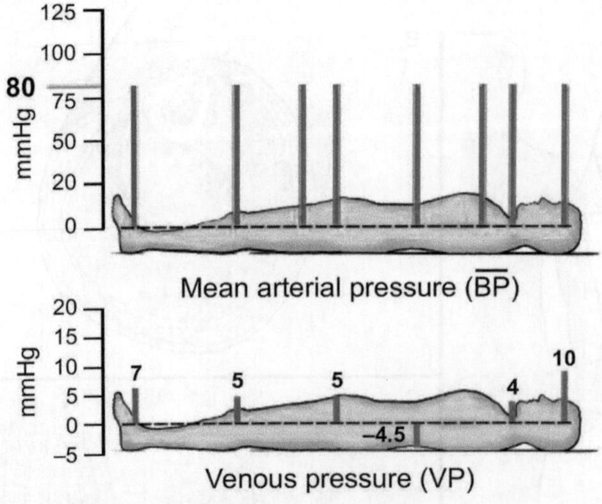

Fig. 4.12 Blood pressures in weightlessness. In weightlessness the cardiovascular pressures are undisturbed by hydrostatic effects of postural change and are more or less constant. Although not a common posture in flight, the subject is shown in an supine position for ease of comparison of pressures to those in 1 g. Mean active blood pressures are typically reduced approximately 10 mmHg or the first few weeks in flight. Only central venous pressure has been measured and the others have been extrapolated

minutes of launch. At main engine cutoff (MECO), the vehicle and crew became weightless. This produced a unique pressure distribution that is not seen in other environments (Fig. 4.12). At this time, completion a rapid shift of some 400 mL of blood from the legs into the body's circulation took place, where, unlike the usual ups and downs on Earth, it remains as an excess volume until corrected by the body's control systems.

In the absence of back pressure produced by abdominal weight on the vena cava, liters of tissue fluids also leave the legs moving towards the torso, in contrast to the fractions of liters seen on Earth when one goes from standing to supine posture (Fig. 4.13). This shift is largely complete in less than 24 h, and the effects are both visible and perceptible to the crew. Legs are smaller, faces are full and even puffy, and veins from head to foot are visibly full but not distended. Shortly after weightlessness onset many perceive a "rush of blood to the head" and most have stuffy noses and altered sensations of taste. Headache may or may not be present. Significant changes in olfaction (smell) and gustation (taste) occur and tend to persist. (See Chapter 9 on metabolism, p 216.)

The following description of such events was made by Ed Gibson, pilot on SL-4.

> "…Early in the flight we experience a sensation of head fullness. …One notices that the eyes turn red, which happened after a day or so. …the face [becomes] a little rounder and a little redder, veins in the neck and forehead become distended, and one's sinuses feel congested. These conditions did not change significantly in-flight, they just tapered off. The eyes gradually cleared, but the congested sinuses, while not too bothersome, were always

there. ...Bill Pogue, ...had the head fullness during docking, [and] experienced some headaches.... After approximately 24 hours, Bill's headaches disappeared. The congestion for all of us remained, although I think it was probably a little more severe for Bill. The Commander and I noticed this feeling of head fullness and the accompanying symptoms for the first two weeks or so.... Several variables were observed to affect the fluid symptoms and the sensation of head fullness. One was exercise. We always felt a lot better for about a half hour to two hours after we exercised on the bicycle.

"We were also able to see the leg volume changes because of the fluid shift. First of all, we could see the muscles shrink when we got up there. It was obvious to the eye, and it could be confirmed by measurements. A couple of times we measured the calf after exercise on the treadmill. It increased about a half inch or so after a reasonable amount of exercise and then it shrank down fairly rapidly (15 to 30 minutes) as soon as we stopped." [16]

4.3.6 *Volume Changes*

Clearly during weightlessness the trunk and head[9] receive an excess of blood and fluid shifted from the legs. Kidneys might be expected to produce large volumes of urine, but in all the subjects whose urine volume was measured inflight, only one had an increase in volume during this period. This is in contrast to the rapid diuresis seen on Earth in response to much smaller fluid volume increases, say after a few bottles of beer. Apparently this volume is lost through decreased intake triggered by decreased thirst from its center, which senses an increased fluid volume and blood pressure Loss of the fluid volume shifted from the legs was followed on Skylab 4 over the next 3–5 days by mass measurement (Fig. 4.13) [4]. Body mass was lost equivalent to the individual's loss of mass from leg fluids. Leg compliance was simultaneously increased [17]. Both of these measurements appeared to remain stable with increasing mission length.

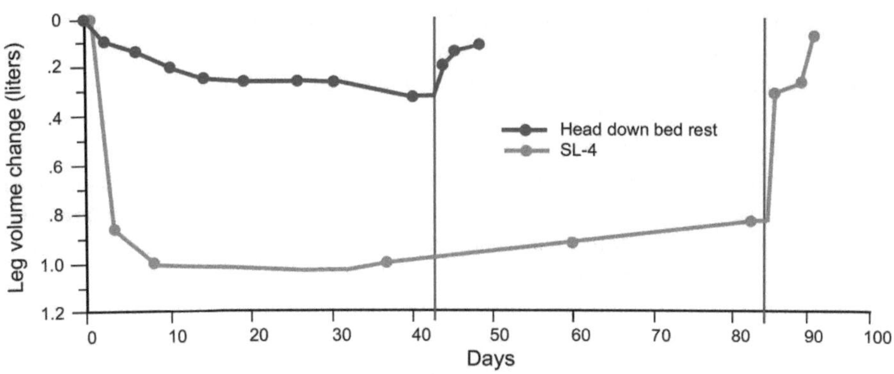

Fig. 4.13 Leg volume changes in BR and weightlessness. Measured differences in one leg comparing spaceflight to a bed rest study. The slow continued loss in bed rest appears to be the loss of muscle

[9] Measurement of arm volumes show no increase.

4.3.7 Cardiovascular Changes

Blood pressure (BP) and heart rate were initially reduced, especially during sleep [18], and central venous pressure is lower than on Earth. Ambulatory blood pressures were still reduced at 3 and 6 months on ISS but 24 h mean values of heart rates were unchanged [19]. The assumed vasodilatation in this paper based on HR and cardiac output was also found in measurements of femoral vascular resistance after flights of up to 6 months duration while cross sections of jugular and femoral veins were increased [20]. These physical changes are indirect products of fluid shift and are in discussed in Chap. 8. Muscle pumping remained effective throughout SL-4.

Gain reduction in control systems: It was noted in the description of cardiovascular controls that most controls adapt to changing conditions. The basic fluid volume controls appear to retain their high gains and to function remarkably well during the unusual large fluid shifts that occur during launch and reentry. However, the gains of the vasomotor and baroreceptor controls appear to be significantly reduced by bedrest and weightlessness, especially in some individuals [21, 22]. These gain reductions are appropriate adaptations to weightlessness and become apparent only on return to Earth.

A study of 30 astronauts, 1–2 h post landing after flights up to 16 days, contained a sub group of eight who could not stand quietly for 10 min. All astronauts had been fluid loaded prior to reentry with 1 L of water and 8 g of Na+. In this subgroup findings included low vascular resistance, lower blood pressure and norepinephrine with increased heart rate. This subgroup also had lower blood pressure and heart rate preflight.

These findings are consistent with blood vessels that did not constrict adequately, presumably from reduced hormonal and nervous signals from the vasomotor and baroreceptor centers. These systems adapt, and readapt more slowly. It had been shown in another postflight group that the baroreceptor neurological control gains had been reduced in as little as 2 days of weightlessness. Another study of 35 astronauts (30 men and five women) found that none of the women could complete a 10-min stand test versus 80% of the men who could. Other findings showed a similar reduced control by the vasomotor/baroreceptor controls.

Two studies examined the effects of short (4–30 days) and long (4–6 months) flights on cosmonauts' cardiovascular control systems. One study compared the results of stand tests by four cosmonauts after 4–30 day flights to the results from ten cosmonauts after 4–6 month flights. A combined total of four did not complete the tests. One woman completed the stand test and showed no effects of flight. No significant difference was noted in baroreceptor response in the two groups. It was concluded that control system changes were complete by 14 days and remained unchanged with increasing mission length. A second study of eight male cosmonauts over 162–196 days reached the same conclusions [22]. These findings are consistent with blood vessels that did not constrict, presumably from reduced hormonal and nervous signals from the vasomotor centers and physical change.

Both bed rest [21] and weightless [23] studies showed that females have reduced orthostatic tolerance, apparently from reduced release of vasopressor hormones.

In some it appears that gain of both hormonal and neurological blood pressure control loops are reduced in weightlessness, especially in women. Significant losses and gains, on return to 1 g, can occur in days but losses continue to a stable minimum in a matter of weeks.

Exercise, or lack thereof, produces significant functional and physical change in the circulatory system, on Earth or in space. These changes produce some of the alterations seen in fluid shifts and are secondary only to them in orthostatic effects. This is discussed in Chap. 8.

4.3.8 Rationale for Fluid Shift and Loss

Blood and fluid shift and loss are produced by the absence of hydrostatic pressure. The less compliant leg tissues force blood and fluid into the more compliant body tissues. A question sometimes heard is, "Why does so much more fluid leave the legs in weightlessness than when going from horizontal to vertical on Earth, as in rising in the morning?" On Earth weight of abdominal contents compress the vena cava and raise leg venous pressure 10–15 cm of water or more depending on gut size. Leg tissue compliance (Box 4.4), volume versus pressure, is large (Fig. 4.14), allowing large volumes of fluid to be trapped in leg tissue from the small increase in venous pressure produced by abdominal weight in 1 g. This back pressure is absent in weightlessness allowing surprisingly large volumes of fluid to leave the legs for body.

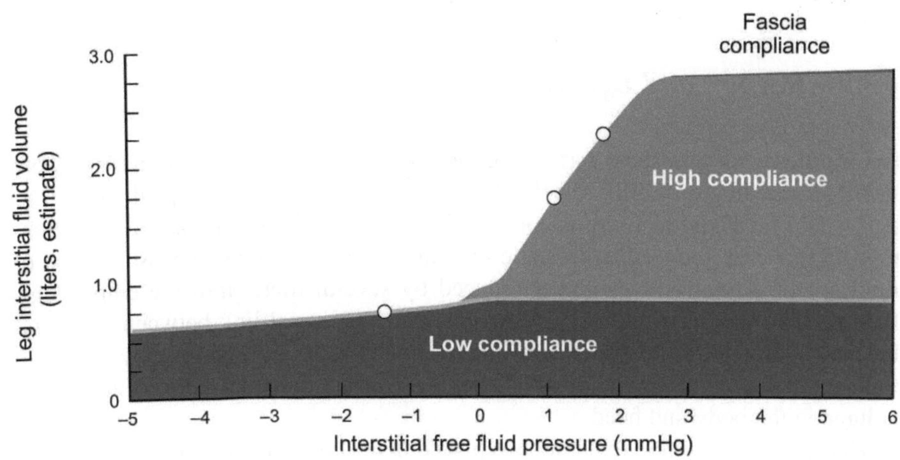

Fig. 4.14 Tissue compliances in the human have not been adequately measured, and the values shown here are estimates. The curve is a composite from several studies. The difference between fluid volumes shifted on Earth versus that in weightlessness is illustrated by the three points on the curve. The upper is standing erect and the mid point is lying supine. In spite of the steep slope only some 0.6 L of fluid are shifted out of the legs. When the weight of the abdomen is removed in flight venous pressure is further reduced and tissue pressure falls below the break point shifting another 1.0 L for a total of 1.6 L

Such a volume shift into the body should increase both venous and arterial pressures but instead decreases them for unknown reasons. One possibility is that compliance of body tissue, which is normally much greater than leg tissue, is increased even more inflight, possibly because of the removal of tissue weight forces. Blood volume is increased in the upper body, but reflexes will reduce this to normal levels by reducing plasma volume. This reduction in plasma leaves an excess concentration of red cells, which is corrected by stopping production of red cells for days to weeks after the start of weightlessness. Increased fluid volumes do not produce the prompt increase in urine production that would result from drinking an equal volume of water on Earth. Instead the thirst reflex and liquid intake is reduced. These reductions may be caused by: (1) the fluids are normal body fluids, and/or (2) decreased arterial blood pressure.

In weightlessness there is none of the usual shifting of volumes under hydrostatic pressures that occurs more or less constantly with changes in posture in 1 g. Because the gains of vasomotor reflex loops are plastic, they are reduced to suit the new situation. This is consistent with the initial reduction in blood pressures and heart rates, and their slow return to normal.

After a few hours in weightlessness the legs have been emptied of several liters of fluid and some 0.4 L of blood. These volumes are discarded over several days. As will be seen on return to 1 g, gains have been reduced in some of the reflex loops. After longer times in weightlessness, say months, the fluid volumes have remained stable, but reflexes continue to adapt to some reduced but stable level. Judged by 1-g standards, the crew is dehydrated, anemic and has increased leg compliance and altered reflexes.

4.3.9 *Reentry and Egress*

On Shuttle reentry, posture was seated, and hydrostatic pressures on the legs were less than standing on Earth. Acceleration loads reached a transient peak at +1.5 gz that would be trivial to Earth adapted individuals but was significant in this situation. In this case reduced hydrostatic pressures were sufficient to cause a flood of fluid out of a total body volume, reduced by several liters, into the empty legs. During this time the circulatory system becomes a crucial link between body and leg blood volumes. Increased arterial and especially venous pressures rapidly shifted fluid out of the blood stream into the "dry" legs, further reducing a diminished blood volume in the body and head.

During reentry, unless this fluid is replaced in the blood stream from body tissues as rapidly as it is lost, blood volume and pressure may fall to critical levels. In spite of the specified preloading of fluid with saline prior to reentry, one author (WT) recalls one crewman who experienced orthostatic hypotension during reentry and relieved it by immediate and continued ingestion of large amounts of water. This is discussed in the summary. As previously noted there have been rumors but no confirmation of orthostatic problems in pilots during reentry.

After landing there is the usual urge to get back on Earth. After struggling through the neuromuscular, Chap. 9, and vestibular problems, Chap. 3, of standing erect, it is here that maximum change in hydrostatic pressures occur, and one is most vulnerable to OS and LOC. There were a small but significant numbers of both types of incidents on Shuttle. In the great majority of such cases lying supine for a short time allows the blood volume to "catch up", and the individual is able to walk off with the rest of the crew. In some rare cases administration of intravenous saline was required. With the increasing diversity of individuals flown a very few had to be carried off. The summary contains a discussion of the incidence and prevalence of such events.

There is controversy over the primary cause of reentry and postflight OSI; i.e., inadequate fluid to keep the circulation and heart sufficiently filled to maintain blood pressure versus loss of control systems gain. Because the standard treatments are intravenous replacement of fluids or by simply allowing time for redistribution to occur, it appears that fluids play a dominant role during reentry and the subsequent egress from the spacecraft. While control gain can play a crucial role, *it is only when fluid has been replaced that one can attribute OSI to control system failure.*

4.4 Countermeasures

There are three major areas of concern in providing countermeasures to the effects of weightlessness on the circulatory system: (1) replacement of fluid volume; (2) control of leg volume, and (3) compensation for reduced gain of control loops. The relative importance of these factors varies with mission phase that for purposes here, are identified as: return to Earth (in spacecraft), exit of spacecraft, early postflight, and late postflight. The order that counter measures have been introduced was one of expediency for the causes of the problem were unknown until Skylab 4.

4.4.1 Control of Fluid Volume

During the Gemini Program, cuffs placed on arms and legs were periodically inflated and released [20, 24]. Part of the concept was valid, for low constricting pressures around a limb will increase venous pressure and rapidly "trap" fluids in empty legs; however, when pressure is released the effects are rapidly reversed.

In an attempt to replace the fluids lost prior to reentry, Russian investigators initially added salty bouillon, a Russian dietary item, prior to reentry. Currently this is reported to be 0.9–1.2 L of water or juice with 2.7–3.6 gm of salt. This was initially followed in the American program by the addition of approximately a liter of normal saline to be ingested over 1–2 h prior to the start of reentry. Fluid loading with a liter of water plus eight grams of salt tablets to retain the water became part of reentry preparation protocol, [25] and as noted, this may have been increased to 1.9 L. Note that this at best replaces only half of the volume lost that can be as high as 4 L. There are reports of individual experiencing problems ingesting this volume of fluid (or less) and salt tablets.

4.4.2 Control of leg Volume

Brazelet was an early Russian attempt to reduce the fluid shifts out of the legs for up to 30 days of flight, and is still used for shorter periods. They are custom fitted elastic bands of unspecified tension worn continuously around the thigh, over other clothing. This will trap some volume of blood and fluid, depending on their tension, and are claimed to eliminate the discomfort of the rush of blood to the head [26]. There was no measured postflight difference in those who had worn them [27].

Refilling the legs and replacing lost fluids prior to reentry could solve much of the OSI problem. This was attempted by using a negative atmospheric pressure chamber on the Shuttle to "suck" fluids back into the legs. Such devices were large, cumbersome and essentially immobilized the user. Equally important, any shifts of fluid into legs were rapidly reversed. American studies of such devices inflight demonstrated the inherent problems in this approach. Russia continued development and their CHIBIS 3 is portable with a pump and computer with a number of programs. Details and results are unknown.

Most of the countermeasures effort has been in constriction of leg volumes during reentry, and some times afterward, by counter pressure garments such as g-suits or some variation thereof. By placing constricting pressures on legs and lower abdomen, compliance and fluid shift into the empty legs is greatly reduced. Such counter pressures will also prevent any refilling of empty legs. Anti-g suits (AGS) were developed in World War II for protecting pilots against LOC during transient high-g maneuvers and subsequently used by NASA and the Russian space programs before the cause of the problem was understood. They became standard equipment during reentry and landing late in Apollo and remain in use today.

American AGS use elastic bags inflated by air pressure to provide constricting tension around lower abdomen, thighs, and lower legs. Initially they were flight qualified military suits with a subject controlled inflation maximum pressure of 78 mmHg. There was nothing published on an operational test of their effectiveness until after the loss of Challenger, but anecdotally they were effective. After the loss of Challenger a partial pressure suit labeled "launch and entry suit" (LES) was added to the flight garments for launch and reentry plus the AGS [28]. This produced excessive heat for the wearer, which further reduced orthostatic tolerance. A liquid cooled garment (LCG) was then placed next to the subject's skin. This ensemble was realistically evaluated during reentry and on standing in 2003 [28]. Without the LCG blood pressure and heart rate were measured in eight subjects who did not inflate the AGS versus 24 who did. There was no significant change in heart rate, but there was significantly increased blood pressure at peak g loads on reentry and on first standing. There was no difference in blood pressure in those with and without LCG during entry and landing, but those with LCGs had significantly lower heart rate. Subsequently a total body pressure suit, the advanced crew escape suit (ACES) accommodate both the AGS and LCG. In addition to interfering with normal postflight locomotion and other activities, the NASA AGS lost its pressure soon after disconnection from its source in Shuttle.

The *Kentavr* is Russia's AGS, with custom fitted elastic garments enclosing lower abdomen and thighs plus separate lower leg "stockings". Counter-pressure exerted in all areas is 33 ± 5 mmHg. The only published evaluation of the Kentavr compared its effectiveness to that of the American AGS in 1 g using subjects whose body water volume was reduced by diuretics, a condition with little relevance to postflight OS. These suits are apparently worn for several days postflight. Not surprisingly, edema of uncovered perineal, knee and ankle areas is a problem.

It is remarkable that American g suits intended to prevent LO during brief loading of up to seven to nine gz or more in a normal body are effective in this very different situation. The best that can be hoped for is that LOC will be prevented during reentry, but the cardiovascular system will be subject to a sudden stress when suit pressure is removed. This is the point where LOC has been most common.

Commercial stockings with graduated counter pressures have been proposed as a replacement for the AGS presumably after landing. Jobst© ankle to groin elastic stockings are sized by measurement to fit the individual and provide increasing pressures of 6 mmHg at the groin to 18 mmHg at the knee to 55 mm at the ankle. Effectiveness of these garments was evaluated in five astronauts by replacing the AGS with the stockings at landing and 2 h later after walk-arounds, greetings etc. Stand/tilt table testing was conducted with an array of measurements. Results of their testing were compared to those of nine astronauts without stockings following the same protocol. While no stocking wearer failed the tilt table test, this was after 2 h of adaption. The only statistically significant difference between this group and the control group who did not wear stockings was a higher cardiac output.

4.4.3 An Improved Countermeasure

It could be reasonably predicted that if countermeasures were based on known biophysics and not expediency, much better results should be obtained. For example, if the concept of shifting fluids back into the legs and replacing lost fluid volumes were integrated and implemented pre-entry, the majority of this problem would be solved, leaving only reduced control gains and a modest anemia to be dealt with on short flights.

The following schema should meet the above criteria. One to one-and-a-half days pre-entry a simple elastic "cummerbund" with 10–20 cm of water pressure would be placed around the abdomen, and the subject would increase fluid intake until urination is increased. This would replace the missing counter pressure of abdominal contents on Earth and should "trap" a leg volume equivalent to lying supine on Earth. Increased urination would indicate that body fluid volume had been replaced.

The abdominal band would not interfere with reentry garments, and when 1-g loads were reached the band pressure would be released. Advantages of this arrangement are that the legs would be partially refilled, and fluid volumes used in refilling, would be replaced. Fluid replacement would need to continue postflight because vertical posture will shift some additional fluid into the legs. This approach has repeatedly been rejected for several decades in spite of an extensive theoretical and experimental basis that supports it.[10]

4.4.4 Compensation for Reduced Gain of Control Loops

In a few rare cases orthostatic symptoms persist for days or even weeks after landing, long after fluid shifts have been completed. In these cases it appears that readaptation of control gain is delayed. There are conditions on Earth that reduce the gain of the vasomotor control systems producing orthostatic intolerance, and such patients may be helped by drugs that directly and continuously stimulate the constriction of arterial and venous vessels that subsequently decrease compliance and increase blood pressure. Miodrine is such a drug and has been investigated in bed rest studies, but it was not used on astronauts because of side effects. Use of such drugs should be reserved for the rare postflight cases in which gain is reduced and water has been fully replaced.

4.4.5 Physical Rehabilitation

Body fluids, their distribution and circulation can be significantly affected by the subject's usual physical activities. Many of the changes seen here were caused by the unknown or uncontrolled pre, in and post-flight exercise. This is addressed in Chap. 9.

4.4.6 Incidence of Orthostatic Intolerance

Data on the incidence of postflight OSI is complex, incomplete, and often conflicting. Brief descriptions of some of the problems encountered follow. Signs and symptoms are individually variable and also variously described. The most common and reliable signs and symptoms follow in order of increasing severity.

An **objective increase in heart rate**, when combined with decreasing blood pressure is an early and definitive sign. It becomes a symptom when perceived by the subject.

Light-headedness is a sensation of losing contact with the environment often accompanied by an intuitive feeling of incipient loss of consciousness.

---

[10] This statement is based on one author's (WT) personal experience.

Vertigo, a false sensation of movement in the environment or of movement of the environment, is often considered a clinical symptom but is more likely to be a vestibular symptom in this situation.

Visual changes such as dimmed vision, loss of contrast or constriction of the visual field is significant.

Loss of consciousness (LOC) is the definitive sign in the absence of other causes. Other complexities include marked differences by gender, duration of flight, and variation in fluid and salt preload. Clearing of symptoms with horizontal posture and an increase with vertical posture is also a definitive sign. In addition, symptoms rapidly vary with time and also vary with mission phase, but this is seldom recorded.

As noted there is a large difference in susceptibility to OSI measured by tilt table or stand tests versus unconstrained physical performance. Stand tests immobilize the muscle pump, a major countermeasure to OSI. This insures a greater failure rate in stand testing than in unrestrained activity.

Stand test failure rate incidence (%)[11]			
	Short mission	Long missions	
	<21 days	>4 months	
All	20	80	
Male	<20	71	
Female	80–100	100	
Functional symptoms reported postflight (%)[12]			
	Short mission	Long missions	
	<21 days	>4 months	
All symptoms	8.2	Disabling symptoms[13] 7.9	
Male	7.0	0.02	
Female	8.0	8.3	
These data are subjective			
	Short	Long	Female
Sample population	659	567	102
Totals: 131 missions, duration (days)	6456	5483	1082

There are no reports of OS incidence comparing it to mission time except for such crude grouping into short and long missions. Two such reports show very different results. The temporal profile of potential disability is crucial for the planning of long missions. A major concern is whether or not there is a constant increase of disability with time. However, some reports suggest there is a rapid increase of OS during the first 2–3 weeks of a mission, followed by a relatively stable plateau with little or no increase in incidence. The one relatively consistent area of reporting is in male versus female susceptibility to OS in normal 1-g testing and in postflight tests.

[11] These data were provided by NASA's LSAH Program.

[12] Current incidence reports of long missions are contradictory. Russian reports for the same periods are approximately 40%.

[13] Disabling = LOC or visual problems.

The most problematic question concerns functional disability. For purposes here, tunnel vision and level of consciousness are two examples. Details from the Shuttle postflight medical debrief data follow.

Symptoms	Male	Female	Total
Fainting/blackout	1	3	4
Lightheadedness	32	13	45
Rapid heart rate	8	4	12
Vision *(Tunnel Vision, Visual Blackout, Visual Grayout)*	0	2	2

These data are underreported especially as to the level of consciousness. For example, the author (WT) observed an unreported male LOC and the incident was not added to the data. The actual number of LOC incidents is probably several times higher than the numbers shown, especially in the male population. A reasonable estimate lies somewhere between the exaggerated 20% "failure" rate in stand tests and the less than 1% shown in this data, say 5% or less. While female functional failure rates are higher and more prolonged, including one LOC in a press interview, a number walked off Shuttle like their male counterparts.

Stand tests may provide more practical data for mission planning in key areas such as the prediction of liability to OSI. There are some convincing data that stand tests coupled with other studies can reliably predict those who will be significantly affected and those who will not.

Other systems have been compromised by the shift and loss of fluids. After only a few days in space approximately 0.4 L of blood, or the amount given as a blood donor, has been removed. This significantly compromises aerobic activity in space and especially on return to Earth. It is addressed in Chaps. 8 and 10. The senses of taste and smell are also affected and will be addressed in Chap. 10. Changes in vision are significant and will be addressed in Chap. 5.

Recovery of orthostatic tolerance is also important, especially in planning missions to other bodies. There is ample experience and data to indicate that recovery of orthostatic tolerance is rapid after flights of 10–12 days or less. On flights of 1–2 weeks, with adequate intake and activity the fluid replacement can be completed in a matter of hours, and normal activities are usually possible after a night's sleep. My neighbor's comment was typical, "He was out as usual the next morning."

Recovery time of control system gains occur in a matter of days, although in a few severely affected individuals it can be a week or more.

4.5 Weightlessness Simulations in 1 g

In an effort to study fluid shifts and their effects three simulations of weightlessness have been employed on Earth: water immersion; bed rest (horizontal and tilted head down); parabolic aircraft flights; and animal studies using tail-suspended rats. Unfortunately, there have been almost no comparisons of effects seen inflight to those revealed during simulations. Far too many investigations simply claim the effects are equivalent but little data seems to back up such claims. Although simulations have value, to assume that they provide equivalent conditions to 'weightlessness' is not correct and some of the differences are discussed in the following paragraphs.

4.5.1 Water Immersion

If differences in respiratory pressures are controlled, counter balancing of hydrostatic pressure by water immersion should produce a close approximation of weightlessness. Experiments using water immersion as a simulation for weightlessness were first conducted as early as 1945. Some studies entailed underwater breathing gear that allowed subjects to be completely immersed. Others experiments entailed immersion up to the neck. Many experiments were conducted using animals and humans as subjects from 1945 and into the mid 1960s (see bibliography by Duddy [29]).

4.5.2 Bed Rest in Horizontal Position

Horizontal bed rest is the equivalent to a usual sleeping posture on a flat bed. The horizontal position of the body alleviates the pull of blood and other fluids away from the head and down towards the feet. However, while the longitudinal gravity forces are removed significant transverse pressures remain. The most significant of these pressures is the weight of abdominal contents on the inferior vena cave, that returns venous blood from legs to heart. Depending on the size and weight of the belly contents this horizontal bed rest produces approximately ten centimeters of water or more of venous pressure on leg tissues keeping liters of fluid in them. (See Compliance, Box 4.4.) In short, while horizontal bed rest can simulate some of the effects of weightlessness on the body, the effects of gravity on a supine body limit this method of simulation.

4.5.3 Bed Rest with Head Down Tilting

In an effort to more closely simulate fluid pressures in weightlessness subjects are often tilted head down, typically 6°–7°. This adds or subtracts a vertical component of 1-g hydrostatic pressure proportional to the sine of tilt angle times distance below or above the heart. In a typical subject this might increase venous pressure by 2.7 cm of water at eye level. Leg venous pressure will be reduced according to distance above the heart, which in a typical subject might be 8 cm of water at foot level. This, however, is still above the point on the compliance curve that maintains large volumes of fluid in tissue (see Fig. 4.14). Moreover, back pressures from the abdominal contents are slightly increased by the geometry of tilting (see Fig. 4.15).

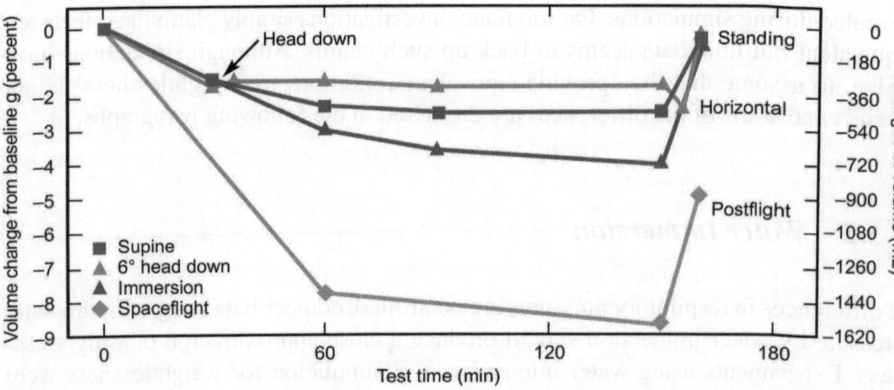

Fig. 4.15 Measurement of leg volume changes. Measured short-term total leg volume changes in six subjects in three 1-g simulations and in three subjects on Shuttle. Note the rapidity of response in flight

Unfortunately, basic measurements such as venous pressure, which requires no more than the venipunctures done millions of times a day on Earth, have rarely been done. Head down tilting was suspended briefly until new safety monitoring could be put in place based on ocular effects discussed in Chap. 5.

4.5.4 Weightlessness in Zero-g Aircraft

Near weightlessness can be achieved briefly in aircraft that follow a parabolic flight path relative to Earth. The airplane flies first on an upward trajectory at a 45° angle. When engine thrust is cut and the nose of the aircraft pitched downward both the plane and its occupants freefall and essentially weightless for a brief period of time. Such flights are good simulations of what occurs when spacecraft engines are cut off and propulsion forces suddenly go to zero. However, differences in gravity environments just prior to the event must be considered. The parabolic flight path of these aircraft produces a change from say 2-g pull-ups to weightlessness and on a typical flight multiple parabolas are flown. Only 25–45 s of weightlessness are possible in the aircraft during each parabola, much too short a time for the vascular system to equilibrate.

4.5.5 Hind Limb Unloading in Rats

Animal studies account for a great deal of what we know of the human body and its treatment. The majority of our DNA is also found in animals. One area of DNA notably different in the species is that which controls size, shape and adaptation needed to produce *Homo sapiens*. In spite of this, large sums of money and other resources continue to be expended on animal studies. A favorite animal model is the rat suspended by its tail to produce a head down tilt angle of 30°. The rat is typically allowed to move about its cage using its front limbs. A glance at the body configuration of a rat versus man shows profound differences in size (10:1) and configuration. Like the majority of quadrupeds most of the rat's body including leg mass is on or near a horizontal line with the heart, equivalent to continuous bed rest in humans. There is no significant tissue mass under large hydrostatic pressures to collect or shift blood and tissue fluid and none of the complexities and problems associated with human legs. The small changes in blood pressure produced by tail suspension can hardly effect body fluid distribution. From a biophysical viewpoint the rat or any other small animal is a most unsuitable model for study of fluid shifts with gravity loading or unloading in humans.

4.5.6 Simulation Versus Flight Results

Figure 4.15 portrays mean results from 2½ h of bed rest and water immersion simulations with six subjects compared to fluid shifts in three astronauts inflight [30]. Baseline mean of leg volumes were 9.3 L for simulation subjects and 8.1 L for the astronauts. At the end of 2.5 h there were total shifts of blood and fluid of 1.5% of leg volume for 6° head down, 2.5% for horizontal and 3.8% for water immersion versus 8.6% for weightlessness. There are no obvious explanations for the unexpectedly small shift produced by immersion or for slightly smaller shifts by head down tilting. There are no other comparable same subject studies, but isolated examples show results similar to these. In sum, one gravity simulations may be useful for studying some effects of weightlessness on the bodies of animals and humans, fluid shifts that occur during simulations are not comparable to those from spaceflight and their use may account for some of the countermeasure failures witnessed during the history of manned spaceflight.

4.5.7 Summary of Fluid Shifts, Effects and Countermeasures

When the human body is placed in weightlessness by spaceflight, a unique condition results that has not been previously encountered. The cardiovascular system adapts to a new state (weightlessness) that cannot be fully simulated on Earth. With removal of hydrostatic and other weight induced pressures, some 0.4 L of blood and 2–4 L of tissue fluid is shifted from the legs to the upper body and head in a matter of hours leaving the legs visibly empty and the head congested. This excess of body fluids is detected by volume control systems and removed over the next several days by reduction in thirst instead of the expected diuresis. The blood then have an excess of red cells that are removed from the circulation in a matter of days. Changes in vascular sizes are reported to occur and to remain stable after that time. Gains of vasomotor and baroreceptor controls are reduced over 10–14 days.

Effects of disuse atrophy can significantly affect fluid volume, circulation and control. The totality of disuse effects may require a month or more to be completely expressed, and are then dependent on subject activities and exercise.

On egress from the spacecraft after return to Earth, and for some hours following, there is a rapid reversal of the previously described fluid changes. Gain of control systems may take several days to recover and in rare cases with severely reduced gains, a week or more. There have been no operational studies of reentry without countermeasures except for a few stand tests. Virtually all reentries used g suits as countermeasures, and most had some amount of fluid preload. Under these circumstances there are no reported incidences of OSI during entry in spite of persistent rumors of rare but significant events. On egress a small number of LOC incidents have occurred. There appears to be documented evidence that woman are much more susceptible to than men.

In a few cases susceptibility to OSI may be prolonged; e.g., one woman had LOC while standing during a press conference several hours after landing. In rare cases this susceptibility may persist for days or more. Available data probably underreport operational incidence while stand testing exaggerates it. Initially the incidence of postflight OSI rapidly increases with flight duration, but it appears to reach a plateau in a matter of weeks with unknown changes after that. There are differences of opinion as to time that stability is reached and rate of increase, if any, of incidence once "stability" is reached on long flights. Current countermeasures, especially drugs, have limitations. However, stand tests done pre- and postflight appear capable of identifying the subpopulation that is particularly susceptible to OSI.

The termination of the Shuttle program drastically changed flight operations and biomedical concerns. These changes include:

- A much reduced population launching and landing
- A shift to what was once considered long duration flights with 6 months now more or less the norm and the first American flight of a year was just completed
- Highly automated entry and landing
- Pilots and crew more or less recumbent with knees flexed on entry of return vehicle (in Soyuz)
- Post landing and egress aid sometimes not immediately available and may be delayed in off nominal situations.

While fluid shift adaptations and reflex changes will presumably have approached their maximum, there may be less orthostatic stress during entry and landing. Change to a smaller population for longer periods will allow the longitudinal studies needed for planned exploratory missions to the Moon, Mars, or asteroid. Many basic studies of tissue and vascular pressures, volume and compliance remain to be done, especially their variation with time in weightlessness.

4.6 The Future

There are current needs for improvement of the following:

- Incidence data that is consistent with the operational environmental operations in addition to tests under controlled circumstances.

- Knowledge of basic physiology both in 1 g and weightlessness; e.g., venous pressure and tissue compliance.
- Better countermeasures based on the biophysics of the circumstance.
- Monitoring and documentation of operational events in addition to stand testing.
- Recognition of the limitation of 1-g bed-rest simulations and the unsuitability of small animals for study of this phenomena in one or reduced g.
- Accurate individual timely pre, in and post flight volume measurements-One size does not fit all

Missions with stays on reduced g bodies (e.g., Mars, Moon) followed by reentry and return to Earth are under consideration. Problems from circulatory adaptations during approach and landing followed by reentry and return to Earth are under consideration. Problems from circulatory adaptations during approach and landing should be much less on the outbound leg. Measurements of such effects, especially of volumes, pressures and compliances, will be needed. The major problem will be return to 1 g, but initially this should be less severe than some long duration Russian flights to date.

At this time only speculation on the effects of landing and living in the reduced-g environment of other celestial bodies is possible.[14] An example of the effects of reduced hydrostatic pressure on cardiovascular parameters is shown in Fig. 4.16. The changes in mean arterial blood pressure in partial 1-g environments are reasonable, but they need to be proven. To make a reasonable extrapolation to the volume changes in the obligatory fluid shifts is not possible in the current absence of compliance data in weightlessness. Such changes should be expected to range between those seen between 1 g and weightlessness. There should be little danger of OSI in a Martian landing even after a long time in flight and much less danger of OSI in a lunar landing. Much more objective and basic data on 1-g landing after long flights are needed before effects of returning to Earth after a Martian or lunar trip mission be reliably predicted.

Values shown for venous pressure in Fig. 4.16 are a conglomerate of measurements of venous pressure in 1 g, some over half a century old. Venous pressures shown for weightlessness are estimates. Venous pressure is zero at the head with erect posture in 1 g and is probably the same at the head in one sixth (the Moon) and one third g (Mars).[15] Absence of valid venous pressure measurements, now a half century overdue, makes the values shown speculative. A limited population in long-duration flights should allow crucial gaps in our basic knowledge of these phenomena to be filled. Related aspects of this chapter are treated in Chaps. 5 and 9.

---

[14] A real challenge will come when a child is born and develops under reduced g of a smaller planetary body or in weightlessness of a long flight.

[15] This may be a relevant consideration in Chap. 5 where ocular effects of weightlessness are addressed.

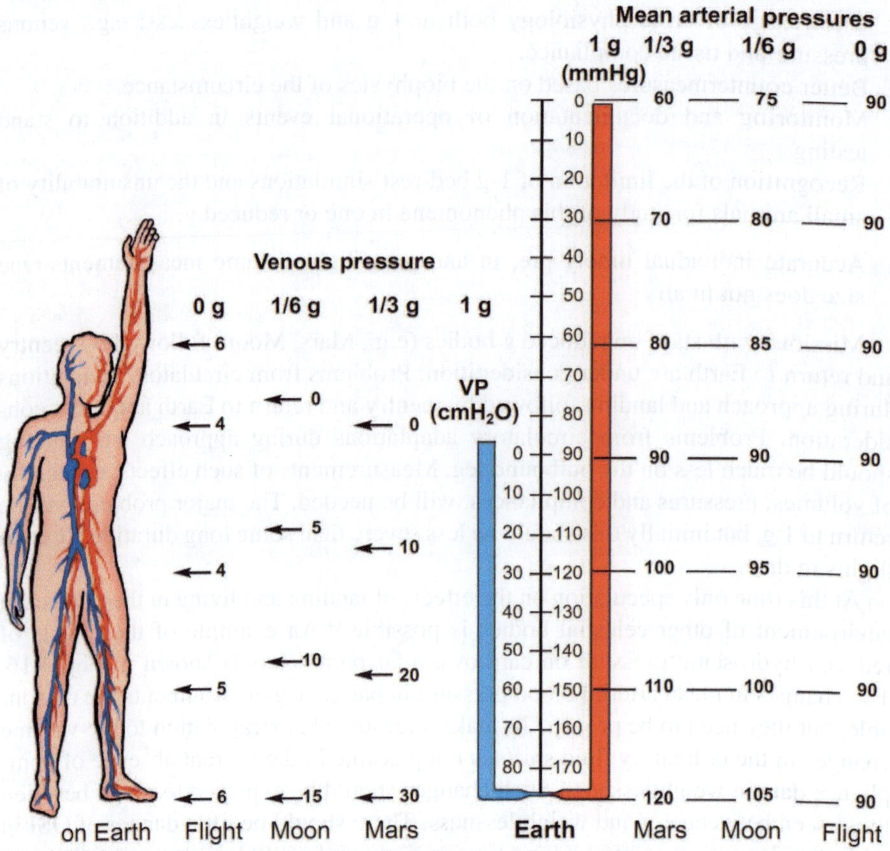

Fig. 4.16 An example of the effects of reduced g and resulting venous and arterial pressures on Earth, in flight and on the Moon and Mars in hydrostatic pressure on cardiovascular parameters

References

1. Berry C, et al. Man's response to long duration flight in Gemini Spacecraft. In: Gemini mid-program conference. NASA SP–191; 1966. p 239, 250.
2. Hoffler G, Johnson RL. Apollo flight crew cardiovascular evaluation. In: Biomedical results of Apollo. Washington, DC: NASA. NASA SP–368, 4:227; 1975.
3. Thornton W, Hoffler G. Anthropometric changes and fluid shifts. In: Biomedical results of Skylab. Washington, DC: NASA. NASA SP–377, 32:330; 1977.
4. Thornton W, Ord J. Physiological mass measurements in Skylab. In: Biomedical results from Skylab. 19:19.
5. Greenleaf J. Deconditioning and reconditioning. Boca Raton: CRC Press; 2004.
6. Anderson, Clayton C. The ordinary spaceman: from boyhood dreams to astronaut. University of Nebraska Press; 2015.
7. Lee S, et al. Orthostatic intolerance after ISS and space shuttle missions. Aerosp Med Hum Perform. 2015;86(12):A54–67.
8. Shubin N. Your inner fish. Sharon: Vintage Books; 2009.

9. Guyton A. Chapter 16: the microcirculation. In: Textbook of medical physiology. 8th ed. Philadelphia: W. B. Saunders; 1994. p. 170–9.
10. Guyton A. Chapter 14: overview of the circulation. In: Textbook of medical physiology. 8th ed. Philadelphia: W. B. Saunders; 1994. p. 150–2.
11. Guyton A. Chapter 15: Vascular distensibility. In: Textbook of medical physiology. 8th ed. Philadelphia: W. B. Saunders; 1994. p. 159–60.
12. Guyton A. Chapter 81: cerebral blood flow. In: Textbook of medical physiology. 8th ed. Philadelphia: W. B. Saunders; 1994. p. 679–81.
13. Guyton A. Chapter 28: renal and associated mechanisms for controlling sodium concentration. In: Textbook of medical physiology. 8th ed. Philadelphia: W. B. Saunders; 1994. p. 308–13.
14. Guyton A. Role of the nervous system for rapid control of arterial pressure. In: Textbook of medical physiology. 8th ed. Philadelphia: W. B. Saunders; 1994. p. 197–200.
15. Lunduvall J, Bjerkhoel P, Quittenbaum S, Lingren P. Rapid plasma volume decline upon quiet standing reflects large filtration capacity in dependent limbs. Acta Physiol Scand. 1966;158:161–7.
16. Gibson E. Skylab 4 crew observations. In: Biomedical results from Skylab. Washington, DC: NASA. NASA SP–377, 4:22; 1977.
17. Thornton W, Hoffler G. Hemodynamic studies of the leg under weightlessness. In: Biomedical results from Skylab. Washington, DC: NASA. NASA SP–377, 31:324; 1977.
18. Bucky J. Cardiovascular changes. In: Space physiology, 7. New York: Oxford University Press; 2006. p. 157.
19. Norsk P, et al. Fluid shifts, vasodilatation and ambulatory blood pressure reduction during long duration spaceflight. J Physiol. 2015;593(3):573–84.
20. Arbeille P, et al. Adaptation of the left heart, cerebral and femoral arteries, and jugular and femoral veins during short-and long-term head-down tilt and spaceflights. Eur J Appl Physiol. 2001;86(2):157–68.
21. Custard MA, de Souza Neto EP, Abry P, Flandrin P, Millet C, Duvareille M. Orthostatic tolerance and spontaneous baroreflex sensitivity in men versus women after 7 days of head-down bed rest. Auton Neurosci. 2002;100:66–7.
22. Baevsky RM, Baranov VM, Funtova II, Diedrich A, Pashenko AV, Chernikiva JD, et al. Autonomic cardiovascular and respiratory control during prolonged spaceflight aboard the International Space Station. J Appl Physiol. 2007;103(1):156–61.
23. Waters WW, Ziegler MG, Meck JV. Postspaceflight orthostatic hypotension occurs mostly in women and is predicted by low vascular resistance. J Appl Physiol. 2002;92:586–94.
24. Dietlein L, Judy W. Cardiovascular conditioning. In: Gemini mid-program conference. NASA SP–191; 1966. p. 381.
25. Buckey J. Fluid loading. In: Space physiology. New York: Oxford Univ Press; 2006. p. 159.
26. Hamilton DR, et al. Cardiac and vascular responses to thigh cuffs and respiratory maneuvers on crewmembers of the International Space Station. J Appl Physiol. 2012;112(3):454–62.
27. Fomina G, et al. Changes in hemodynamic and postflight orthostatic tolerance of cosmonauts under application of the preventive device—thigh cuffs bracelets in short-term flights. J Gravit Physiol. 2004;11(2):229–30.
28. Perez SA, Charles JB, Fortner GW, Hurst VIV, Meck JV. Cardiovascular effects of anti-g suit and cooling garment during space shuttle re-entry and landing. Aviat Space Environ Med. 2003;74(7):753–7.
29. Duddy, John H., et al. Weightless simulation using water immersion techniques: an annotated bibliography. No. LMSD-5-24-65-3. Sunnyvale: Lockheed Missiles and Space Co Inc.; 1967.
30. Thornton W, Edge V, Coleman E, Uri JJ, Moore TP. Changes in leg volume during microgravity simulation. Aviat Space Environ Med. 1992;63(9):789–94.

Chapter 5
Cephalic Fluid Dynamics and Ocular Changes in Weightlessness

5.1 Introduction

Permanent ocular changes in a group of astronauts after long flights is a relatively new finding in space medicine. Here we refer to these changes as Microgravity Ocular Syndrome (MOS) [1] although other names have been suggested.[1] Given that the cause of these ocular changes are as of now unknown, MOS seems the most appropriate name—it does not imply a cause but instead refers to a known set of symptoms that currently attract considerable attention in the space medicine community. MOS is different from other effects of weightlessness on the body in terms of its duration and possible permanency and hence requires a different approach in this chapter. Previously addressed effects of weightlessness produce appropriate reversible somatic adaptations.[2] Ocular changes revealed during spaceflight are not obviously appropriate and may be damaging and irreversible in some individuals. Vision is complex and only the mechanisms directly affected by MOS can be described here. Currently, the origin of MOS remains unproven and the problem remains unsolved. Here we suggest an alternative hypothesis as to its cause that may

[1] Authors have assigned various names to this syndrome including Visual Impairment and Intracranial Pressure (VIIP) syndrome [2], Space Obstructive Syndrome (SOS) [24], and Microgravity Ocular Syndrome (MOS) suggested by Marshall-Bowman, Barratt and Gibson [1]. The causes of changes in vision due to weightlessness are currently very much in question. MOS does not assume any one cause, in contrast to Visual Impairment and Intracranial Pressure (VIIP) syndrome, a name that implies the cause is elevated intracranial pressure (ICP). History has many examples of long and expensive pursuit of misapplied nomenclature from, Miasmas to Space Motion Sickness.

[2] Some investigators might take issue with this. but there does not seem to be a record of functional disability after an adequate time in suitable conditions for rehabilitation in one g.

W. Thornton, F. Bonato, *The Human Body and Weightlessness*,
DOI 10.1007/978-3-319-32829-4_5

point the way to fruitful research pathways that could lead to a better understanding of MOS and how it might be prevented or treated.

Vision provides the most important sensory inputs during human aerospace flight. This is not surprising given that vision is arguably the dominant sense for our species. Vision has also been one of the most intensely studied physical features in flight medicine. Military organizations have lost more potential pilots to visual requirements than any other cause. For many years NASA used military visual requirements in astronaut selection and for astronaut retention visual function has always been closely monitored.

Shifts from normal to distant vision (hyperopia) became a more or less expected and accepted experience for some individuals during spaceflight. These shifts were reversed post flight and so little attention was given to the phenomena until a few astronauts on ISS flights of six months developed visual change that persisted post-flight [2]. During detailed studies of their eyes, an initial seven astronauts were found to have significant functional and anatomical eye changes. These appeared to be the first pathological changes induced by spaceflight in its history and produced an eruption of media interest, additional studies, a summit meeting to address the problem (The Visual Impairment Intracranial Pressure Summit) [2], and a series of other large meetings and numerous contracts aimed at defining and studying the MOS problem [3]. For a time, it may have been the largest and most dominant effort by Johnson Space Center flight medicine and NASA life sciences.

On post flight functional examination, all of the seven astronauts initially studied had increased the focal distance of their eyes and become 'farsighted' or hyperopic. Anatomic changes included flattening of the rear of the eyeball, changes in the retina and optic disc, and swelling and distortion of the optic nerve and its sheath where they enter the eyeball. Some of these changes have persisted for years post flight. They have been extensively studied and clinical findings well characterized and organized. NASA and its consultants hypothesized that fluid changes inflight cause increased cerebrospinal fluid pressure (CSF), which in turn produces the ocular problems. Hence, MOS is now often referred to as Visual Impairment and Intracranial Pressure (VIIP) despite the lack of conclusive evidence that MOS is definitively caused by increased CSF pressure. The VIIP hypothesis and with it, means of measuring intracranial pressure, continue to be extensively pursued. The number of reported individuals significantly affected by these phenomena is so far small, but they could be denied approval to fly additional missions because of potential harm to them. The possibility of even longer missions makes MOS even more problematic. If ocular damage can occur during a 6-month ISS mission it stands to reason that an even longer mission to Mars for example could lead to even more damage. For these reasons, MOS has consumed, and will continue to consume, major amounts of time and resources for the foreseeable future.

After a brief history of the problem, a history of how the present position on MOS was reached follows. Background begins with a biophysical description of the known mechanisms directly involved, relevant experience in 1 g, and experience in weightlessness. A review of theory is given. Suggestions for an etiology of MOS that incorporate relevant recent findings and discoveries are made.

5.2 Brief History

Visual performance attracted attention in the Mercury program when Astronaut Gordon Cooper reported identification of automobiles on roads and aircraft on runways from an altitude of 100 miles. Both near and far visual acuities were tested in 1965 on Gemini V and VII. Four million square feet of backgrounds for the visual acuity 'charts' were plowed into desert earth [4]. No changes in acuity were found. Relatively complex crew-dependent land contrast studies were attempted on Gemini VII and then on Gemini X in 1966, but equipment failure on Gemini VII and mission operational problems on Gemini X precluded data collection. As flight time increased, some astronauts had difficulties with near vision on flights as short as two weeks. No lasting changes were noted and +1.4-diopter glasses to accommodate in flight hyperopia became an inflight option.

In 2011 Mader and colleagues reported ocular changes in astronauts studied over the previous 'several years' [5]. Seven astronauts with ISS flights of approximately 6 months who had complaints of visual changes inflight were intensively studied, at varying times postflight. The total population is not stated so the number of unaffected is unknown. Significant ocular changes, some apparently permanent, were found in all seven examined. Based on these findings a postflight survey of approximately 300 astronauts was made in search of inflight changes in vision. On short Shuttle missions 29% reported mild to moderate changes in near vision. For longer duration ISS missions 60% reported mild to severe changes. Postflight persistence of these changes is unreported. A pre- and postflight magnetic resonance imaging study of eye and nearby regions of the skull by Kramer was published in 2012 [6]. A presumably male and female sample of 27 astronauts, eight of whom had flown twice was studied with Nuclear Magnetic Resonance (NMR). Their average flight duration was 108 days. Changes in the posterior of the eye and associated optic nerve were found in 26% of this sample. An additional four ISS astronauts were subsequently added to the first seven who had persistent changes. This group of 11 currently appears to be the total of those considered to have sustained permanent ocular changes in flight. This is still a work in progress, and the incidence and permanence of changes is still to be established.

The Human Research Program Evidence Report [3] released in July 2012 was an extensive and in depth review of the problem by selected experts in various aspects of the problem, from NASA, its contractors, and various institutions. Procedures for studying the eye pre- and postflight were increased to include all that the state-of-the-art allows and are appropriate. In-flight measurements now include advanced photography, ultrasonic studies of structure and fluid flows, including Cerebrospinal Fluid (CSF), and optical studies that allow identification and measurement of the retina, its choroid and their distribution.

Noninvasive means for measuring intracranial pressure (ICP) have been sought for decades. A description of these efforts would make a sizable monograph. A description of the claims, including patents, and attempted usage of various means attempted for this would fill volumes. This search was escalated by NASA for investigation of these ocular problems. Apparently a group at Johnson Space Center had been actively investigating noninvasive means of measuring ICP a decade before MOS was discovered and they had an experiment aboard ISS using Doppler ultrasound and cranial compliance to derive ICP. A commercial instrument that had been developed and tested in England was flown and scheduled for inflight use at least twice but both times appears to have been cancelled. On the appearance of MOS an additional instrument was flown but the status of this measurement remains unreported at this time.

Although increased intracranial venous pressure is the most commonly postulated cause of increased CSF pressures, measurement of this type of pressure currently requires invasive means and these tests have not been conducted inflight. NASA JSC has also announced that it had initiated a 4-year program on ISS to study fluid shifts but as of now the status of these investigations remains unreported. The small number of astronauts and their long duration expeditions will now make statistical significance for such a study difficult.

Other unanswered concerns about MOS are: What effect does it have inflight in addition to hyperopia? How does it change with flight duration? What is its course on return to Earth? In flight the operational impact appears to have been small to date. There are no reported changes in corrected visual acuity or other significant changes in visual performance on longer missions, and its long-term functional impact remains to be seen. From the perspective of individuals affected, the nature and duration of such effects are most important. Some of the changes seen have resolved, albeit some very slowly. However, structural or functional change such as abnormal CSF pressure could be significant. Initially there was such concern over this that head-down bed rest studies were briefly interrupted but have now been resumed.

There has been a rush to judgment in several aspects of MOS. CSF pressure remains a tenuous postulate but there is little evidence to support increased CSF pressure as the causal factor associated with MOS. The finding of high normal or slightly increased CSF pressure postflight in four of the initial seven subjects was based on lumbar punctures and manometric measurements without pre-flight baselines. It is not unreasonable to assume that some or all of the four subjects had slightly increased CSF pressure pre-flight. If the usual criteria for scientific measurement had been applied here, a diagnosis of elevated CSF might be problematic. A second deficiency in the approach to this problem was the use of the old and now incomplete model of CSF flow, which did not include the most important function of CSF, support of the central nervous system neurons by a flow over all of its neurons and not just those near the surface. Possibly more important was failure to consider the third leg of the vascular system, the lymphatic system.

A number of alternative causes of MOS have been proffered including increased atmospheric CO_2 [7], vitamin deficiencies [8] and excessive forces in exercise [7]. At this time no definitive cause for MOS has been identified and no effective treatment has been reported. Conversely, there has been a continuing series of discoveries in CSF and lymphatic circulation on Earth. There have also been recent reports that claim the incidence of MOS for ISS astronauts is as high as 50% [9] or 60% VIIP [10]. More work is clearly needed to more accurately determine the incidence of MOS, what causes it, and how it can be prevented or reduced.

5.3 Background

The human eye is the most complex and comprehensive of the human senses with approximately 125 million sensors arranged to detect a range of wavelengths coded as colors while accommodating a range of intensities from bright sunlight to a few photons. It does this over an angular coverage of approximately 180° horizontally and can focus over a range of a few inches to infinity [11]. Support, feeding and collecting waste from such an array of sensors and neurons requires a complexity of circulation of blood and fluids unseen elsewhere in the body. Such complexity requires that this chapter be devoted to only those items that are affected by weightlessness and may be related to MOS. Even this restriction requires simplification. It begins with a partial schema of the region of the eye and optic nerve affected by MOS. Continuity of eye and brain is shown, with circulation of CSF over the brain and into and out of brain and optic nerve tissue. Also described is the collection and return of used CSF to the blood's circulatory system by lymphatic vessels. Known relevant effects on these systems by 1 g, simulations on Earth and spaceflight follow. Current MOS hypotheses are discussed and an alternative approach to its understanding based on recent discoveries is proposed. To appreciate changes seen in MOS requires elemental knowledge of the complex circulation of the eye and brain. In addition to the usual vascular circulation by arteries, capillaries and veins, circulation of fluids filtered from and returned to the bloodstream are also essential to the life of neurons in the brain and eye [12]. This circulation of fluids is important in understanding the alternative hypothesis presented here.

Vascular circulation in the eye may be seen directly by means of an ophthalmoscope as a network of arteries and veins against a normal reddish background of the retina which is comprised of a multilayer sheet of optical sensors and their capillary support. Abnormal changes in areas of the retina such as cotton wool exudates and small hemorrhages have been found in MOS. Such exudates occur when fluid circulation is inadequate and metabolic products accumulate. A great variety of other conditions in the eyes, nerves and vessels may be seen, especially with more advanced optical devices and procedures.

Eyes are a direct extension of the support structures of the brain. Millions of optic nerve fibers leave the brain, mix and match as they cross in the optic chiasm and carry on in two bundles of about a million fibers each to each eye (Fig. 5.1). An artery and veins that feed these nerves and neurons in the retina are in the center of the fibers. CSF must also flow around these fibers for effective for maintenance of

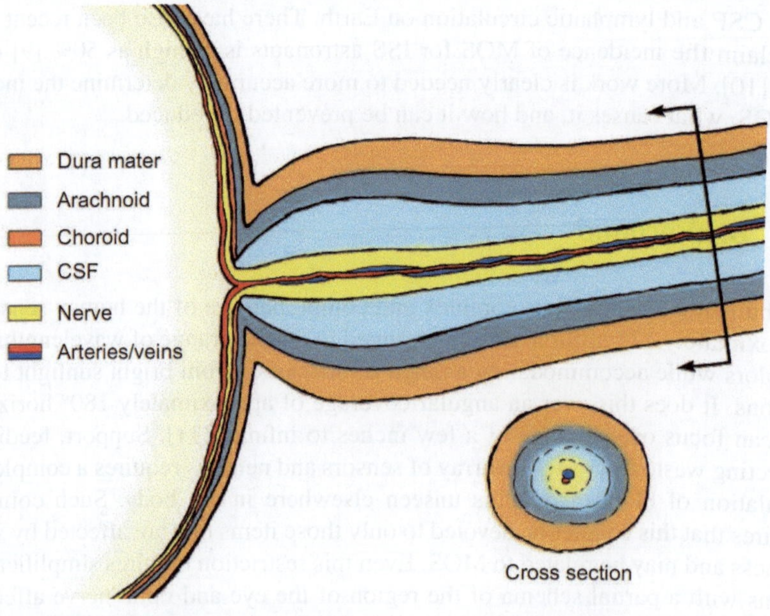

Fig. 5.1 Diagram of structures of the eye where changes in MOS occur. The collective optic nerve is a continuation of brain tissue and supporting structure including blood and fluids. Dura and arachnoid tissues form a nerve sheath that surrounds a layer of CSF (*blue*), which flows from brain through glia and around nerve fibres into a lymphatic collection system in the optic nerve (*yellow*) and its sheath and is returned to the venous system by large lymph vessels along the nerve and sheath. The nerve proper contains some million fibers each of which on entering the eye divide into some 130 fibers that spread over the inner surface of the eye to support the complex light sensors forming the retina. Most of the inner surface of the eye is covered by the retina. Life support for this mechanism is provided by the choroid, a vascular layer which is perfused by the retinal artery and veins. Note that the dura of the nerve and brain becomes the sclera of the eye. All of these features are affected by VIIP, Glia and lymphatics are not shown and some features are exaggerated for emphasis

them. Surrounding this nerve bundle is a cylinder of CSF enclosed in tubular extensions of *arachnoid* and *dural* membranes from the brain. This complex coaxial assemblage is commonly designated as the *optic nerve*.

Just prior to connecting to the eye, the dura and arachnoid expand and then spread into membranes to become the arachnoid and sclera of the eye. Figure 5.1 depicts a simplified typical diagram of this anatomy that does not show the *lymphatics* that drain the CSF. To enter the eye the nerve fibers and blood vessels pass through a grid of tough fibers that mechanically anchor and stabilize them. Once past this *cribriform grid*, nerve fibers, arteries and veins bend sharply to spread over the choroid to form the retina, a layered sheet of light sensors and supporting cells. The pale circular area at the point of nerve entry is easily seen

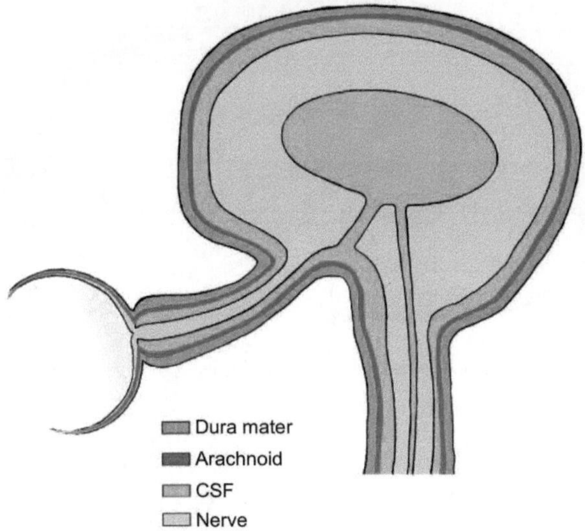

Fig. 5.2 This is an exaggerated conceptual illustration of distribution of CSF in the central nervous system (CNS). CSF is contained around the brain by two membranes, the tough dura and a more compliant nonporous arachnoid. This envelope is continued around the spinal cord and the cranial nerves. One of these, the optic nerve, is shown in exaggerated form to illustrate the basic arrangement of fluid containment and nerve. It is a continuation of the arrangement in the brain and spinal cord. To be effective, fresh CSF must flow through the interstitial spaces of all the nervous tissue. This is addressed in Fig. 5.3

with an ophthalmoscope and is called the optic disc. In the process of entry and spreading, each fiber of the optic nerve—the axon of a retinal ganglion cell—gives rise to on average one hundred plus synaptic connections with individual light sensing cells (*rods* and *cones*) in the retina.

Brain, spinal cord and many of their nerves are surrounded by a shallow sea of fluid filtered from the blood. An understanding of this fluid and its circulation is necessary to understanding of its circulation in the optic nerves where changes with MOS occur. Figure 5.2 depicts a distorted conceptual view of CSF volume distribution of concern here. Brain and nerve tissues, are surrounded by a layer of CSF, which moves slowly over them as conceptually diagrammed in Fig. 5.3. This fluid probably contains materials essential to maintenance of the nervous tissue. A layer of CSF is contained within an envelope of two sheets of the meningeal membranes, the *pia mater* and *arachnoid*, forming the *subarachnoid space* between them that is filled with CSF. This assemblage is covered by the much tougher and thicker *dura*. These membranes become cylinders containing a nerve surrounded by fluid as the nerves leave the brain. An overview of the traditional understanding of CSF generation, circulation and clearance is shown in Fig. 5.3. This is the model on which some prevailing current theories of MOS are apparently based.

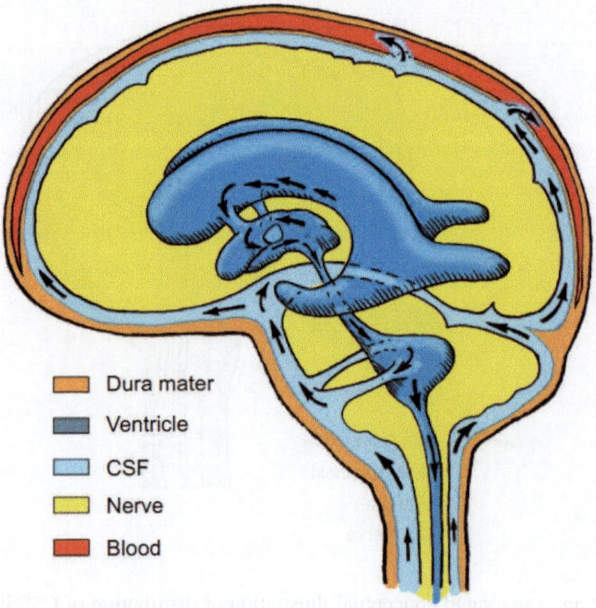

Fig. 5.3 Diagram of conventional theory of CSF flow. There are four volumes, ventricles, of fluid in the brain. Note that ventricles (*dark blue*) are in three dimensions while flow is in cross section. In each of these there is a choroid plexus which produces CSF. The large paired right and left ventricles empty into a third that empties into a fourth. Openings in the fourth let the fluid flow into the spinal canal and also into a space between brain and arachnoid covering and dura. The arachnoid membrane that is closely applied to the dura is softer and impermeable to C.S.F. It is not depicted here but is shown in Figs. 5.1 and 5.2. In an area along the top of the brain the dura splits to form a blood filled sinus. Arachnoid granulations two of which are shown here as openings between CSF and blood, are folds of the membrane that allow CSF to pass into the venous sinus, which empties into the internal jugular veins. While this is a major pathway for CSF and still often treated as the only pathway, the pathways into the space around the nervous tissue are diffuse and not shown. Drainage of this region is now being discovered and is discussed in a later portion of this chapter. The flow pattern shown is a supply line to the nerve tissues, which interchange materials with CSF as it flows over the brain, spinal cord and nerves. Choroid tissue in each of the ventricles generate several times the static volume of CSF each day thus ensuring an adequate supply through the nerve tissue

Generation of CSF appears to be independent of arterial and venous blood pressures; however, it may be varied by other factors. When the generation of fluid is constant, and in the absence of other inputs to the CSF space, pressure becomes dependent upon the rate of fluid removal. It was recently reported by Kramer [13] that CSF fluid generation in flight is unchanged from that on Earth. This makes any change in pressure dependent on its rate of removal and on cerebral venous pressure into which it must pass. As previously noted, pressure in this critical fluid circulation has not been directly measured.

All of the central nervous system tissue of brain and spinal cord lie between pools of CSF. There are four CSF filled *ventricles*, fluid filled spaces, within the brain. Each of these ventricles has a *choroid plexus* producing CSF that empties from the first and second into the third and thence into the fourth, which connects to *subdural space* around the brain and to the central canal in the spinal cord by a hole or *foramen* through which fluid also flows into the space enclosed by two continuous sheets of tissue, the *arachnoid* and thick, tough *dura*. On the upper surface of the brain the dura is split into two sheets, which form a sinus of venous blood in the central furrow at the top of the brain. CSF exits through outpouchings or granulations of the arachnoid through openings in the dura into venous blood of the central sinus. The sinus is drained by a series of veins emptying directly into the internal jugular veins. This is normally an unobstructed path for catheters that can be inserted into the sinus by way of the neck veins thus providing a direct means of measurement of sinus pressure. If sinus pressure becomes higher than that of CSF, then the arachnoid granulations collapse forming a valve that blocks any reverse flow of blood into the CSF. With this arrangement CSF pressure must be higher than cerebral venous sinus pressure to exit. In weightlessness there is an obligatory rise in cephalic fluid pressures from loss of 1 g hydrostatic pressures (Fig. 4.15) and this may be the basis of some MOS hypotheses and will be discussed. When CSF flow is blocked by tumors or other causes, pressure will rise and brain function above the blockage will be affected producing characteristic signs and symptoms. Again, the CSF flow pattern shown in Fig. 5.3 and described here has been and still appears to be the model generally accepted by MOS investigators.

The presence of another essential component of the Central Nervous System (CNS) fluid circulations, the *lymphatic system*, has been unrecognized for many years. Barring rare individual exceptions, whose voices were lost among general claims of the establishment[3] that there are no lymphatics present in the CNS. This was accepted by most as medical, anatomical and physiological fact when MOS was first discovered and defined. A previously accepted function of the then known lymphatic system was its role as a one-way collection and drainage system. It differs from the venous system in that it generally collects and deals with material directly from interstitial fluid. To paraphrase Guyton, we would be dead in 24 h without it [12].

5.3.1 *The Traditional Lymphatic System*

Chapter 4 describes the exchange of water and other materials across blood capillary walls, but protein molecules are retained in the blood circulation by the fine mesh of capillary walls. Some leakage of protein into the interstitial fluid occurs. Also microbes and other cells may grow in or be shed into the interstitial fluid including neoplastic (cancer) cells. Normally such material, plus excess water, is picked up by a pervasive lymphatic network in the interstitial spaces. Lymphatic networks begin with a closed-end lymph capillary, a tissue with holes large enough to collect protein and cells. These capillaries coalesce into three progressively larger

[3] Existence of CNS lymphatics was still generally denied until recently.

networks of lymph vessels of increasing diameter, which pass through lymph nodes whose functions include filtration of lymph for microbes and cancer cells and any other material foreign to the body. Lymph is then returned to the bloodstream through veins. Only the 'cleaning function' of lymphatic fluid flow will be discussed here.

Our interest here is not in the immune function but in the lymphatic system's ability to remove the protein and other osmotically active substances, which will otherwise attract water and produce swollen edematous tissue. When a major lymphatic trunk is blocked, one may see a grossly swollen tissue such as an arm when its lymphatics are disrupted by surgery. Worse still are the horror pictures from tropical countries where insect larvae permanently block lymphatics and grotesque anatomical distortions occur. As visualization and identification techniques improved, increasing areas of the CNS were found to be rich in lymphatics including the cranial nerves. A series of animal experiments [14–16] showed that these nerves played a significant role in absorption of CSF, especially the olfactory nerves. Blockage of the olfactory cribriform plate through which these nerves pass can lead to up to 50% of CSF absorption being blocked. Lymphatics are also now posited to play significant roles in drainage of the spinal cord and the eye. In 1999 lymphatics were demonstrated in the optic nerve and these collect CSF fluid and maintain the flow around it [17]. In the cat's eye [18] there are lymphatics along the entire length of the optic nerve and they are more abundant as the nerve approaches the eye's globe.

There has long been anatomical evidence provided by animal studies that lymphatic drainage can play a significant role in CSF drainage. Although it was and sometimes still is claimed that CNS nerves do not have accompanying lymph vessels, this is no longer supported by evidence. When the optic nerve leaves the parenchyma of the brain it is covered by compliant tubes of dura and arachnoid and these contain a layer of CSF that must be collected by lymphatics to maintain a flow over the nerve. (See Figs. 5.1 and 5.2.) Using sheep olfactory nerves as a model, a series of studies from 1997 to 2005 demonstrated that up to 50% of CSF drainage is facilitated by lymphatics. Thus CSF has two outlets, one by way of the brain's arachnoid granulations into the sagittal sinus, which flows into the internal jugular. The second outlet is by way of the lymphatics, which accompany the cranial and spinal nerves. In 2012, several publications described the demonstration of a unique and complex combination of glia and lymphatics[4] in the mouse brain at the University of Rochester Medical Center [19]. This complex was designated *glymphatics*. Other investigators have subsequently demonstrated a lymphatics system in the brain's dura [20]. After known findings relevant to MOS in 1g, bed-rest and spaceflight are reviewed, a critique of current MOS theories is made and suggestions offered.

[4] *Glymphatic* is the name given to the discovery of the CSF flow pathways through the brain. Glial cells and lymph vessels are key components, hence the term: *Glymphatic.*

5.3.2 *One-g Phenomena Relevant to MOS*

When increased CSF pressure became the prime suspect as the cause of MOS, examples of known effects of increased CSF pressures were sought and easily found. These effects ranged from the distorted skulls and other disastrous effects of hydrocephalus to known cases of increased pressure in which no negative effects could be found. Idiopathic Intracranial Hypertension (IIH) [21] entails increased CSF pressures of more than 25 mmHg and produces a global array of signs and symptoms. Few of these symptoms are found in MOS and major symptoms such as headache, aural changes, eye and cerebral dysfunction are not present. Hence, MOS appears to be a unique problem and this is reflected in the number of new names for it.

A major 1-g consideration relative to MOS are bed rest simulation studies. Head down bed rest would be a reasonable analog in this situation. One study to date documented a single individual for 30 days [22]. Results could be fairly summarized by saying no significant changes were found. However, one would have to be optimistic to expect any in a single young subject in only 30 days. Also, invasive means of CSF pressure measurements are unsuitable for this situation and there are no reports of noninvasive or indirect measurements here. Measurements of pressures and other quantities are also needed for a baseline in flight studies.

5.4 One-g Technology

A number of devices used to study aspects of MOS are in standard practice on Earth and some have been flown including ophthalmoscopes, functional test charts, and scanning and imaging devices. A major effort to find and demonstrate noninvasive means of measuring CSF pressure was made and commercial devices are said to be on board. Some results from current standard procedures on Earth are described here.

Measures of quantities of interest are given in Table 5.1. For anyone interested in such pressure measurements, a lesson from Cambridge, Intracranial Pressure Measurements and Monitoring CSF Pressure [23], is strongly recommended. Direct measurement of CSF, which is now conducted and sometimes monitored in appropriate cases, requires surgical access through the skull into a CSF volume and sometimes through the brain into a ventricle. Electronic transducers display and then monitor pressure.

When an otherwise normal subject's CSF was directly recorded from a patient in horizontal position, there was a mean pressure of 12.5 mmHg (16.5 cmH_2O) with combined respiratory and cardiac variations of 5 mmHg (6.5 cmH_2O). On sitting, the patient's mean pressure becomes 1.5 mmHg (19.5 cmH_2O) with a respiratory/cardiac variation of 4 mmHg (5.2 cmH_2O). The mean difference of 11 mmHg (14.3 cmH_2O) is well within the range of expected difference in hydrostatic venous pressure change on going from 1 g to weightlessness for an average person. In a group

Table 5.1 Cerebrospinal fluid (CSF) and related pressures

One g							Weightlessness		
	Upright			Horizontal					
	Mean	(Pulse/respiration)	[Max]	Mean	(Pulse/respiration)	[Max]	Mean	(Pulse/respiration)	[Max]
Subject	Measured peak to peak (mmHg)								
CSF ventricle	2	(5)	[20]	13	(5)	[20]	–	–	–
Parenchyma, brain	2	(–)	[–] [–]	13	(–)	[–]	–	–	–
Intraocular	10–20	(–)	[20]	[a]			10–20	(–)	[20]
Venous	0	(–)	[10]	10	(–)	[20]	–	–	–

[a]Little or no correlation with other pressures – = Unknown

of 61 patients the mean seated ICP was 1.0 mmHg (1.3 cmH_2O) with a range of −7.5 (−9.8) to 15 (19.5) mmHg (cmH_2O) and when horizontal the mean pressures became 10 mmHg (13.0 cmH_2O) with a range of +2 (2.6) to + 20 (26.0) mmHg (cmH_2O) for a mean difference of 9 mmHg (11.7 cmH_2O) between upright and supine posture. The changes in this larger sample are also consistent with the expected difference in normal CSF pressures on Earth and space [23].

Because of difficulties and risks in direct measurement, most CSF pressures are approximated by measuring CSF pressure at the lumbar level with the subjects on their side, using a spinal needle and fluid manometer, a technique devised in the 1800s. One must agree with the authors of the Cambridge measurement review that "manometric lumbar pressure measurement may be misleading". Other normal baseline pressures listed in Table 5.1 include intra-ocular and lumbar ICF pressures.

Other significant normal 1-g findings are that ICF and brain parenchymal pressures are essentially equal. When horizontal CSF pressures are simultaneously measured with equivalent transducers at brain and lumbar sites, at pressures up to 40 mmHg, CSF and cerebral venous pressures are equal. Vascular pressures and volumes in 1 g and weightlessness are described in Chap. 4 on Fluid Shifts. These and other 1-g measures are significant as the normal points of departure for changes seen in MOS.

5.4.1 In-flight Changes and MOS

Table 5.2 lists the major signs and symptoms seen in MOS and their incidence in the study of the 11 aforementioned cases [5, 6]. Their order and identification numbers are arbitrary. First to be noted was the change in focal range (item 1, Table 5.2). The apparent cause of this and other changes follow. Eyeballs, or ocular globes, were shown to be flattened in Kramer's study (item 7, Table 5.2). These and other changes are illustrated in Fig. 5.4. Swelling of the nerve sheath, especially at its

Table 5.2. Symptoms and signs in eleven astronauts with MOS

Symptoms	Incidence
1. Focal distance increased, hyperopic, +1.5 to +1.75 diopters	11
2. Scotoma	2
Signs (six subjects examined)	
Visualized	
3. Retinal exudate, cotton wool spot	4
4. Hemorrhage	1
5. Choroidal folds	7
6. Optic disc edema	8
Scanned	
7. Flattened posterior globe	8[a]
8. Thickened retinal nerve layer	8[a]
9. Enlarged optic nerve bundle and sheath	8[a]

A number of changes were unilateral in the right eye and in cases that were bilateral, the changes were usually predominant in the right eye except in one case

Note: Some of these exams were conducted up to two years postflight and may not represent true incidence

[a]Of nine examined

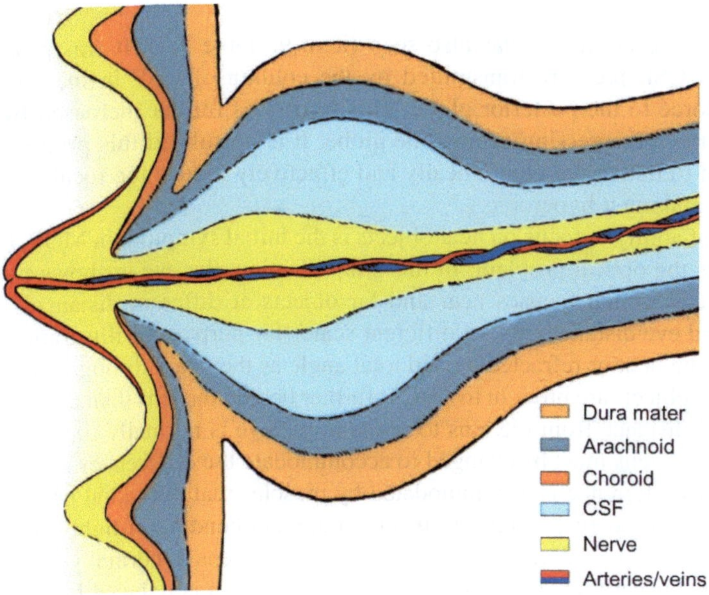

Fig. 5.4 Conceptual Diagram of nerve/eye junction in VIIP. Changes illustrated here include flattening of the posterior globe from pressure of the hypertrophied optic nerve complex on it. This flattening plus pressure on the veins cause thickening of the choroid layer beneath the retina. These pressures are relieved by ripples or folding if the choroid and retina. Flattening of the eye also reduces its focal length (Fig. 5.5). Tissues of this portion of the optic nerve are swollen by edema. This causes the nerve fibers to protrude into the eye producing "disc edema". The nerve fibers and their sheath are swollen and bulge outward. Lymphatics and some other details are not shown and features are distorted for clarity

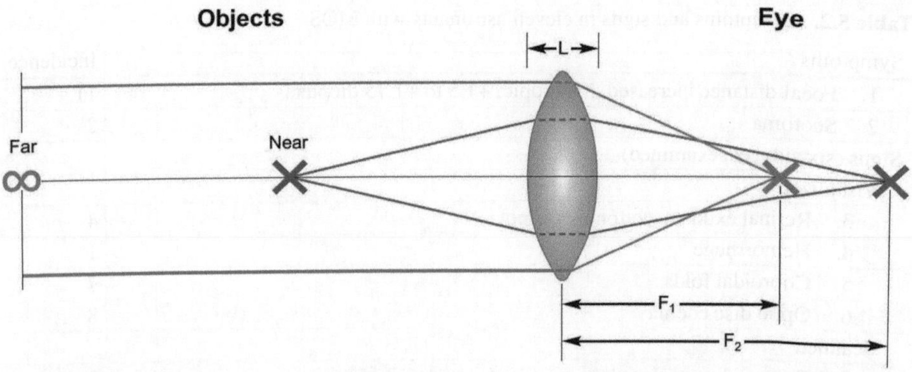

Fig. 5.5 Simplified schema of eye to illustrate focussing of objects at different distances. L is thickness of the lens, F_1 is the focal distance for distant objects, shown in *green*, and F_2 for near objects, shown in *red*. In a normal eye at rest there is passive radial tension on the lens, which flattens it. To bring near objects into focus ciliary muscles around the lens contract allowing it to increase its curvature and move the focal point from F_2 to F_1 where the retina is located. When the retina is pushed near the lens in hyperopia the lens must further increase its curvature to bring both far and near objects into focus on the retina, and the eye is called hyperoptic. Note: All of Guyton's Text Books of Physiology have full sections on eye focus and accommodation (color figure online)

junction with the globe, was also seen (item 9, Table 5.2). It was proposed that increased CSF pressure transmitted by the column of CSF in the nerve sheath applies force to the posterior globe. This force was further increased by enlargement of the area pressing against the globe. It is postulated this produced a force sufficient to flatten the globe locally and effectively reduce the focal length of the eye thus making it hyperopic.

Loss of ability to focus on near objects is the initial symptom in MOS. Figure 5.5 illustrates the optical problem. In this simplification, the eye is shown as a simple convex lens, which focuses near and far objects at different distances. Note the object and eye distances are on different scales for purpose of illustration. Rays of light will be bent or refracted a fixed total angle as they pass through the lens. Thus the nearer objects are brought to a focus further behind the lens than the distant ones. The focal distance from the lens to retina in the eye is normally fixed so the focal length of the lens must be changed to accommodate the differences in viewing distance. This difference is accommodated by muscles that put radial pressure on the elastic lens of the eye, increasing its curvature and bending of light rays such that near objects are now focused at F_1. This is what happens in normal younger eyes.

F_1 is the usual focal length of the eye. If this length is reduced by pressure and flattening, nearer objects will be out of focus and the eye is hyperopic. The lens must now accommodate both distant and near objects. Most subjects will require reading glasses for near vision. In at least one case the eye was so flattened that reading glasses were required to put distant objects in focus.

The additional effects and number seen (in bold type) and presumed to be caused by increased fluid pressures and flattening are; folding of the choroid, 5, and elevation of the optic disc, 6. Increased pressure on the retinal veins will increase fluid volume in both choroidal and retinal nerve tissue resulting in reduced area from loss of curvature in flattening that can produce the wrinkling or folding often seen in MOS. Pressure on veins entering the eye may also explain the cotton wool exudates, 3, and hemorrhage, 4, seen in the retina. Scotomas, 2, experienced by several astronauts, may have been produced by edematous pressure on nerve fibers or by vascular problems in local areas of the retina. Optic disc edema, 6, and nerve and nerve sheath swelling, 9, are currently assumed to be direct effects of increased CSF pressure.

5.4.2 Cause of MOS

As previously stated, the cause of MOS is currently debated. Initially, the theoretical and qualitatively observed increase in venous pressure caused by the cephalad shift of fluids in weightlessness was immediately invoked as the cause of an unproven increase in CSF to abnormal levels. However, no calculations or estimates of the magnitude of such shifts have been published or presented to support the magnitude of such changes. A number of other possible causes or exacerbating factors have been discussed including increased atmospheric CO_2 [7], diet [8], and even exercise in weightlessness [7]. The small absolute increase in venous pressure produced by weightlessness was recognized by Wiener who proposed that larger venous pressure increases might be caused by entrapment of the internal jugular vein between the clavicle and sternocleidomastoid muscle reducing or abolishing flow and increasing back pressure. Wiener hence coined the term "space obstructive syndrome" [24]. Currently the primary emphasis and efforts apparently remain focused on increased venous pressure caused by weightlessness that in turn is hypothesized to produce increased CSF pressure.

5.5 Evidence Based CSF Flow

Here we add what is currently known about lymphatics to the traditional model of CSF hydrodynamics and then explore the implications of a more complete model that is still a work in progress. Animal models have demonstrated evidence of lymphatic drainage from the brain for many years. For example, in 2012 the flow of CSF through the brains of mice was demonstrated [20].

It was accepted that CSF flow is essential to normal function of the central nervous system. However, circulation by the accepted arachnoid venous sinus route that provides CSF flow over brain surfaces allows only a superficial layer of brain cells to be exposed to CSF. This situation becomes worse in the optic nerve, and presumably some other cranial nerves, for while the optic nerve fibers are surrounded by CSF, it comes to a dead end as shown in Figs. 5.1 and 5.3. The situation is more untenable in the spinal column with its long central dead end canal surrounded by a thick column of nerve tissue. Openings in the brain's fourth ventricle will not allow the development of a pressure gradient to force fluid through the spinal cord's nervous tissue. In sum, the traditional model of CSF circulation is inadequate.

Findings of lymphatics in the brain plus their known presence in cranial nerves allow for a new conceptual model of CSF circulation consistent with its function and known behavior. This model has two major flow paths. One is that of the classic model in which CSF is produced by choroidal tissue in the series of connected ventricles that generate flow over the surfaces of the brain and empties into the sagittal venous sinus (Fig. 5.3). Note that the thick brain and spinal cord have access to CSF from two sides: ventricular and peripheral surfaces of the brain, and central canal and peripheral surfaces in the cord. This is the global source of CSF for brain, cranial and spinal nerves.

In the new model this macroscopic flow over the brain and cord draining into the venous sinus is the source for a second diffuse circulation of CSF through the brain, spinal cord and many nerves. Vascular capillaries may remove some portion of the interstitial CSF water but cannot remove larger particles. The CSF path through through the brain tissue consists of the perivascular spaces surrounding innumerable small arteries and veins entering and leaving the brain, plus coaxial channels beneath the sheaths of cranial and spinal nerves. Presence of these CSF channels have probably been known for more than a century. As the small arteries penetrate the brain the cylindrical surrounding their perivascular spaces are completely covered by foot plates of *glial* cells which transfer CSF into the interstitial fluid surrounding neurons. This fluid picks up and carries waste into venous perivascular spaces leaving the brain. Waste and fluid are progressively collected by lymphatics that coalesce and finally empty into the venous system. This system was described by Iliff and colleagues [20] in 2013 and named the *glymphatic* system.

In 2015 another team reported on further work with these CNS lymphatics and their emptying into veins in the dura covering of the brain [19]. It has now been demonstrated that there is a river of CSF flowing over the CNS whose level and pressure is dependent on its input and out flow. This river provides untold numbers of tributaries into and out of the brain and CNS nerves to support their function.

5.5.1 CSF Flow and Lymphatic Drainage in MOS

To date two possible causes for the postulated increased intra cranial pressure in MOS have been pursued or at least considered: (1) increased CSF pressure secondary to fluid shifts in weightlessness and (2) increased venous pressure from shift of body components in weightlessness (Fig. 5.6). To date, neither of these hypotheses have produced increased objective understanding of the problem or useful physical finings. MOS Symptoms do not appear to be global with headaches and other signs and symptoms expected with increased CSF or cerebral venous pressure. Studies in Rhesus monkeys with ICP of up to 400 mmHg did not produce distension of the optic nerve sheath [25].

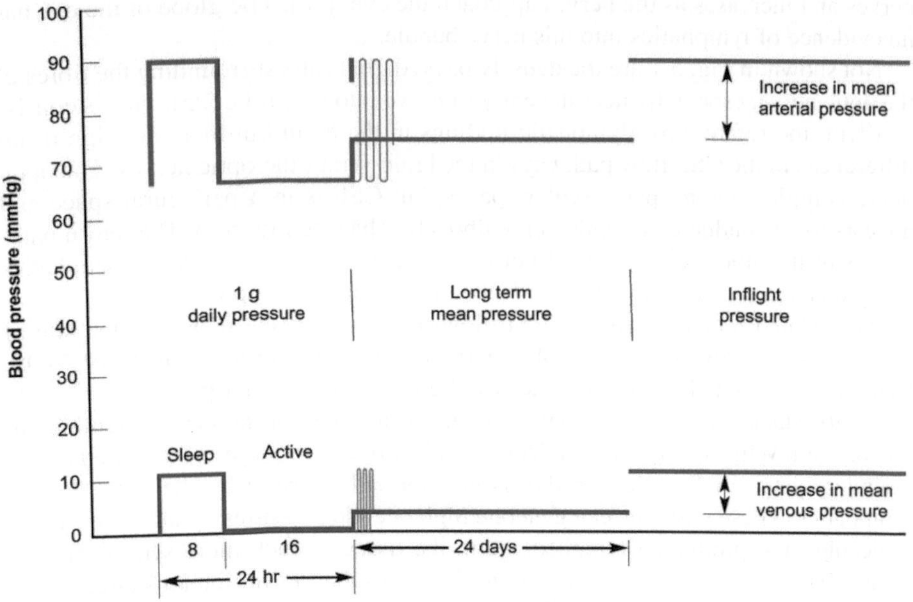

Fig. 5.6 In the absence of measured cephalic blood pressures in flight, calculated pressures become the best approximation. Values for arterial (*red*) and venous (*blue*) pressures in weightlessness are obtained by removing known hydrostatic pressures in 1 g. Since the effects of VIIP are slow to develop, a time average is more meaningful and is also shown. Based on this, all of the changes are definite but small. If CSF pressure is closely related to venous pressure, the calculated on-orbit venous pressure should not be increased above established 1-g limits (color figure online)

There have been increasingly noted differences between the symptoms of proven cases of increased ICP and MOS. In spite of a decade or more of attempts to measure, and reported shipments of instruments for this being shipped to ISS, no inflight measurements of ICP have been reported. A number of inflight imaging studies have been conducted and those capable of CSF flow measurements indicate normal global flows. All of this and more argue against ICP as a direct cause of MOS.

Instead symptoms are limited to the eyes, or sometimes to one eye, and abnormal findings are clustered in and around the junction of optic nerve and eye. This region deserves close scrutiny, especially the optic nerve, Fig. 5.1. This nerve is a relatively complex structure with the artery, vein, and nerve fibers. Since it is an extension of the brain, this nerve is surrounded by a column of CSF that is an extension of the CSF surrounding the brain and is contained in a tube of arachnoid and dura. This column of CSF comes to an end at the juncture of nerve and eye, however it must have a drainage path for a flow over the nerve fibers. Capillaries in the sheath and nerve fibres can absorb only water and small molecules. Large molecules, detritus and most of the fluid must be moved out of the nerve by lymphatics with their larger pores and gaps. Animal studies show that lymphatic drainage is prominent in optic nerves and increases as the nerves approach the eye[5] [18]. The globe of the eye has no evidence of lymphatics into this nerve bundle.

Not shown in Fig. 5.1 are the densely packed glial cells surrounding the fibres of the optic nerve, especially near the entry of nerve into the globe [25]. This is consistent with the referenced glymphatic findings in the brain but there are significant differences in the CSF flow pathways in the brain versus the optic nerves. The optic nerve complex has no perivascular spaces, but CSF is in a perineural space and appears to be conducted to optic nerve fibres by glia (see Fig. 5.1). The return pathway here has not been described but the presence of numerous large lymphatics along this nerve complex makes a connection to a dural vein probable.

An updated understanding of CSF flow plus an analysis of flow at the site of ocular change provides a plausible alternative or contributing etiology of MOS. Lymphatic flow can be easily blocked by slight external counter pressure especially in the first stages of the system that has no intrinsic valves or demonstrated pumping mechanism. When this occurs, molecules and particles too large to be cleared by the vascular system will collect in the tissues normally cleared by the now blocked lymphatics. These particles can generate high oncotic pressures by attracting water molecules and producing lymphedema in the tissues. Much more serious are the tissue changes produced by obstruction. Some of the most horrendous effects may be seen in chronic Lymphatic Filariasis, a tropical parasitic infestation, said by some to be the world's second most prevalent disability. If obstruction occurs in an optic nerve's lymphatics, then swelling of the nerve and nerve sheath is to be expected. Such swelling and expansion is regularly seen in the scans from MOS (see Table 5.2). Pressures from such tissue expansion could produce higher longitudinal forces on the globe than that postulated from the column of CSF under pressure, flattening and wrinkling the adjacent globe. The so called kinking of the nerve appears to be a product of interference with ocular motion. Intrinsic expansion of dura, arachnoid and nerve are more consistent with lymphatic dysfunction than increased CSF pressure. Unilateral effects, usually right-sided, are frequently seen and this seems unlikely in a global process such as increased CSF pressure. Conversely there are very frequently large differences in the length of the intraorbital portion of the optic nerve and its lymphatics in the same individual. Length of the lymphatics may play a role in their blockage. In sum the symptoms and signs seen in VIIP, especially their localization to a small region, are inconsistent with ICP as a sole cause.

We propose that the apparently permanent ocular changes seen in a relatively small number of older male astronauts after six months or so of weightlessness is a product of lymphatic obstruction. This obstruction could be mechanical or fluid pressures, or a combination of the two. There is an obligatory increase of mean cephalic fluid pressures in orbital space flight. See Fig. 5.6. This estimated increase in arterial pressure is 13 mmHg and 6 mmHg venous. How these pressures affect other cephalic fluid pressures are unknown. External mechanical pressures on the orbital portion of optic nerve lymphatics may be exerted by other intraorbital tissue,

[5] A thinning of arachnoid tissue occurs in parallel with this.

especially by the varying positions of the eye. Lymphatic Occlusive pressures and the nature of these lymphatics are currently unknown as are the associated tissue counter pressures and biomechanics. Investigation of these questions have started to take place on Earth, but not in space.

5.5.2 *Summary of Current Status of MOS*

MOS is an unexpected effect of weightlessness. Investigators of the problem have viewed this as a dangerous pathological change which is beyond the capacity of accommodation or adaptation. Apparently the body counteracts the stress on ocular components by more or less permanent, and by 1-g standards, abnormal structural change. Such changes degrade focal capacity of the eye, but apparently maintain visual acuity. Presbyopia regularly produces such functional changes but this is usually considered a normal aspect of aging, albeit in another ocular component.

Key follow-up questions include:

- Are the changes progressive postflight?
- Are the changes progressive in flight or do they reach a stable plateau?
- How are they related to similar but reversible changes seen in a large percentage of flown individuals?

The first two questions will take years to answer, even if all individuals on the Space Station are involved. If the changes are progressive they will adversely compromise acuity or other essential function. If they have reached a stable plateau, this is arguably an extreme adaptation. Much should be learned from answering the third question.

A large clinically-oriented effort has resulted in adequately characterizing the problem, determining demographics, organization and dissemination of available data. Such efforts however have been hampered by lack of essential, difficult to obtain data and lack of a current and comprehensive approach, but there should be enough data, understanding and experience with this problem to make the necessary decisions to allow exploratory missions.

If a number of questions on missing knowledge concerning the human eye are properly answered, clinical medicine as well as space medicine will benefit.

Glossary

Arachnoid A fine membrane, the middle one of the three membranes or meninges that surround the brain and spinal cord, situated between the dura mater and the pia mater. The sub arachnoid space contains cerebrospinal fluid.

Arachnoid granulation (Or arachnoid villi) are small protrusions of the arachnoid (the thin second layer covering the brain) through the dura mater (the thick outer layer). They protrude into the venous sinuses of the brain, and allow cerebrospinal fluid (CSF) to exit the sub-arachnoid space and enter the venous system.

Cerebrospinal fluid (CSF) The fluid contained within the ventricles of the brain, the subarachnoid space, and the central canal of the spinal cord and spinal canal surrounding the cord.

Chiasm The crossing of two lines or tracts, as of the optic nerves at the optic chiasm where nerves from the right and left side of the brain meet, mix and continue to each eye.

Choroid The vascular layer of the eyeball between the retina and the sclera which metabolically supports the retina.

Choroid plexus A network of blood vessels in each ventricle of the brain. It is derived from the pia mater and produces the cerebrospinal fluid.

Cotton wool exudates A white, fluffy-appearing lesion, an infarction of the nerve fiber layer, observed on the retina of patients with certain systemic conditions, such as diabetes, acquired immunodeficiency syndrome, hypertension, and lupus. It can also be observed in retinal infections.

Cribriform Perforated like a sieve

Diopter A unit for refractive power of lenses, being the reciprocal of the focal length expressed in meters.

Dura (mater) The tough outermost membrane enveloping the brain and spinal cord.

Foramen An opening, hole, or passage, especially in a bone.

Hydrocephalus A condition in which increased fluid pressure enlarges the head sometimes causing brain damage. It, typically occurs in babies and very young children.

Hyperopia Or farsightedness, is the condition of the eye where incoming rays of light reach the retina before they converge into a focused image.

Idiopathic intracranial hypertension (IIH) A disorder of elevated cerebrospinal fluid pressure of unknown cause previously called Pseudo Tumor Cerebi.

Microgravity ocular syndrome (MOS) One of several terms given the complex of signs, symptoms and physical changes affecting the eyes of some subjects in weightlessness.

Optic chiasma The X-shaped structure formed at the point below the brain where the two optic nerves cross over each other.

Optic disc The intraocular part of the optic nerve formed by fibers converging from the retina and appearing as a pink to white disk; because there are no sensory receptors in the region, it is not sensitive to stimuli.

Optic nerve Each of the second pair of cranial nerves, transmitting impulses to the brain from the retina at the back of the eye.

Parenchyma The functional tissue of an organ as distinguished from the connective and supporting tissue.

Sclera The white outer layer of the eyeball. At the front of the eye it is continuous with the cornea.

Scotoma An area of diminished vision within the visual field.

Subarachnoid space In the central nervous system, the subarachnoid cavity (subarachnoid space) is the interval between the arachnoid membrane and pia mater.

Ventricle A set of volumes with choroidal tissue which produce and contain cerebrospinal fluid in the brain.

References

1. Marshall-Bowman K, Barratt MR, Gibson CR. Ophthalmic changes and increased intracranial pressure associated with long duration spaceflight: an emerging understanding. Acta Astronaut. 2013;87:77–87.
2. Fogarty J, Otto C, Kerstman E, Oubre C, Wu J. The visual impairment intracranial pressure summit report. National Aeronautics and Space Administration. TP–2011–216160; 2011.
3. Alexander DJ, Gibson CR, Hamilton DR, Lee SM, Mader TH, Otto C, et al. Evidence report: risk of spaceflight-induced intracranial hypertension and vision alterations. In: Human research program human health countermeasures element. National Aeronautics and Space Administration Lyndon B. Johnson Space Center. Approved for public release July 12, 2012. p. 85–6.
4. Duntley SQ, Austin RW, Taylor JH, Harris JL. Experiment S-8/S-13, Visual acuity and astronaut visibility. In: Gemini midprogram conference including experiment results. National Aeronautics and Space Administration, NASA SP–121; 1966.
5. Mader TH, Gibson CR, Pass AF, Kramer LA, Lee AG, Fogarty J, Tarver WJ, Dervay JP, Hamilton DR, Sargsyan A, Phillips JL. Optic disc edema, globe flattening, choroidal folds, and hyperopic shifts observed in astronauts after long-duration space flight. Ophthalmology. 2011;118(10):2058–69.
6. Kramer LA, Sargsyan AE, Hasan KM, Polk JD, Hamilton DR. Orbital and intracranial effects of microgravity: findings at 3-T MR imaging. Radiology. 2012;263(3):819–27.
7. Alexander D, et al. Alterations in carbon dioxide (CO_2) metabolism in: evidence report: risk of spaceflight-Induced intracranial hypertension and vision alterations. Johnson Space Center; 2012. p. 78.
8. Zwart SR, Gibson CR, Mader TH, Ericson K, Ploutz-Snyder R, Heer M, Smith SM. Vision changes after spaceflight are related to alterations in folate–and vitamin B-12–dependent one-carbon metabolism. J Nutr. 2012;142(3):427–31.
9. Marshall-Bowman K, Rittweger J, Suarez JI, Venkatasubba Rao C, Donoviel D, Mulder E, Bershad EM. Effects of head down tilt with and without 0.5% carbon dioxide on intracranial and intraocular pressure. Aerosp Med Hum Perform. 2016;87(3):251.
10. Ivkovic V, Spielmann G, DiPasquale D, Hu G, Zhang Q, Strangman GE. Cerebral pulsatility in prefrontal cortex correlates with acute exposure to microgravity. Aerosp Med Hum Perform. 2016;87(3):251.
11. Purves D, Augustine GJ, Fitzpatrick D, Hall WC, LaMantia A-S, McNamara JO, Wiliams SM, editors. Vision: the eye. In: Neuroscience, 3rd ed. Sunderland, MA: Sinauer Associates; 2004.
12. Guyton AC. Chapter 16: The microcirculation and the lymphatic system: capillary fluid exchange, interstitial fluid, and lymph flow. In:Textbook of medical physiology. 8th ed. Philadelphia: W. B. Saunders; 1990. p. 170.
13. Kramer LA, Sargsyan A, Hamilton D, Tarver W, Powers E, Hasan K, Polk J, Straube U, Shimada K. Magnetic resonance derived CSF production rate as a predictor of orbital abnormalities after microgravity exposure. Aerosp Med Hum Perform. 2013;84(4):326.
14. Mollanji R, Bozanovic-Sosic R, Silver I, Li B, Kim C, Midha R, Johnston M. Intracranial pressure accommodation is impaired by blocking pathways leading to extracranial lymphatics. Am J Physiol Regul Integr Comp Physiol. 2001;280(5):R1573–81.

15. Boulton M, Armstrong D, Flessner M, Hay J, Szalai JP, Johnston M. Raised intracranial pressure increases CSF drainage through arachnoid villi and extracranial lymphatics. Am J Physiol Regul Integr Comp Physiol. 1998;275(3):R889–96.
16. Silver I, Kim C, Mollanji R, Johnston M. Cerebrospinal fluid outflow resistance in sleep: Impact of blocking cerebrospinal fluid transport through the cribriform plate. J Neuropathol Appl Neurobiol. 2002;28:67–74.
17. Killer HE, Laeng HR, Groscurth P. Lymphatic capillaries in the meninges of the human optic nerve. J Neuroophthalmol. 1999;19(4):222–8.
18. Lüdemann W, von Rautenfeld DB, Samii M, Brinker T. Ultrastructure of the cerebrospinal fluid outflow along the optic nerve into the lymphatic system. Childs Nerv Syst. 2005;21(2):96–103.
19. Louveau A, Smirnov I, Keyes TJ, Eccles JD, Rouhani SJ, Peske JD, Derecki NC, Castle D, Mandell JW, Lee KS, Harris TH. Structural and functional features of central nervous system lymphatic vessels. Nature. 2015;523(7560):337–41.
20. Iliff JJ, Wang M, Liao Y, Plogg BA, Peng W, Gundersen GA, Benveniste H, Vates GE, Deane R,Goldman SA, Nagelhus EA. A paravascular pathway facilitates CSF flow through the brain parenchyma and the clearance of interstitial solutes, including amyloid β. Sci Transl Med. 2012;4(147):147ra111.
21. Biousse V, Bruce BB, Newman NJ. Update on the pathophysiology and management of idiopathic intracranial hypertension. J Neurol Neurosurg Psychiatry. 2012;83(5):488–94.
22. Taibbi G, Kaplowitz K, Cromwell RL, Godley BF, Zanello SB, Vizzeri G. Effects of 30-day head-down bed rest on ocular structures and visual function in a healthy subject. Aviat Space Environ Med. 2013;84(2):148–54.
23. Czosnyka M. Brain physics in 30 short lectures (not only for engineers). Neurosurgery Unit, Department of Clinical Neurosciences, University of Cambridge. 2011. http://www.neurosurg.cam.ac.uk/pages/contact.php.
24. Wiener TC. Space obstructive syndrome: intracranial hypertension, intraocular pressure, and papilledema in space. Aviat Space Environ Med. 2012;83(1):64–6.
25. Hayreh SS. Management of non-arteritic anterior ischemic optic neuropathy. Berlin: Springer; 2011.

Chapter 6
Loss of Muscle and Bone During Spaceflight

6.1 Introduction

The next five chapters deal with the MusculoSkeletal System (MSkS). It is the largest system of the body and both its structure and function are involved. While previous chapters addressed issues that have significant impact, weightlessness effects such as Space Motion Sickness (SMS) are transient—SMS lasting only 2–3 days—quite short when one considers a long mission such as a 6-month stay aboard ISS. There is also a defined time period when fluid shifts have limited in-flight impact and postflight readaptation is also relatively short. In contrast, musculoskeletal adaptation can be potentially disabling for significant periods of time after long missions, and countermeasures can require major amounts of time and resources during spaceflight. Two chapters are given to the musculoskeletal effects; one (the current chapter) is devoted to cause and effects of bone and muscle loss. The second chapter (Chap. 7) addresses the theory and practice of countermeasures and the prevention of adaptation with its loss of muscle and bone. Adaptation of the MSkS also has major effects on the CardioVascular System (CVS), which are described in a separate chapter (Chap. 8). A neuromuscular adaptation plays a significant role in transient strength loss postflight and may have in-flight effects as well. This is addressed in Chap. 9. In addition, Chap. 10 that covers loss of body mass, also addresses a MSkS problem. Inadequate diets have played a significant role in loss of muscle and bone. A summation in the form of a suggested integration of the content of these chapters into a more effective countermeasure program ends this collation.

Musculoskeletal problems were not unfamiliar to the early pilot astronauts. There is a long tradition of pulled muscles and ligaments and more serious spinal problems associated with flying high performance aircraft. Some Houston neuro-

W. Thornton, F. Bonato, *The Human Body and Weightlessness*,
DOI 10.1007/978-3-319-32829-4_6

surgeons' walls are covered with astronaut pictures presented in gratitude for saving a career. There were also mishaps during off duty hours.[1] This tradition does not entirely disappear on selection as a pilot astronaut. Ironically, many if not most of the musculoskeletal problems related to spaceflight are the results of too little, not too much, physical activity.

It soon became obvious in the Mercury program that spaceflight acutely reduced g tolerance. Without strength measurements there was no way of knowing how much of the postflight instability was a result of strength loss. Had measurements been made after the 14-day Gemini 7 flight we can predict that a significant loss would have been found in leg strength. Only cardiovascular stress testing was done on Apollo missions. Vehicular and EVA Apollo space suits restricted motion and opposed virtually any motion when pressurized, especially finger motion. Crews spent hundreds of hours in the suits during training and developed skills and strength that enabled them to function effectively on the Moon in spite of severely restricted range of motion. They were aided by the reduced lunar gravity that resulted in an 83% reduction in weight. Hence, the suit backpack that weighed 134-lb on Earth became 22 lb on the Moon. During the time of the Apollo missions the Russian Soyuz missions included a 17-day mission after which the crew was reportedly too weak to walk upon landing.

During the Skylab Program, the SL-2 crew flew for 28 days with only the bicycle ergometer for exercise. Postflight, leg strength was significantly reduced and arm strength less so[2]. Arm strength exercise was added to the SL-3 crew's (59-day) flight plus leg exercise for the SL-4 crew (84 days) [1]. Crews were all able to walk following each mission although with obvious difficulty. Measurements documented significant flight time dependent losses in muscle and strength in arms and legs, in endurance, and in leg volumes, all of which could be attenuated by exercise.

No specific recovery program from these effects was implemented for these missions. It was left to the individual[3]. A typical return to normal activities might take a week following an Apollo mission. However, Ed Gibson, an avid fitness advocate, jogged several miles 4 days after his SL-4 84-day flight, but he didn't repeat the feat for several days because of sore legs. Yet by the 11th day postflight the entire SL-4 crew had regained leg volume lost inflight.

With Apollo and Skylab experience, the 7-day flights of Shuttle caused no operational concerns about muscles and strength. On Shuttle's first flight Commander John Young discovered another aspect of strength loss in weightlessness, and its solution. His verbal debriefing was succinct, "The brakes on that thing (the Shuttle) are really heavy. You better practice on them on the way down." Braking forces were

[1] Chuck Yeager was flying with undisclosed broken ribs when he broke the sound barrier in 1947. The ribs were broken in a collision with a closed gate on a wild horse ride two nights before the flight.

[2] As it was perfected, lean body mass measurement was added to later selection physicals.

[3] Traditionally preflight physical workouts were voluntary, individual and greatly increased.

normal as were John's legs after a 2-day flight. The weakness was produced by transient neuromuscular strength inhibition (see Chap. 9).[4]

Ken Mattingly warned one of the current authors (WT) of another aspect of this strength inhibition. "When you first try to stand up (after landing) you might as well have left the straps on." He was correct because the first attempt to stand was futile. A maximum effort with arm and legs produced an unstable stance for 15–20 s. Repeated efforts produced a remarkable increase in strength, and in some tens of minutes it was possible to exit the spacecraft for the walk around inspection. An Astronaut Office survey showed that this was a typical experience. By the next morning my neighbor observed, "He was putting the garbage out as usual." This was the usual experience following Shuttle flights of 7 days.

Some individuals had difficulties as the diversity of fliers increased, and some Shuttle flights were extended to 16 days, and equipment weight was increased. A large series of muscle and strength studies was done over a 6-year period (1989–1995) [2]. Results were inconsistent. The investigators concluded that "muscles are less able to maintain endurance—after spaceflight—and exercise may avert decrements in aspects of endurance." A treadmill was flown on all Shuttles from STS-3 (1982) to STS-50 (1992) before it was replaced with a cycle ergometer. Other exercise devices were flown for evaluation including a rowing machine, a redesigned cycle ergometer and a motorized treadmill. Russia continued its development of vehicles and missions of increasing duration until the current longest duration flight of 435 days was made in 1995.

In 1995 a series of flights by astronauts on the Russian space station Mir was initiated by Norm Thagard with a 115-day stay. There were a total of seven such flights with an average length of 140 days. The only published report on the functional effects of these flights is Jerry Linenger's autobiography [3]. The participants were returned from Mir by Shuttle in a horizontal couch and then carried by stretcher into the elevated mobile medical facility adjacent to the Shuttle. This minimized effects of the 1.5-g reentry and subsequent 1 g on the ground, allowing investigators to better assess the subject's condition in weightlessness. According to Linenger, before this arrangement, astronauts had been unable to stand unaided after returning from Mir, and some had to be carried off by stretcher. He described his own extreme efforts to walk and stand giving the impression of a debilitated person who needed the subsequent 3-month recovery administered by a clinically trained physical therapist. His recall of cosmonaut stories and pictures of their condition on landing after long duration flight was that they were more or less physically incapacitated. Even the few who were able to stand and take a few steps were carried away on stretchers.

International Space Station (ISS) began the first of its "expeditions" in 2000. There have been 50 expeditions as of September 2016 with durations of up to 340 days. Until recently NASA medical privacy agreements have effectively suppressed reports of postflight crew condition. Available reports on ISS expeditions and those

[4]It became standard practice for pilots to exercise leg muscles by actuating the brakes during Shuttle reentry.

from the Mir-astronaut missions are discussed in the history, inflight, countermeasure, and postflight segments of this and the next chapter.

There has been a series of empirical studies and efforts on ISS aimed at reducing losses of bone and muscle strength. Currently there are cycle ergometers, a simulated weight machine and a treadmill in the American areas. Verbal reports are that after a typical 6 months on ISS with effective use of a reasonable amount of the daily schedule of 2 h for exercise, crews can walk and are able to return to normal work after a 3-week readaptation schedule. An astronaut returned from a 6 month expedition reported that he was able to walk after 2 h in 1 g. In spite of progress, it is reasonable to question the adequacy of the current knowledge and technology to support long duration Lunar missions much less a Martian mission.

6.2 Brief History

Unlike the problems described in the two previous chapters, loss of muscle and bone in space is not a direct effect of weightlessness. Instead, bone and muscle loss are the product of disuse caused by the inability to walk, jog or engage in many other 1-g physical activities in weightlessness. Effects of such inactivity on bone and muscle have been observed throughout history. In 1892, a German physician Julius Wolff expressed these effects in what has become Wolff's law, largely based on changes in the MSkS [4, 5]. Simply stated, form is determined by function. Undesirable effects of bed rest were noted in the 1850s. Bone mass and strength loses that occurred during bed rest were extensively studied by calcium balance and other means during World War II and thereafter. Investigators who conducted these studies predicted large bone losses in weightlessness when the space program began.

In 1965, plain film X-ray studies on Gemini 4 and 5 [6] showed large bone losses in the heel and hand. These were short missions of 4 days and 8 days respectively. Although the extraordinarily large losses were later shown to be erroneous, they triggered a rush of bone studies that continue to the present. At that time, the concept of muscle forces on bone controlling bone strength was not generally accepted. Incidents involving EVA on Gemini demonstrated operational needs for strength that were met by the construction of an astronaut gym with strength training facilities. Early plans for a cycle ergrometer on Apollo were dropped, and a capstan resistance device (Exergenie®) was flown, but only occasionally used. Skylab initially had only a cycle ergometer for exercise in spite of being the longest flights attempted during their time.

The Astronaut Office was sufficiently concerned with the possibility of crewmembers being unable to walk after such flights that for the Skylab missions one of the current authors (WT)[5] was allowed to measure strength pre- and postflight [1]. Skylab's study of mineral loss from bone by measuring all intake, output, and body mass on all crewmen before, during and after flight remains the largest and most

[5] This was a Detailed Supplementary Objective—a procedure added to mission plans.

complex life sciences study in space [7]. Loss of calcium in the heel and hand were measured pre- and postflight by tissue absorption of radiation [8]. During the first manned 28-day Skylab mission (SL-2) the crew complained of lack of adequate means of exercise and upon landing walked with obvious difficulty. Strength measurements 5 days after landing showed significant losses in arms and large losses in the legs. For the second manned Skylab mission (SL-3) a set of extension springs with handles and an isokinetic loading device with a cable and handles were added for trunk and arm exercise [1]. On return to Earth the crew appeared to be at their limit in walking, and it was agreed that without some locomotor exercise the 84-day crew of the third Skylab mission (SL-4) would not be ambulatory on return. This was reflected in the 59-day crews' measured loss of leg strength. Conversely, arm strength losses had been eliminated by their use of the added strength exercise.

A Teflon plate and bungees added to an onboard harness made crude approximation of walking possible on the 84-day SL-4 mission. The crew returned in better condition than the crew from the 56-day mission. Strength, body mass and leg volume measurements confirmed this [1, 9, 10]. Metabolic studies of bone minerals showed a significant loss on all Skylab flights and this finding was supported by density studies [8, 11]. In-flight findings and their significance are discussed in the sections that follow.

Following Skylab there were major changes in management in which the Astronaut Office lost much of its influence and autonomy. It was no longer possible to initiate life sciences studies. Exercise hardware development was allowed but not supported. It was obvious that locomotor exercise was essential in flights of more than a couple of weeks and desirable on shorter flights. In cooperation with Henry Whitmore, a mechanical genius in San Antonio, a series of exercise devices were developed and demonstrated including treadmills, rowing machines and strength exercise devices adapted for spaceflight. (See Chap. 7.) Unknown to the Americans at that time, the Soviets were also developing exercise devices including a treadmill to support their flights of increasing duration.

As Shuttle operations approached, a prototype treadmill was built that met Shuttle's severe size and weight constraints. On the second Shuttle flight in 1982 (STS-2) Joe Engle demonstrated jogging in place loaded by bungees and a harness. In 1982 on STS-3 the treadmill was demonstrated. Instead of using an improved design allowed by relaxed Shuttle specifications, life sciences managers ordered a number of copies of the prototype with its limitations and assumed control of their use. They were then flown over the next decade until replaced by a stationary bicycle. Other devices were flown on Shuttle in preparation for ISS.

During the Shuttle era the Soviets (and then the Russians) developed a variety of strength countermeasures. In the clinical world a variety of systems based on absorption of radiation by tissue were developed that could measure loss. Such devices when coupled to computers provide automated identification and measurement of tissue, shapes and volumes of muscles and bones and now dominate such studies. They supported some of the earliest musculoskeletal studies on long duration Mir studies including the astronauts on the Shuttle-Mir missions.

ISS was initially flown with the Interim Resistance Exercise Device (IRED) and a Johnson Space Center life sciences treadmill. A commercial cycle ergometer was also flown. The early years of musculoskeletal exercise and studies were delayed by failures of the IRED, treadmill and cycle. Scheduling of exercise was inadequate and poorly adhered to. One study showed maximum foot ground forces when running on the treadmill was considerably less than on Earth with predictable losses in leg strength and bone density [12]. Some individuals who understood some essentials of exercise hardware and design and use made marked advances with a pseudo weight station named the Advanced Resistance Exercise Device (ARED), and with modified commercial treadmills and cycle ergometers. Two hours plus per day for exercise is now part of the basic schedule (see Chap. 7).

A large number and variety of bed rest and in-flight studies have been [13] and are being conducted during the ISS era. Animals have also become a significant part of muscle and bone studies. One-g studies of mice and rats with hindquarters supported by their tails have become a common method of research for animal investigators. ISS now has cage facilities for rats and mice. A significant change in ISS protocol is the use of functional performance as a strength criterion. All ISS astronauts have been scanned for radiographic density, pre- and postflight, creating a virtual treasure trove of information. However, access to these data has been limited and publications and hence publications based on these date is also limited. Participation by U.S. Astronauts in ISS was reduced by termination of the Shuttle program. NASA's attention is currently focused on exploratory flights of long duration, a very different situation from orbital flight that will require very different countermeasure protocols and hardware.

6.3 Background

Bone and muscle are usually treated as separate, very different entities. In this chapter, wherever possible, they are treated as a unit for in this situation they are a functional interdependent unit and treating them separately has caused difficulties and delays, and this digression.

Phylogenetically muscle is much older than bone. There were many successful animal species for eons of time before mineralization of cartilage to produce bone appeared and unnumbered boneless species still exist. Functionally muscle is more trouble free than bone and especially bony articulations or joints. This may account for some of the earliest ancient physicians' records of treatments of bone while muscle was relegated to military and sports trainers. This situation is still seen today to some degree.

In the space program this dichotomy resulted in bone studies being conducted early in Gemini [6], years before the first ad hoc study of muscle on Skylab whose major, and still most complex study in space, was of bone mineral metabolism. Most researchers in NASA bone studies have been trained and based in clinical medicine which deals with a variety of pathological bone problems whose treatment is primarily medical.

Logically, the physiological countermeasure for loss of muscle and bone in space is replacement of the forces that maintain muscle strength and mass and consequently bone strength and mass on Earth; i.e., physical activity. Initially this was not accepted by the bone community who sought more esoteric causes directly related to microgravity that was postulated to produce some pathological condition. Effects of totally inadequate inflight exercise were regularly taken as proof that exercise was ineffective with a mantra, "X% of bone was lost in spite of vigorous exercise throughout the mission"[6]. A few bone researchers pursued application of force in unusual form such as static loading or vibration[7]. Predictably, efforts were made to use the medications used in pathological atrophy on Earth and this is now a major countermeasure effort. While it is increasingly accepted that forces from muscle activity is the natural stimulus to bone strength, the attraction of a simple medication will remain for decades to come, probably as an adjunct to exercise.

This emphasis on bone in contrast to muscle has in a way put the cart before the horse in both studies and countermeasures to date. The following is focused on somatic adaption of muscle then bone as the primary cause of their mutual losses in spaceflight.

6.3.1 *Nature of Adaptations of the Musculoskeletal System*

The human body is in a constant state of change even when activities and environment are stable. Every component of the body is being constantly removed and replaced (turned over), usually a cell at a time[8] [14]. Some examples of the turnover time are:

Skin cells	~14 days
Red blood cells	120 days
Liver cell	300–500 days

Normally the rate of replacement equals that of removal such that size, shape and function of the tissue remains relatively unchanged. However, a change in environment often causes changes in rate of removal and replacement of a tissue's cells producing a net gain or loss of tissue and function. A simple example is the case of energy consumed versus energy used. Any excess of energy intake from food and drink that is not used by the body is stored as fat (see Chap. 10). In a person with normal fat stores, if the energy used is greater than that available from food, it will be obtained from the stored fat. If body mass or weight is used as an indication of energy balance, then the relationship can be expressed by a simple first order linear equation.

Weight ~ Energy in ± Energy out

Mass = K(Energy in − Energy out)

where K is mass energy^{-1}

[6] A survey of the literature on this topic reveals typical bone loss rates of 1–2% per month with the load bearing bones (hips and below) suffering the most loss.

[7] This is still pursued as a countermeasure by some.

[8] An exception to this is the carbon in DNA and some brain cells.

The resulting graphic plot will be a straight line whose slope is an indication of energy balance. This may be seen in the in-flight body mass curves from Skylab in Chap. 10 (see Fig. 10.8, p. 282).

Loss of other tissues and functions from disuse and other causes is complex, on Earth and in space., for example, when a person is placed at bed rest. In the normal turnover of muscle or bone the body continually removes and replaces a more or less fixed and equal amount of tissue. This equilibrium is modified by changes in environmental stresses (Chap. 2, pp. 2.3–2.5). Under increased stress the rate of replenishment is increased; therefore, muscle and bone strength will increase. The effects of increased stress on bones and muscles can be seen in athletes. Exercise and training can result in large increases in bone and muscle strength. If stress is reduced, as in bed rest or weightlessness, the rate of removal may increase and replacement decrease reducing mass and strength.

Quantitatively tissue will be decreased as a percentage of tissue available for removal. This is analogous to a bank that charges 10% per month for keeping an account. Deposit $10 into the bank, and at the end of the first month the account will be worth $9. At the end of the second month it will be worth $8.10 and so on. Although there will be decreasing amounts at the end of each month, less is removed each month such that the time for total loss[9] becomes very long. This is the equivalent of part of what happens in bone or muscle loss. The mathematical expression for this is shown in Eq. 6.1.

$$X = X_0 e^{-at} \qquad \text{(Eq. 6.1)}$$

where: X = money or muscle remaining

X_0 = initial amount of money or adaptable muscle

a = rate at which money or muscle is removed; i.e., amount per unit of time

t = time from start of removal of money or muscle

e = natural logarithm

While the bank account calculations are not as simple as those of the linear equation of weight loss or gain, given the known original balance X_0, with the time t and rate of loss a, the remaining amount of money or strength can be calculated. Or if the amount remaining at a given time and loss rate were known, the time to reach a given value could be determined.[10]

It is not quite that simple with muscle and bone because the human body does not depend only on the bone or muscle's environment not to lose all its muscle and bone. Some additional significant amount of both tissue and function are constantly maintained by DNA. It is not difficult to imagine situations where a fixed reserve of tissue has survival value. It has always been known that even with complete loss of stimulus, such as when the spinal cord is transected, appreciable bone will remain

[9] Mathematically the time to reach zero will be infinite, but in practice some amount will be reached which for practical purposes will be zero.

[10] Some arbitrarily small number for practical zero can be used.

as long as the subjects live and that local stimulation of muscle can restore tissue production of strength and motion. For accurate calculation this fixed reserve requires that a constant *C* be added to the equation (Eq. 6.1) that completes the relation for muscle or bone loss (Eq. 6.2).

$$X = X_0 e^{-at} + C \qquad \text{(Eq. 6.2)}$$

This is equivalent to having a fixed untouchable reserve in the bank.

Although there is a single homogenous mass of tissue, it is a mathematical necessity and a conceptual advantage to treat the single mass of tissue as if it has two components: one variable in response to the environment and the second fixed.

Determination of the individual limits of adaptation needed to calculate the true value of X_0 is difficult to impossible for upper limits of muscle strength are reached by only a few athletes after years of extraordinary effort. Even then it is arguable that bone strength will have reached its adaptable maximum. In this situation the upper limit is taken as the maximum muscle strength or existing bone density at the time losses begin. Only a Swiss group have made a complete measurement of bone in this situation to date [15, 16] using matched surrogates for X_0 max and measured X_0 min while other European investigators have applied this concept to muscle measurements in a bed rest study [17].

Under these circumstances a reasonable question is why should there be concern with such a concept? The concept itself is of great significance to a correct understanding of the problem. The current practice of treating exponential processes as a linear process without a limit leads to egregious errors. With graphics and other measurement based estimates, far more realistic views of the situation can be obtained. A pragmatic solution to this difficulty in measurement is to reduce the loss rate to zero by adequate countermeasures, and this may have happened recently in a few individuals.

Unfortunately, misleading interpretations of the results obtained has become standard practice. A linear rather than the correct logarithmic relationship to describe loss has been and continues to be used for muscle [18] and bone [19]. This practice has produced frequent and significant misinterpretations of incomplete data. Measured, limited losses of bone are being extrapolated into fractures when attempting to walk on Mars after the long flight time required to get there. Such stories are welcomed by media and by individuals and groups seeking funding for investigations in this problem. All of these issues are discussed in the following sections except countermeasures which are discussed in the following chapter.

Changes seen in the MSkS in weightlessness illustrate classic examples of species and somatic adaptation (Fig. 6.1) as previously described in Chap. 2. Transformation of our ancestral layers of body muscle into our present complex of specialized muscle groups was a remarkable species adaptation as shown in Fig. 6.2a. An even more remarkable species adaptation was transformation of the

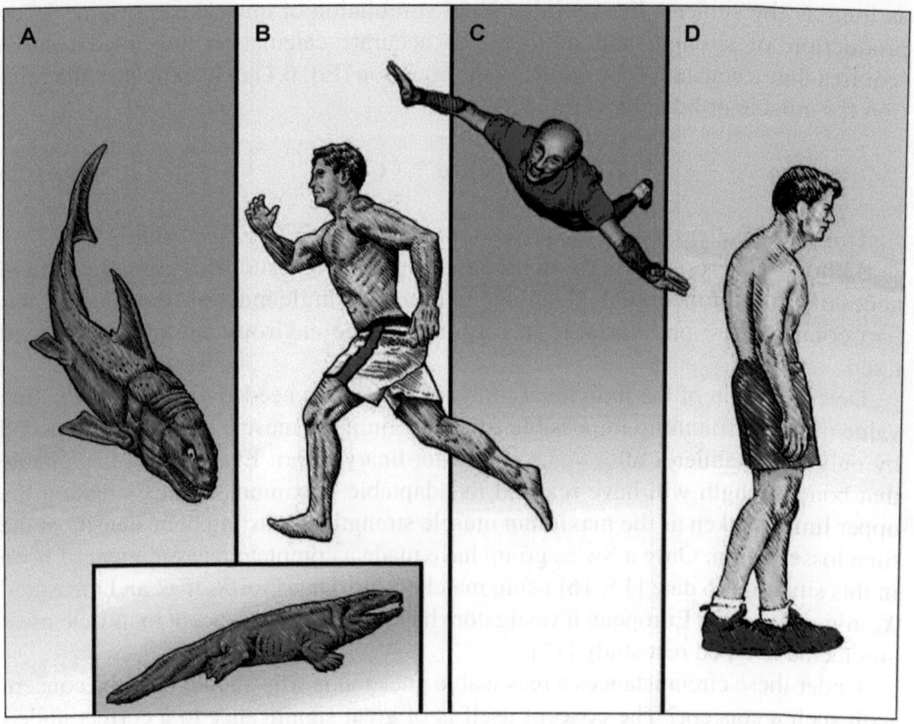

Fig. 6.1 As our ancestral fish *left* its original habitat and moved onto dry land, its body became species adapted by evolution to weight and to other features of terrestrial life (**a**, **b**). The capacity to somatically adapt to varying demands of the 1-g environment became embedded in our DNA (**b**). When weight is removed as in spaceflight (**c**), somatic adaptation sheds unneeded capacity such that muscle and strength become inadequate for normal 1-g function (**d**). Loss of unneeded bone will also be in progress

bones shown in Fig. 6.2b, especially the transformation of each of the bones associated with the fins into an essential bone in the array forming the skeleton of our hands and feet[11] [20].

In our DNA there is another remarkable capacity: the ability, within limits, to adapt our body systems to whatever stresses the environment imposes, especially the MSkS (see Chap. 2, Fig. 2.1, p.21). If work or exercise requires larger than usual forces muscle will increase its size and strength to meet the increased loads. Bones have a safety margin of 8–10 times their usual maximum load. Since muscle and bone are interdependent, bone will also increase its strength to safely accommodate the increased muscle forces. When loads are reduced, or removed in weightlessness, unneeded muscle and strength is removed, which is followed by loss of bone and its strength.

[11] Dubin's description of this in his "Inner Fish" is good reading. It was recently featured in a TV series by PBS.

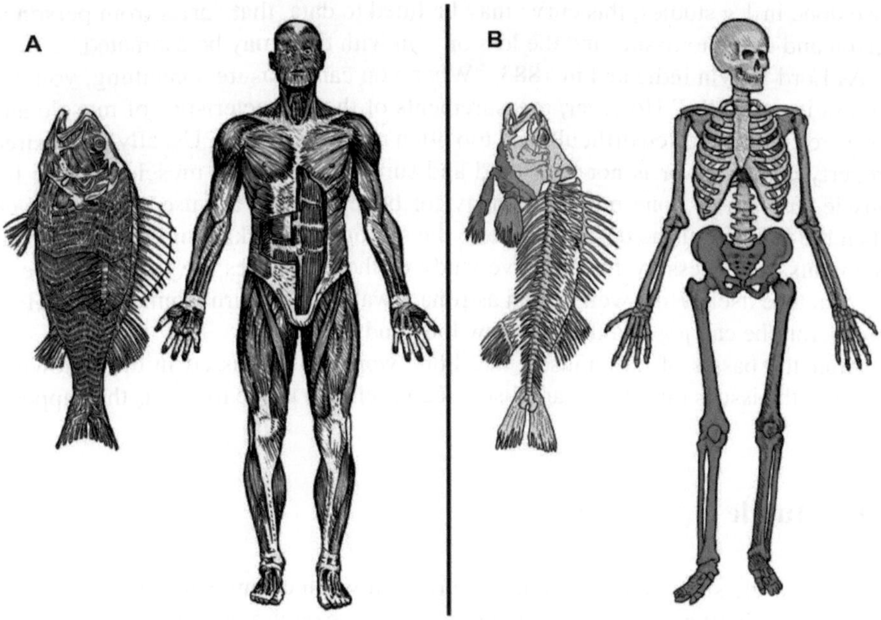

Fig. 6.2 The magnitude of species adaptation may be seen in this image of muscle (**a**) and bone (**b**), in fish and man. In (**a**) both bodies primarily appear to be muscle. Up to 70% of the fish's body is muscle, most of which is used to bend the body and propel it rapidly through water. Except for the shoulder girdle, the human trunk muscle is in thin sheets covering a bony skeleton. Some 40% of a typical human body is muscle, and most of that is in the legs. Transformation of the flexible fish skeleton into the complex rigid human skeleton is even more remarkable, especially the fins (*orange*) and their tiny bones that became feet and leg bones in humans (color figure online)

Somatic adaptation to loss of the coordinated forces of many muscles in walking or jogging, in 1-g locomotion, can be prevented by applying the same forces in weightlessness, as those on Earth. However, somatic adaptation is a surprisingly exacting process, and when the countermeasures applied are not exact, muscle and bone do not respond. All too often one hears "but exercise didn't stop the muscle loss" when the exercise did not replicate the forces of locomotion on Earth, for example when a treadmill generated only half the forces usually seen on Earth.

Another principle of adaptation that is frequently misunderstood is how the changes produced vary with time. They are not linear. All too frequently one hears, "Bone or muscle is lost at say 3% per month, and the time it will take to go to Mars is 12 months," which is then applied to estimate bone or muscle remaining on reaching that planet. If adaptation were linear with time, in the above case there would be no bone in 34 months. It is known that such losses or gains slow with time and reach a limit. The rate of adaptation is exponential with time and can be expressed mathematically by Eq. 6.2. By correctly using this relation, as some European investigators have done in 1 g studies, this curve may be fitted to data, that varies from person to person and tissue to tissue, and the loss or gain with time may be estimated.

As Lord Kelvin indicated in 1883, “When you can measure something, you can discuss it rationally.” However, measurements of the characteristics of muscle and bone are complex, often difficult, and too often made confusing. Usually the desired property cannot be or is not measured and surrogates such as muscle volume for muscle strength or bone mineral density for bone strength are used instead, even when the relation of the desired value to the surrogate is unknown. Although measurements are necessary for effective study of these changes the final measure is performance itself. How well can an astronaut walk after return from a flight? How long before he can jog and then for how long and how fast?

After the basics of how muscle and bone work are discussed in the following sections, the issues raised here are discussed in relation to the topics as they appear.

6.4 Muscle

Loss of muscle strength can be a major problem when cosmonauts and astronauts return to Earth after a few weeks without adequate exercise countermeasures [18]. It could become the primary problem when astronauts try to walk on Mars carrying 150-Earth-pound backpacks after months in flight if exercise has not been adequate. While bone strength will have been lost from the lack of muscle force imposed upon it, without proper exercise en route, many will be too weak to walk. Conversely, if muscles are properly exercised bone will also be protected, bone and muscle being an integral functional unit.

Muscle can produce a unidirectional force that usually results in a reduction in its length [22, 23]. This action is generated at the molecular level by interaction of two twisted chains of protein molecules of Actin and Myosin (Fig. 6.3a), arranged as in Fig. 6.3b [21, 22]. This is the basic motor unit (sarcomere) of human motion. When triggered by a nerve impulse the action diagrammed in Fig. 6.3c occurs; myosin molecules attach to sites on the actin chain and forcefully rotate, moving the chains a microscopic distance with respect to each other. This action is analogous to that in a ratchet. If a stimulus is continued the chains will eventually overlap and reach their limit. More or less force will be generated as more or less fiber connections are made (Fig. 6.3d). This force versus length relation will be reflected in whole muscles and becomes significant in the biomechanics of exercise.

6.4.1 Motion

Maximum change in length in a single unit during contraction is microscopic, say 0.002 mm. Since molecular action occurs in microseconds the motor unit will shorten to its minimum length in milliseconds if the stimulus is continued.[12] To

[12] In the absence of a stimulus the fibrils disconnect and slip into their relaxed position.

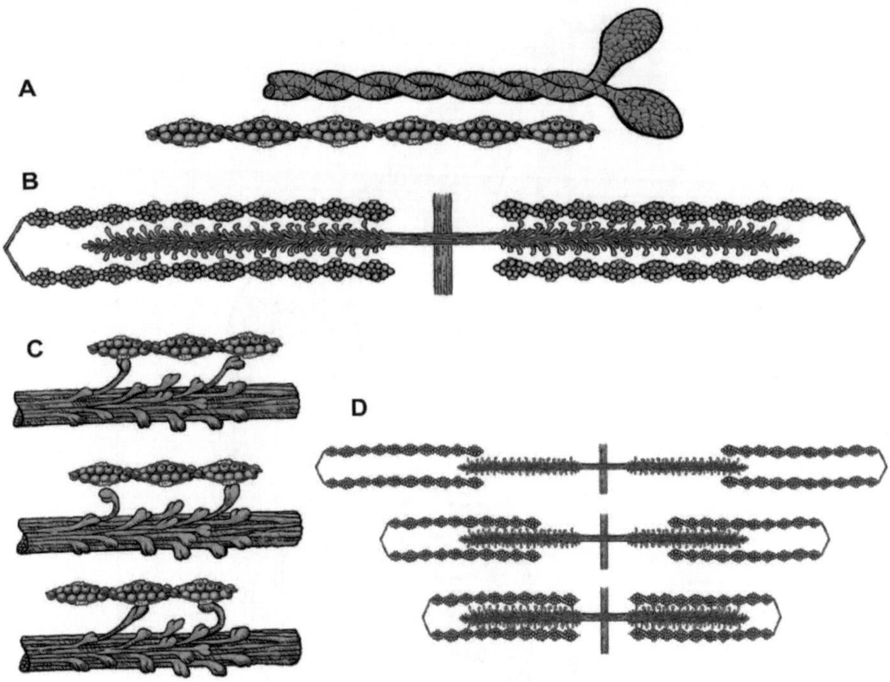

Fig. 6.3 All muscle motion and force are generated by untold millions of microscopic motor units called sarcomeres. The basic driving elements are chains of molecules of actin (*blue*) and myosin (*red*) (**a**), arranged as in (**b**). When actuated by Ca^{2+} molecules released by neurological triggers, myosin bends and brings paddles into contact with actin. Attaching to points on the actin (**c**), the paddles move forcefully propelling the myosin chain along the actin in ratchet like action. At this level action is rapid. Under continued nervous stimulation the chains are moved past each other (**d**), and a sarcomere 0.004 mm long can shorten until the rods overlap, and the unit is only 0.002 mm long. Forces are correspondingly small. Motion times are measured in micro- and milliseconds (color figure online)

generate useful motion, motor units are connected in series to multiply the action of a single motor unit to form a *myofilament* (Fig. 6.4a). To produce a maximum motion of 1 mm would require 500 motor units in this case. Some filaments are tens of centimeters long.

6.4.2 Force

To create useful forces, the myofilaments are collected into parallel, progressively larger bundles that directly multiply the myofilaments's force (Fig. 6.4a). The first such bundle containing thousands of filaments is called a *myofibril* (Fig. 6.4b). Many fibrils are then bound into a *muscle fiber* (Fig. 6.4c). The binding of large numbers of fibers into *fascicles* (Fig. 6.4d) continues the multiplication of force. Varying numbers of fasicles are bound into a *complete muscle* (Fig. 6.4e) some of which are

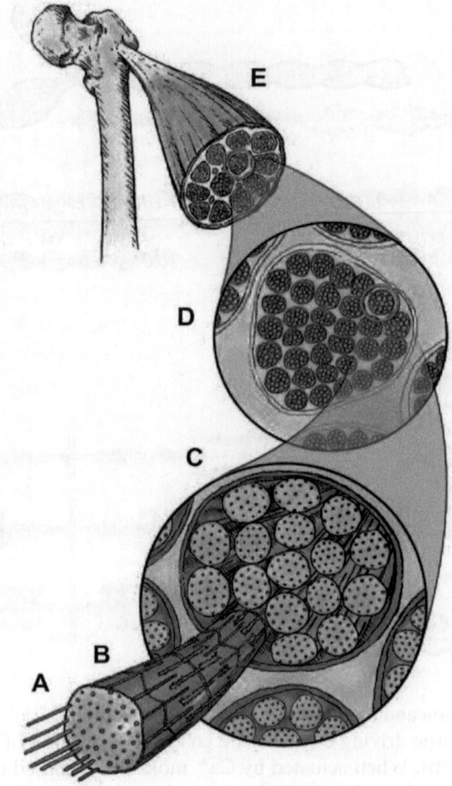

Fig. 6.4 Multiplication of molecular forces: production of useful muscle force from the almost infinitesimal force of the actin/myosin molecules in the ultramicroscopic sarcomeres is achieved by bundling or placing them in parallel. The process begins with untold numbers of the myofibrils (*a*), shown as thin projections in the *bottom left* into fibrils (*b*), thousands of which are bundled into muscle fibers (*c*), that are then bundled into fascicles (*d*) (*upper right circle*). Each bundling multiplies the force thousands of times. The necessary number of fascicles, easily visible in muscles, is then incorporated into a muscle (*e*), to produce its force, often hundreds of pounds. Note that increases in actin and myosin produce an increase in force and size of the myofibrils that are then multiplied by bundling, sometimes into the huge muscles of body builders. Cross section of the muscles is roughly proportional to their strength

capable of producing maximum forces of hundreds of pounds. Maximum muscle force is proportional to the number and size of fibers, which also determines the cross sectional area (CSA) of a muscle. CSA is frequently used as a surrogate for strength. All too frequently this is not done with the necessary rigor and errors are common.

6.4.3 Muscle Fibers

The *muscle fiber* is a complete cell and is the basic complete operating unit of a muscle. It is a large specialized cell with multiple nuclei (Fig. 6.5). Muscle contraction requires energy and muscle fibers are surrounded by an extensive capillary network to

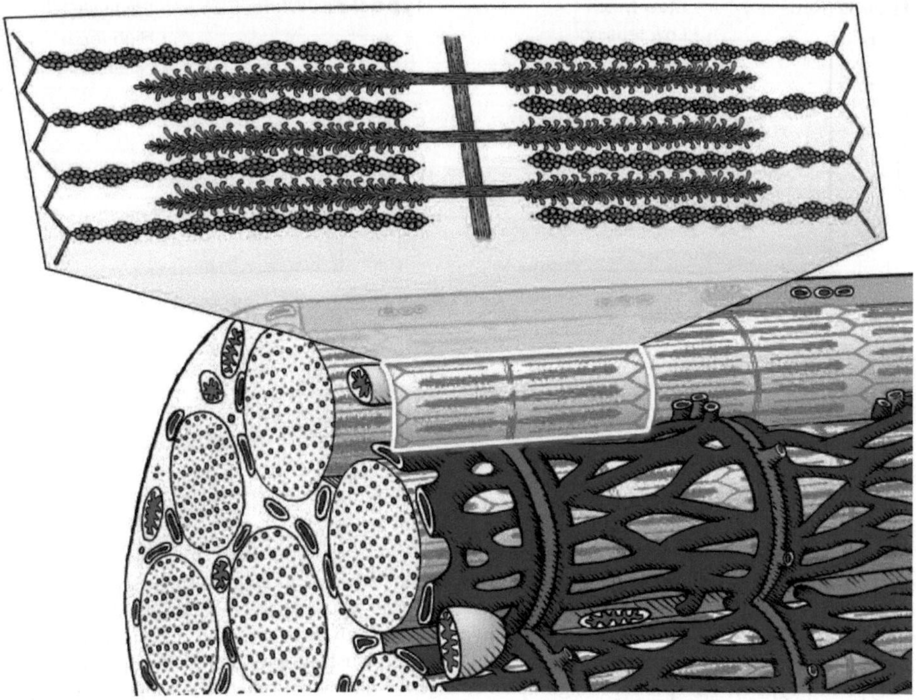

Fig. 6.5 Muscle fibres, one of which is depicted here, are the basic functional units of muscle. It begins with the actin/myosin molecular chains bundled into thousands of fibrils rather than the few shown here. These are then gathered in great numbers, in contrast to the few shown here, into muscle fibers. A complete functional unit is laced with capillaries, which are omitted for clarity here, and give Type I fibres their color and endurance. There are multiple nuclei with DNA to control the many support and operation functions. Mitochondria, depicted as *light blue* packets here, convert energy substrates from the interstitial fluid into the chemicals required to support contraction. Segmented sarcolemma networks (*blue*) allow the nervous system to activate the necessary number of actin/myosin contractile elements to produce virtually any degree of total force required (color figure online)

refresh the interstitial fluid surrounding thousands of fibrils with many mitochondria for energy conversion. (Capillaries are not shown.) There are also extensive networks of *sarcoplasmic reticulum* surrounding the fibers that trigger simultaneous contractions by segments of the fibers when stimulated by individual neurons.

A key to understanding many aspects of muscle performance and its adaptation lies in understanding the muscle fibers. A fiber's purpose is conversion of energy into force and motion, but there are fiber types that do this in different ways. Three types are usually recognized but only two are described in Fig. 6.6 and discussed here.

Type I or slow twitch fibers produce a relatively low force over several hundred milliseconds time. Their energy is derived from aerobic or oxidative processes continuously supplied by blood flow, hence they can make repeated contractions for long periods. This adds a red color to these fibers. They do not increase in size with

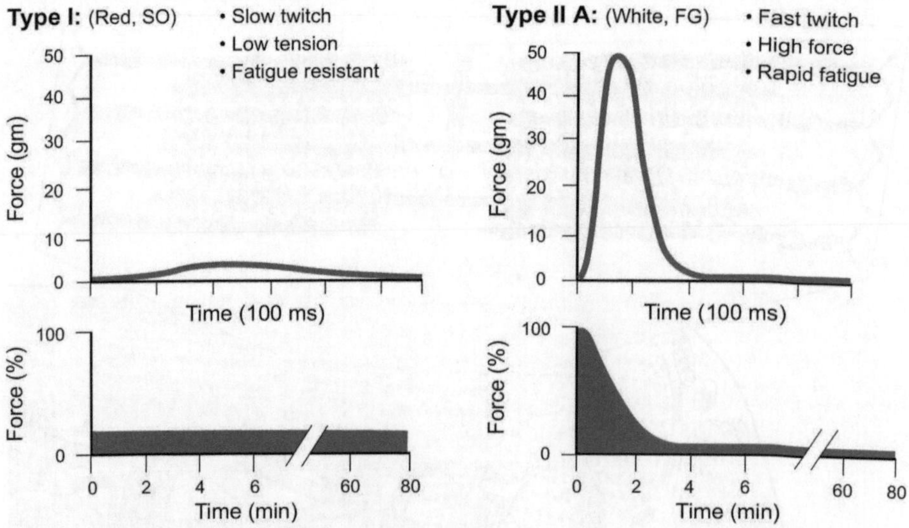

Fig. 6.6 Muscle fibres are divided into several types based on their form and function. Characteristics of the two basic types are shown here, but there are a range of subtypes with more or less of the characteristics of these two fundamental ones whose characteristics are shown here. Although the numbers of fibres remains relatively fixed, both their size and characteristics can change to adapt to differing environments. In all cases of disuse atrophy there is a transformation of Type I to Type II. This produces significant changes in individual functional capacity that will be examined in the next sections

exercise, but their associated blood vessels, which are not shown in the figures, proliferate with endurance exercise. There is a corresponding increase in the number of mitochondria required to transfer energy from blood to fiber.

Type II or fast twitch fibers produce a short but powerful twitch. Their energy source is from glycogen sources stored in the cell. This can be rapidly depleted in seconds to minutes by repeated contractions. Without the red blood supply in Type I fibers, these appear pale or white. Unlike Type I fibers, Type II can hypertrophy under exercise stimuli producing the muscle mass seen in weight lifters. This hypertrophy is produced by increased amounts of protein in the motor units, especially actin and myosin. It occurs when the muscle is regularly stressed by greater than normal force loads increasing the units and the muscle's size and strength. Under stress of very large force loads the number of motor units and myofibrils can also increase. There is controversy over whether the number of muscle fibers can increase. When force loads are reduced as in weightlessness, these processes are reversed producing atrophy. Some details of mechanisms that cause these changes are unknown.

Moreover, fiber types under continued appropriate exercise stimuli can be transformed at the muscle fiber level from one type to another. This results in significant differences often easily seen in individuals' bodies such as a professional linebacker versus a distance runner. The runner's body is seldom heavy, and the all important legs may be no larger or stronger than those of an untrained counterpart. Yet the muscles in those legs can use all the oxygen a well-conditioned circulatory system can provide for an hour or more.[13]

Conversely the linebacker's body can produce bursts of force and power plus sprints of surprising speed, but on an open road he will soon be left in the dust of the distance runner. As might be expected muscle fibers and their distribution are very different in these two examples. Thick, pale, fast twitch Type II fibers predominate in the linebacker's muscles while red Type I fibers are predominant in the distance runner's legs. As also might be expected, the exercise required to produce and maintain these very different bodies is also very different.

In contrast to the runner's miles on the track to develop *endurance*, the linebacker will spend some of his time lifting weights so heavy that only a few repetitions are possible to develop strength. An understanding of these two generic exercises is enough to give basic insight into exercise countermeasures and their results. Extended use of the cycle ergometer, an endurance exercise on used on Skylab for the SL-2 mission, maintained endurance and aerobic capacity of legs and cardiovascular system, but the low maximum forces could not stress legs to maintain strength. Conversely, a dedicated 1-g jogger on SL-3 increased his arm strength inflight [1] using isokinetic strength exercise. Exercise and its effects will be pursued in the section on Somatic Adaptation in Chap. 7 on countermeasures.

Fascicles are collected in varying numbers and characteristics and organized in various ways into muscles with which we are more or less familiar, ranging from tiny extraocular muscles that react faster than an eye's blink to massive leg muscles that may take 100 ms to project a body on Earth into a weightless trajectory for a second or more. All muscles share some generic characteristics that are significant when combined with bone as will be seen in subsequent segments.

6.5 Bone

Our need for a rigid skeleton of bones can be seen by comparing a human skeleton to that of our ancestral fish (See Fig. 6.2). Fish are effectively weightless in the sea. Up to 70% of the fish's body is in sheets of muscle arranged to generate waves of muscle down the body that react directly with the water to propel it forward, much faster than humans can run. These forces are diffuse, and the flexible small skeletal bones are largely continuous tie points for the muscles.

In humans, bones are large and rigid to support the weight of the body and to transmit concentrated muscle forces several times the force of body weight to produce locomotion and other actions. Bone is some 12% of body mass in a human female and 15% in the male. It is 12–15% collagen, 70–80% calcium and phospho-

[13] The circulatory system's ability to supply blood is the limiting factor in performance in endurance events. Blood doping or auto transfusion is a means of increasing blood and energy to the muscles and performance.

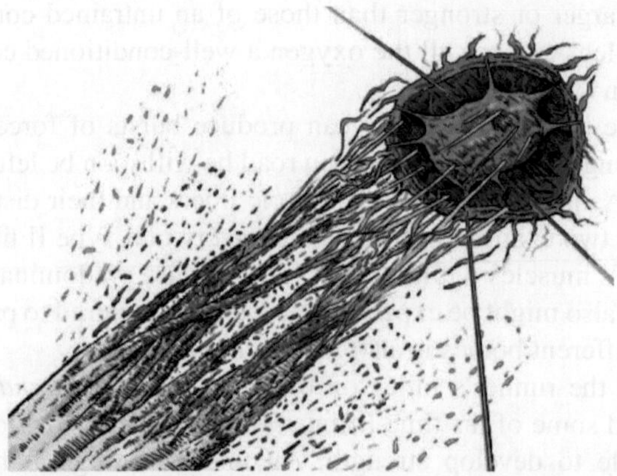

Fig. 6.7 An ultramicroscopic image of an osteoblast is copied here in a diagram to show its bone production. Normally it would be completely surrounded by the processes shown. Segments of collagen (*yellow*) are extruded within the cell where they attach end to end to form collagen fibers and are pushed from the cell into an interconnected mesh. This mesh is then covered by a material (*green*) to form osteoid, which attracts and attaches calcium and phosphorous compounds from surrounding fluid to the collagen. Over the next weeks or months this is transformed into bone (color figure online)

rous, and a small percentage of water. In adults some 1% of old bone is being constantly removed and replaced with new. This process can be greatly accelerated if stresses on the bone are increased producing larger, stronger bones. If loads are decreased, bones become smaller and weaker. This metabolic activity in bone is slow, a tiny fraction of that in muscle, so the bones' circulatory system is correspondingly small with a multitude of tiny canals and passages in the "solid" bone to insure that every cell is supplied with fresh blood or interstitial fluid.

Bone has many functions other than transmitting forces; in a recent long paragraph on bone in space an expert placed bones' loss of strength in a single sentence at the end of a long list of other potential problems such as kidney stones. The reverse of this approach is taken here, where cause and prevention of muscle loss is taken as the problem whose solution will also prevent the other potential problems.

Bone's generic structure may be seen in reinforced concrete in which flexible rods of steel "rebars" of high tensile strength are embedded in cement of high compressive strength. Bone consists of an extensive flexible skeleton of intertwined networks of collagen fibers, some 30% of total bone mass[14] [23]. The remainder is a compound of calcium and phosphorus deposited around the skeleton to complete the bone.

Fabrication of bone begins with cells called osteoblasts that assemble proteins from surrounding interstitial fluid into three chains that form segments of *collagen* fibers, shown in yellow in Fig. 6.7. They connect into long inter-meshed fibers that

[14] A memorable childhood "chemistry set" experiment was dissolving the minerals in a bone such that the remaining collagen skeleton could be tied in a knot.

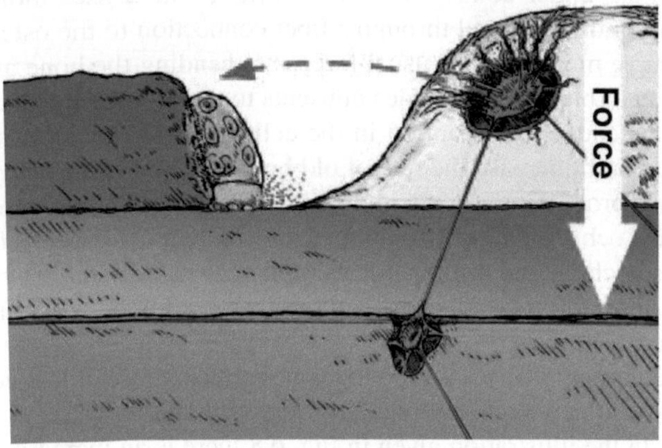

Fig. 6.8 Removal and replacement of bone through a combination of three cell types: osteoblasts that replace bone, osteoclasts that remove bone, and osteocytes, that are osteoblasts trapped in bone and modified to control the process. By this process osteocytes are provided sustenance through tiny canals in the bone (*blue*). Other canals (*red*) provide signal pathways to osteoblasts. If force on the bone is increased the nutrient supplying canal can be constricted reducing flow and producing abnormal byproducts that become signals conveyed to the osteoblast by a canal and indirectly to the osteoclast. There is a complex interconnecting network of osteocytes and osteoblasts. With increased bone stress the numbers and activity of osteoblasts are increased and bone production is increased. Bone is removed by two actions of large osteoclasts. Mineral compounds are dissolved into their calcium and phosphorous components. Cartilage is broken down by enzymes and removed by the blood stream. The number and actions of osteoclasts are controlled by the osteocytes and osteoblasts. If loads on the bone are increased the osteoclastic activity (removal) is reduced and osteoblastic activity (new bone) is increased until bone strength matches its loads. The reverse occurs when bone loads are decreased. It should be noted that single cells shown here represent the army of cells in practice (color figure online)

are coated by the cell with materials, shown in green, that on being extruded by the osteoblast form a material called *osteoid*, which attracts and binds calcium and phosphorous compounds to the collagen[15] and becomes bone over several weeks.

As noted, in mature normal bone approximately 1% of older bone is being removed and replaced constantly by new bone produced by osteoblasts. It is removed by secretions from cells called *osteoclasts*. Rate of removal versus rate of replacement determines size, density and strength of bone. This process is under direct and continuous control of modified osteoblasts that sense the stresses on bone surrounding it and direct the osteoclast and osteoblasts to remove and replace bone such that its strength is commensurate with applied force. This is the mechanism of bone loss in space.

Figure 6.8 uses the three cell types to illustrate this elemental process based on current basic research. An *osteoclast* is dissolving a layer of old bone (brown) leaving a microscopic canyon that is being filled with fresh bone by an *osteoblast*

[15]This phase is analogous to cement being poured around rebars in building construction.

(orange). Both are under control of an *osteocyte*[16], which uses molecular signals plus some form of direct signal through a fiber connection to the osteoblast [24].

In this example more than the usual forces are bending the bone and closing the microscopic canal (blue) that provides nutrients to the osteocyte via a flow of interstitial fluid. This causes a disruption in the cell's normal metabolism, which produces signals to both increase the rate of old bone removal and greatly increase the rate of new bone production, shown in this illustration by the increased thickness of the layer (yellow ochre). These adaptations are slow, requiring weeks for removal of old bone by osteoclasts and months for its replacement by osteoblasts. In this case of larger than normal forces the result will be increased bone mass and strength to support the increased force.

In normal situations where loads on the bones remain more or less constant, equal amounts of old and new bone are regularly removed and replaced to maintain size and condition. In the illustration given in Fig. 6.8 there is an increase in force loads produced by a change in the subject's activity. If an active person retreats to a Lazy Boy™ and TV, more bone is removed than replaced resulting in a net loss. In the case of a human in weightlessness there is maximal unloading of the legs and lower trunk with rapid rates of bone losses and minimal replacement in the unloaded bones [19].

6.5.1 Human Skeletal Bones

Differences in bones were undoubtedly recognized in prehistoric times when they became weapons and tools. In 1683 Galileo noted a relation between applied load and bone morphology. More recently, the standard example was the increased size of bones in the playing arm of tennis players, when it was a one-arm game. In 1892, German physician Julious Wolff (1836–1902) expressed this in a form that has come to be known as Wolff's Law: "Every change in the function of a bone is followed by certain definite changes in internal architecture and external conformation [4, 5]." This may be seen at several points in any given bone.

Figure 6.9 illustrates a right thighbone or femur, the largest and strongest bone in the body and one significantly affected by weightlessness. It has most of the features of the other bones of interest here. Its major portion is a hollow tube, which is a strong and rigid structure. While a primary source of strength in a bone is the density of its minerals, half its strength is also determined by its form; e.g., the strength of a tube varies as the fourth power of its radius. The walls are almost solid cortical bone composed of layers of microscopic columns whose gaps are filled with flat rectangular sheets (lamella). At the ends of the cylinder where a bone meets another bone in motion, the cylinder flares into large surfaces and volumes to allow motion over a large area and to reduce contact pressures and stress. To reduce mass the cortical bone is thinned but reinforced by trabecular bone, which follows the lines of stress.[17]

[16] Osteocytes are osterblasts that are buried (embedded) in the bone they produce and later become sensors that control the amount of bone removed and replaced.

[17] Structure for high performance aircraft frequently also follow this pattern with sheets of metal connected by a light honeycomb.

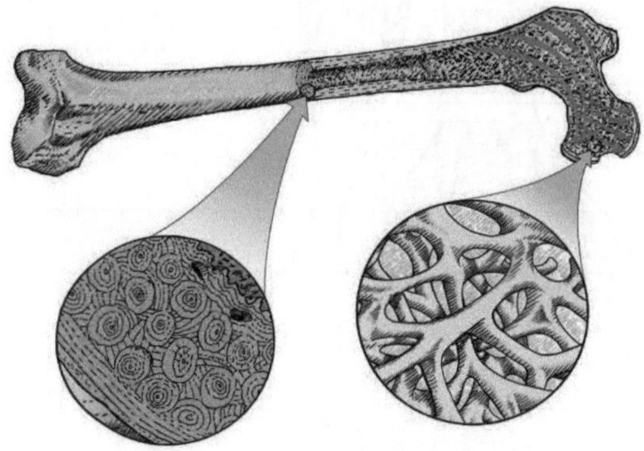

Fig. 6.9 A typical long bone, the femur, is diagrammed here. It is a thick walled tube, the diaphysis, which is expanded at each end, the epiphysis, which provides a larger contact surface and lower contact pressures for the moving surfaces. The wall of the shaft is formed by dense compact or cortical bone while the volume of the epiphysis is filled by an interconnecting network of trabecular bone, a network of branching interconnected arches. Cortical bone is formed by successive circular, solid layers, interconnected by solid plates. There are tiny canals throughout the "solid" bone to carry fluids and blood vessels

Evidence of Wolff's law[18] may be seen again and again in this and any other bone. In the shape just discussed, two "bumps" (*tubercules*) at each end of this bone increase surface area and reduce stress where tendons are attached to transmit major forces to the bone, and in the hemispherical head of the femur, which allows rotation of the thigh versus the dual joint surface of the lower end that constrains the lower leg to move in a single plane. These are also examples of species adaptation. Our concern with Wolff's law is in somatic adaptation of bone to meet changes in stresses. This takes months or years to become obvious on plain X-rays or ultrasonic scanning (techniques used to track and study loss or gain of muscle and bone are given later in this chapter).

In understanding loss of muscle and bone in weightlessness, separate knowledge of muscle and bone only becomes fully meaningful when they are combined into the functional unity they evolved to be. Simple biomechanics provide essential insight into this process. This will be a major aspect of the remainder of this chapter.

6.5.2 *Bone and Muscle Unity*

Bone and muscle are interdependent and function as a unit. Muscle creates motion and force. Even though a large muscle's motion is limited to a few inches of contraction, its force can be hundreds of pounds or more. By the addition of a bone and joint arranged as a lever, muscle motion can be multiplied, from inches to feet while muscle force is reduced accordingly. By such mechanical arrangements of bones

[18] Form follows function.

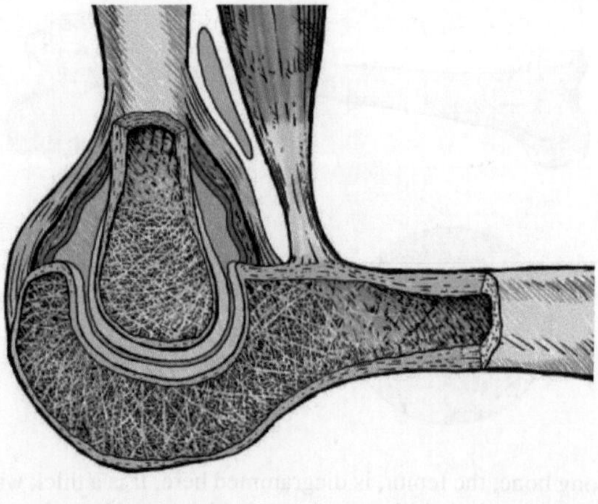

Fig. 6.10 This representation of a joint shows key features of all joints. This bone is shaped to allow motion in a single plane. A second view would show the bones shaped to further constrain it to this motion. A thick cuff of ligaments encloses the joint space and limits and protects the bone against excessive motion. Not shown are a more extensive array of ligaments for additional protection and stability. Contact areas are covered in collagen (*yellow*), which also secretes a lubricant to further reduce friction. A vascular synovial membrane (*red*) covers the cuff and produces synovial fluid, which exchanges nutrients and other materials with the fluid that supports the avascular cartilage. One of several tendons that transmits force and motion for muscle to bone is shown (color figure online)

and joints and with the addition of several muscles, a remarkable range and variety of motions are made possible. There are many other bone/muscle combinations in the body, but we will concentrate on the muscle/bone leverage systems in the limbs and back where major muscle and bone strength losses occur in weightlessness.

Connections of muscle and jointed bone are complex and subject to injury and failure, as most individuals know from personal experience. Broken bones are expected in emergency departments, but far more "turned ankles", sprains, dislocations, and "pulled" muscles and tendons are seen. There have been few such problems in- or postflight, so this subject has received relatively little attention.[19]

Joints are too varied and complex to be pursued here beyond a generic description [25]. Figure 6.10 is a diagram of the major joint components. Bone on bone erodes rapidly and painfully. Plates of collagen are attached to the bone ends to provide a viable low friction surface. Fibroblasts are cells embedded in the collagen plates in order to generate replacement collagen fibers analogous to osteoblasts in bone and to produce a viscous fluid that is trapped in the collagen fibers as a means of hydraulic shock absorption. There are cells in the surrounding synovial fluid that remove damaged fibers. Since joint metabolism is even lower than in bone, circulation to transport energy and nutrients and remove waste is provided by the viscous synovial fluid, which also contains materials to make it a lubricant for the moving

[19] During SL-2, the first manned Skylab mission, Pete Conrad managed to dislocate a thumb, which was reduced inflight by Dr. Joe Kerwin without sequelae.

surfaces. A vascular synovial membrane produces the fluid, interchanges oxygen, CO_2, nutrient and waste products with the blood stream and constrains it to the joint area. The joints are configured to allow only constrained motions of the joined bones. In normal circumstances a neuro-sensory feedback system limits muscle motion to a safe range of motion. To further protect the joint from excessive motion, a series of tough limiting ligaments are arranged around each joint. Without these ligaments, bones in the joint could be fractured if their range of motion were exceeded. These ligaments are frequently overstressed in accidents and stretched or torn producing "strained" ankles and the like.

Muscles are attached to bone by tendons that are strong and have dense collagen fibers with embedded fibroblasts that produce new collagen as needed. Muscle fiber collagen blends into those of the tendons. In the leg, connecting tendons are particularly large and long. Conventional wisdom was that all of these "connective tissues" react to stresses placed on them by becoming larger and stronger under increased loads and smaller and weaker under reduced force loads. This is now being questioned.

After 4 days of weightlessness there were no joint mishaps on the Moon in spite of vigorous activity in a strange, reduced-force environment. The constraints of space suits, and especially boots, may protect against such injuries on landing on Mars following long spaceflights that lacked adequate countermeasures, but the probability of tears or sprains seems more likely than a fracture.

6.6 Biomechanics of Force

Contents of this and the next segment are essential to an understanding of the loss of muscle and bone in space. It is deliberately presented in the most elementary fashion beginning with leverage.

Figure 6.11 shows the forces on bone and muscle in an ankle and foot supporting the body's weight with the heel just clear of the ground. In this case the fulcrum or pivot of the lever is the ankle joint that supports two forces. One force is bodyweight (BW) shown by the red vector at the end of a lever arm 2.7 times the length of the second lever arm to which are attached the muscle forces that balance the lever.

A question that must be asked is, "How valid are calculations of force, and are they supported by measured forces?" Direct measurement of forces in humans is very difficult. Only a few measurements from instrumented hip replacement joints could be found. A British study [26] of a man, measured forces on the femur while walking, jogging and jumping. In a German [27] study of four subjects, two men and two women with an average age of 61 years, were studied while standing, walking, and climbing steps.

Using pooled data from this small sample the ratios of BW to measured bone (femur) forces for five activities were calculated to be as follows: 1) standing on one leg with no significant activity, 1.0 BW; 2) walking; 2.85 BW; 3) jogging, 3.8 BW; 4) jumping, 1.6 BW; 5) climbing steps, 2.6 BW. Mean is 2.7:1. This mean ratio for the thigh is significantly different from the calculated ratio of 3.7 for the lower leg

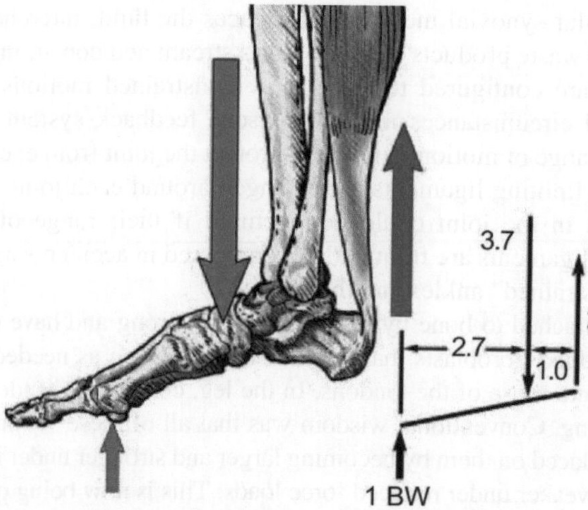

Fig. 6.11 This ankle joint and foot illustrate the effects of leverage on bone and muscle forces. The ratio of lever arms is shown in the diagram. Leg bones are the fulcrum that must support the total force from foot ground contact with the heel raised plus muscle force to support it. The magnitude of resulting forces is given in the drawing. Note that body weight is a fraction of the forces that the muscle must generate and bone must support. When foot ground forces increase two to three times in jogging, one body weight becomes even less important. "Weight bearing" is a small fraction of forces on bone (See Box 6.1)

Box 6.1 Weight Bearing Bones

Weight bearing bones is a term that evolved naturally before the biomechanics involved were appreciated.

It should be used with care. Too often it is misconstrued to imply that body weight is the critical force on such bones instead of muscle forces that may be ten or more times body weight, and it is this that determines bone mass and strength.

Since the ratio of arm length is 2.7 to 1.0 with a ground force of 1 BW pushing against the longer arm there is a torque of 2.7 BW that must be balanced by the shorter lever arm with a relative length of 1.0. This requires a force of 2.7 BW to be produced by the gastrocnemius and other calf muscles. This puts a total force on the bone of 3.7 BW. In this example, less than one third of the force on bone is weight, and more than two thirds is muscle force. Looking at the classic example of leverage in the forearm gives an even larger ratio of muscle to weight forces. Similar arrangements are present throughout the body for arms and legs.

bones, but it could be a difference in muscle/bone geometry or in the assumptions made for calculations. There are much more sophisticated calculations available. In any event there is no question that muscle force, not weight, is the primary stimulus for bone strength.

In the next section magnitude and nature of the forces required to maintain strength in 1 g are shown.

6.6.1 Measurements in Muscle and Bone Evaluation

Measurements of the characteristics of muscle and especially bone have become increasingly complex and the following defines and attempts to simplify them. The simplest and most direct measures are used here wherever possible.

6.6.2 Measurement of Muscle Function

Strength is defined by the forces that a body can produce. These forces vary with the load conditions (Fig. 6.12). Traditionally it is the maximum weight that can be lifted one time. It is now measured under a variety of conditions.

1) *Isotonic Strength*, maximum—the maximum force that can be exerted in a given movement with maximum effort. It is the peak of the force curve in Fig. 6.12.
2) *Isokinetic Strength* is the force that can be exerted at a fixed velocity of motion. It is usually expressed as the maximum force at a given velocity. It should be noted that to be meaningful the velocity must also be expressed (Fig. 6.12).
3) *Isometric Strength* is the maximum force that can be exerted against a fixed object at a known position.

Forces are given here in pounds force (lb), Newtons (0.225 lb force), and body weight (BW) for the weight of the individual being measured. Increasing use of scanning and computing devices has resulted in an increase in the use of cross sectional areas (CSA), volumes and lean tissue masses as surrogate measurements for strength, usually resulting in a significance loss of accuracy in assessing strength. Radiation absorption methods are most frequently used but magnetic resonance imaging (MRI) may also be employed. Cross sectional area in cm^2 of one or more muscles is frequently used and in some cases has fair correlation with strength. Muscle volume (cm^3) and lean body mass (LBM) are also used as indicators of muscle strength, but in this author's experience they provide only poor approximations.

Another significant aspect of strength is endurance, or its reciprocal, fatigue. Endurance is defined as the ability to make continued forceful motions. Endurance can be measured only by direct means and is the reduction in peak strength after a number of cycles expressed as a percentage of peak strength at the start of the exercise.

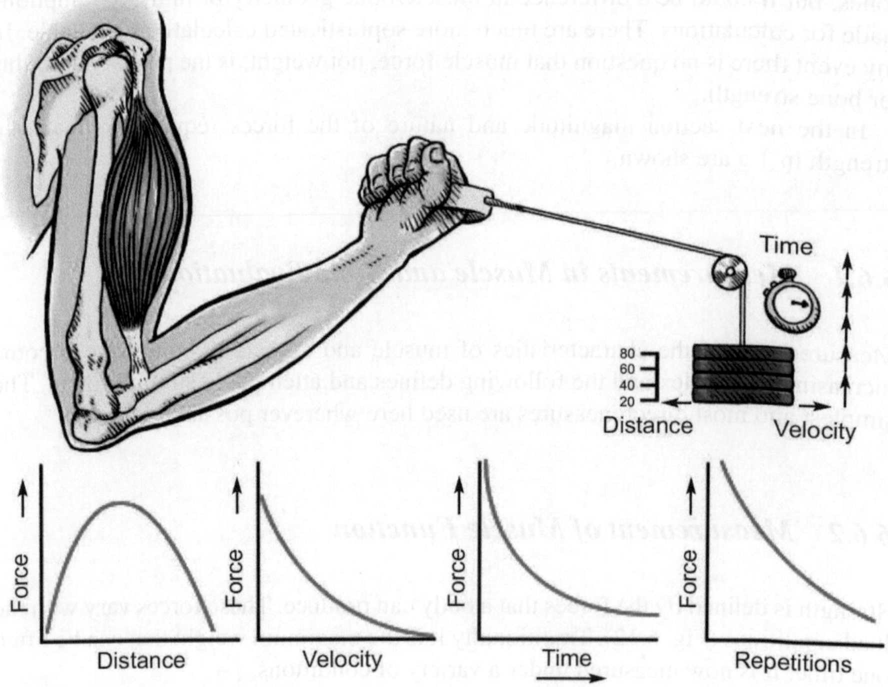

Fig. 6.12 A typical combination of muscle and bones is shown here. The resulting force at the hand is not constant. The various characteristics of the hand force vary over its range of motion as shown. These characteristics are measured in various ways. A classic isotonic arrangement for isotonic motion of a constant load measurement is shown. Strength in this measurement is defined as the maximum load that can be carried through some motion. Strength or maximum force generated varies inversely with velocity $\Delta X/\Delta T$ and acceleration $\Delta X^2/\Delta T^2$. Measured force at a fixed velocity is called isokinetic. Force against an immovable load at zero velocity is isometric. Muscles will Fatigue and their force will decrease with repetitive motion. This diagram uses a single muscle for illustration

6.6.3 Measurement of bone function

Essential characteristics of bone are:

1) *Strength*—Maximum amount of force that can be sustained without failure. May be measured in compression, tension or deflection.
2) *Flexibility*—Amount of deflection produced by a force in a given direction
3) *Fatigue resistance*—Ability to sustain repeated stress.

None of these qualities can be measured without removal from the body. Bone strength is the quality most often sought but can only be estimated from current scans of bone mineral density. Such estimates are complex for strength is dependent on a bone's complex shape and arrangement and on its collagen skeleton and its minerals. If the shape and distribution of material in three dimensions is known, complex calculations of strength can be done with fair accuracy, but the time, complexity and cost practically limit these procedures to scientific investigation. In the absence of a direct measure of bone strength in living tissue, current practice is to make use of some surrogate measurement that may be related to strength. This measurement typically assesses the amount of mineral in the bone. In short, investigators measure what they can and designate it bone strength. Virtually all current quantitative bone measurements are based on radiation absorption as it passes through tissue. The oldest and most widely used of the clinical tools is dual energy X-ray absorptionometry (DEXA). The selected body part is scanned in two dimensions producing a map of the tissue absorption. In the case of bone this is called bone density absorption (BDA) with dimensions of grams per cm^2. Clinical medicine has invested a great deal of time, effort and resources in this technique and has used clinical and statistical measures to estimate bone strength and the probability of fractures. Such estimates have been established as standards by the World Health Organization and are widely used in prediction and preventive treatment in populations at risk for fracture. An increasing number of reports show the poor predictive value of this work. Currently this technique remains the standard for comparison in clinical work but the poor correlation of such estimate as T values with reality and results is a current topic in orthopedics.

Quantitative computed tomography (QCT) is a more exact method of bone measurement. In this technique an X-ray beam makes successive transverse scans or a continuous spiral scan of the body from which a great deal of information about tissues, especially bones, can be derived. This includes composition, densities, shapes and volumes. Such three dimensional scans obviously offer advantages over the two dimensions of DEXA studies. Mineral content of bone now becomes a volumetric measurement expressed as bone mineral density (BMD) in gm/cm^3 versus the area estimates of bone density in gm/cm^2 of DEXA measures.

Peripheral quantitive computed tomography (pQTC) scans smaller volumes such as limbs with increased resolution and greater accuracy of dimensions and densities.

6.6.4 Measurement Interpretation and Application

The ways measurements are interpreted are as important as the measurements themselves, especially in how they are used to understand and document muscle and bone loss. Losses are never total, an obvious fact supported by the significant amounts of muscle and bone that remain 50 years or more after a person is totally paralyzed from the neck down. It has long been observed that the rate of muscle and bone loss decreases with time. There are many analogous situations in medicine, physics, engineering and banking that are described by analogous exponential relations.

If an accurate series of measurements over sufficient time for changes to occur is available, plotting bone and muscle loss over time often provides a semi quantitative view of the situation.

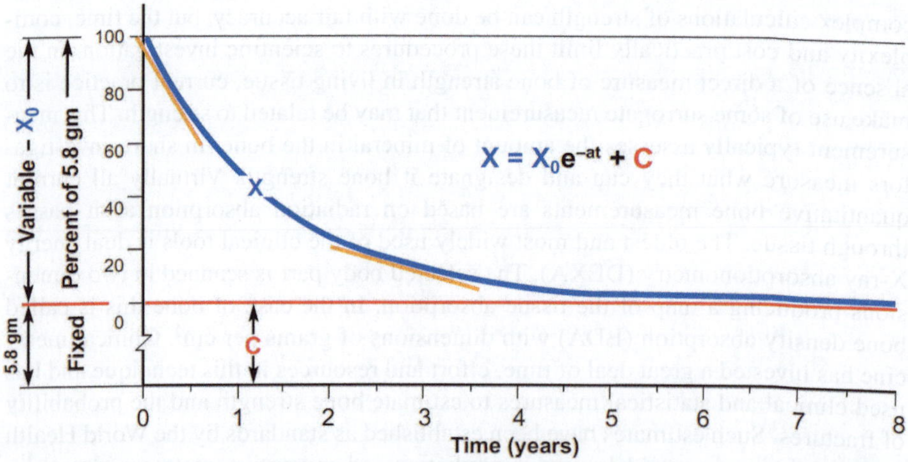

Fig. 6.13 The loss of bone from the joint (epiphysis) region of an SCI patient whose loss process is complete is illustrated. Bone has two components; a fixed mass, indicated by C of 5.8 g in this bone, and a second mass that can vary with load conditions. At the time of the accident the variable mass was 5.6 g and total mass, $C + X_0$, was 11.4 g. Over the years variable mass, X_0e^{-at}, was essentially reduced to zero. The scale for fixed mass is collapsed to allow the variable region of the graph more visibility. An *upper reference line* indicates the mass when loss began. The exponent a, or the rate at which loss occurs for a particular bone, was determined from measured loss levels of similar patients at various times, T, after their SCI by curve fitting. Once the constants are known the amount of bone may be determined at any time. By assigning a level for zero the time to reach a stable fixed mass C may be calculated

A group of Swiss investigators [15, 16] conducted a cross-section study analysis by measuring bone loss at four anatomical sites in a large number of patients paralyzed by spinal cord injury (SCI) for durations of 2 months to 50 years. They also measured a matched cohort to estimate initial amounts of bone originally present. Results for one bone site are illustrated in Fig. 6.13. Other [15] studies analyzed results from muscle lost during bed rest using only the adaptation component of loss, Eq. 6.1. These analytical procedures, which require time and a number of subjects and measurements, have not been reported in other bed rest and flight studies to date.

Losses to date have been expressed as quantity lost during a specific interval of time often in terms of percent lost. Current practice is to provide only a loss rate normalized to a month (e.g., percent/month). This approach has obvious limitations and offers opportunities for error or misinterpretation, a not uncommon event. Figure 6.14 illustrates some of these limitations.

Loss of strength is often given as a measured loss of muscle area or volume while bone loss is expressed using a measure of bone density. Only the amount of tissue in muscle or the mineral in bone lost over a given time is produced by such studies. This means that the loss rate is calculated on the basis of total bone mass, not on the much smaller maximum loss that could occur or the even smaller actual losses in

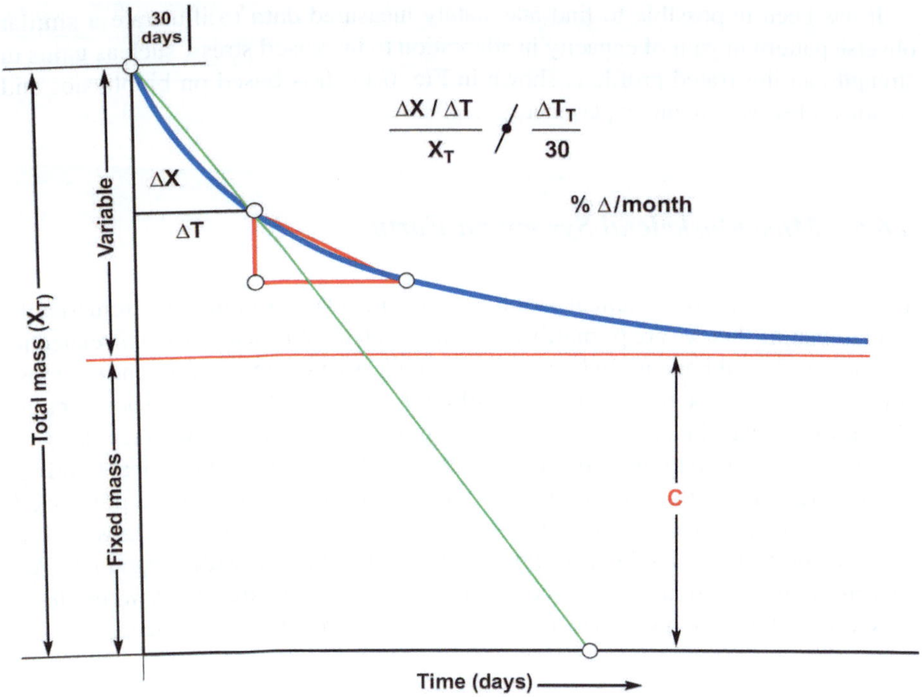

Fig. 6.14 A derivation of percentage lost per month plotted on a representation of actual loss is shown. The measured loss, ΔX/ΔT, is divided by the total mass for a fractional rate. This is converted to rate per month by the fractional total time divided by a month's time. In this case the time is in days. The fractional rate is converted to percentage by multiplying by 100 (not shown). The *red triangle* illustrates the constantly decreasing rate with time that will approach zero

spaceflight or bed rest. This may produce numbers useful for comparison, such as comparison of the relative effectiveness of countermeasures. It cannot be used, however, in any analytical way because it indicates a much smaller loss rate rather than one based on possible actual loss, among other problems.

A serious misapplication of this number is to extend it to long periods of time. An inspection of the extension line shown in Fig. 6.14 shows that not only will the maximum possible loss be exceeded, but the point will soon be reached where neither bone nor muscle are left. An error in the rate of this number will occur if it is taken over long periods such as now is almost routine in ISS expeditions. Note the difference in the two rates, ΔX/ΔT, shown in Fig. 6.14. If the time, ΔT, is long enough for the rate to change, the resulting rate per month is wrong, especially when the implication is that it represents the first month of the process. Examples of such problems are not uncommon in publications.

Although the example shown in Fig. 6.13 is bone the same approach should be used in muscle. In fact, there are many if not most of the body's systems that adapt in the same time-dependent way as shown.

It has been impossible to find adequately measured data to illustrate a similar obverse pattern in gain of capacity in adaptation to increased stress, such as gains in strength, an illustrated profile is shown in Fig. 6.17. It is based on biophysics and personal observation and experience.

6.6.5 *Musculoskeletal System on Earth*

The range and variety of human MSkS activity and adaptation in 1 g is remarkable and fascinating, but we are primarily concerned with the physical activities related to loss of muscle and bone in flight crews in weightlessness. Jogging remains the most common astronaut exercise with weight lifting next. A variety of other sports activities may be engaged in, and a number of this group also cycle and swim regularly.

Two years prior to flight a crew is scheduled for 4 h per week of physical training. Aerobic training is left to the individual's choice, but strength training is prescribed with free weights. This combination of personal and prescribed activities with a wide age range in a mixed gender population must result in a wide range of MSkS capacities, but attention here is focused on those properties affected by weightlessness, especially those properties that have been measured or manipulated.

6.6.6 *Walking, Jogging and Running*

Bipedal locomotion is the activity of primary significance in muscle/bone loss in weightlessness. Some 30% of total body mass is dedicated to muscle and bone that enables bipedal locomotion. Knowledge of the stresses in locomotion in 1 g is essential to understanding the effects of its absence in spaceflight. Bipedal locomotion requires forceful leg motion. A large number of muscles of varying size, shape and strength in legs, pelvis and lower back are actuated by control areas in the brain to produce a coordinated pattern of repetitive forceful movements to generate one of three modes of bipedal locomotion; 1) walking, 2) jogging, and 3) running.

A single step in walking is shown in Fig. 6.15. In this figure (Fig. 6.15a), the right leg is supporting the body as the left contacts the ground and supports the body as it pivots forward on the hip while swinging the right leg forward for its heel strike and repetition of the cycle by the left leg. Note that the body is always supported by one foot on the ground.

Forward speed is a product of step length and step rate and can be increased by increasing either or both. Maximum step length is anatomically limited in walking by the necessity of ground contact. Throwing the body up and forward and clear of ground contact can increase step length such as when jogging or running occurs as in Fig. 6.15b. In these modes the body goes through repeated small weightless parabolas clear of ground contact. This differentiates running and jogging and walking. The difference between jogging and running is usually based on a speed, but

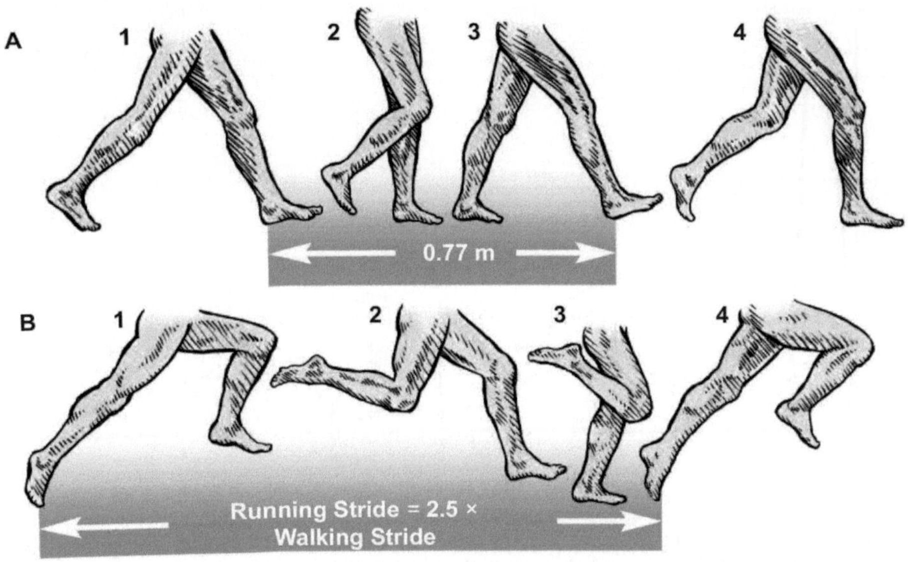

Fig. 6.15 Kinesiology actions in locomotion are illustrated in this figure. Note the hip and foot flexions that project the runner into the air for one-third of a second during running. Foot ground forces (FG) in walking are essentially all body weight and primarily supported by bone. Maximum foot pedal forces from a trained cyclist are shown for comparison in Fig. 6.16

this is not always agreed upon. Six miles per hour is frequently used as the speed at which jogging gives way to running.

Typical vertical *foot to ground forces* (FGF) generated during locomotion on Earth are shown in Fig. 6.16. Horizontal forces, except in sudden changes in speed or direction, are small and can be neglected here. Walking at usual rates produces a double undulation of approximately 1 BW. In this case the force is virtually all weight, mass × 1 g. To project the body clear of ground support, such as when jogging or running, requires forces >1 BW. This added force is inertial, mass × acceleration, and is reflected in the FGF recorded in jogging. All of these forces vary significantly with the individual and all increase with speed.

As steps are shortened in time and extended in length the body must be projected higher until in the case of sprinting, as in Fig. 6.15b, the FGF can reach levels of 5 BW, four of which are from acceleration of body mass. This force of 4 BW is inertial, a point too often forgotten in exercise countermeasures. Also remember that muscle forces are multiplied by leverage of bone, which must also sustain FGF plus muscle forces.

In the example in the previous section, bone sustained 4.7 BW force walking but in a slow jog FGF are doubled and muscle must generate 2 × 2.7 BW force, say 918 lb, and the bone must support 1250 lb or half a ton. Sprinting can make it a ton. But force generation is only one characteristic of locomotion. In addition, speed, number of steps (endurance) and energy costs are significant to what occurs in spaceflight, on the Moon or on another planet.

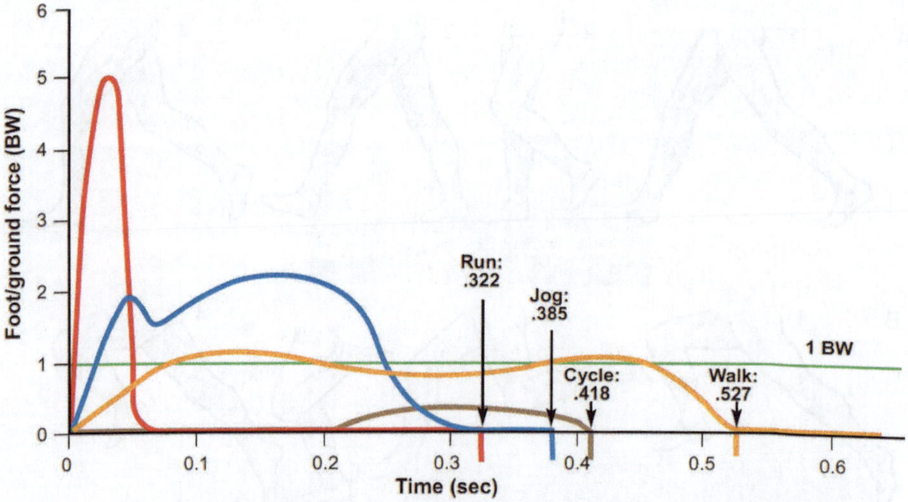

Fig. 6.16 Forces in walking (*orange*), jogging (*blue*), running (*red*), and cycling (*brown*). In jogging and running FGF force magnitude is multiplied by the addition of muscle forces required to overcome the inertial force of projecting the body into the air

Large muscle masses and their rates of energy consumption make controlled locomotion the usual and preferred mode of measuring an individual's capacity to convert energy into motion. This aspect of locomotion and its measurement is treated in Chap. 9 Circulatory System, and Chap. 10, Metabolism. Speed and distance of locomotion are easily measured, but measurement of strength and strength fatigue are more complex.

Currently exercise is classed as: Strength; now usually called resistance, that increases the capacity to produce force; Endurance, that increases the ability to sustain an activity: in some exercises such as bipedal locomotion, aerobic capacity may be increased and they are called Aerobic. Some exercises such as bipedal locomotion can increase strength, endurance and aerobic capacity.

6.6.7 Cycling

The bicycle was invented to reduce the effort and decrease the work of bipedal locomotion. It extends the duration of the required leg force to a half pedal rotation but greatly reduces it. A typical rate is 72 rotations per minute. Forces are much less than in walking and maximum pedal forces for trained riders are on the order of 65 lb. In contrast to walking, jogging, and running, cycling is not a strength exercise. It is a good endurance and aerobic exercise and trained users can reach their near maximum O_2 and energy uptake using cycling.

6.6.8 Weight Lifting and Resistance Exercises

Weight lifting is the classic exercise for increasing strength and an example to illustrate several essential principles of adaptation follows. Exercise machines also exist that mimic weightlifting. Figure 6.17 is a simplified graph of a muscle's adaptation to changes in its usual maximum load. This example begins with a regular usual load, let's say of an arm during everyday activity. Maximum strength capacity is significantly greater than the usual maximum loads. If the usual maximum loads are increased by a regular weight lifting program, such as the classic De Lorme [28] protocol, maximum strength can be increased, even to its DNA limit. Consider starting with a weight that can be lifted only 6–8 times. If this is repeated three times in daily exercise sessions, in days to weeks the number of repetitions can be increased as strength is gained. Afterwards, the load may be increased and the cycle repeated. However, there comes a time, usually after several months, when the weight increases must be reduced and it takes longer for strength to increase; i.e., gains are smaller and take longer. The response curve appears to be exponential.

Only a few athletes reach their maximum limit, for this takes many months to years of training to achieve that level. It is also difficult and risky to work at maxi-

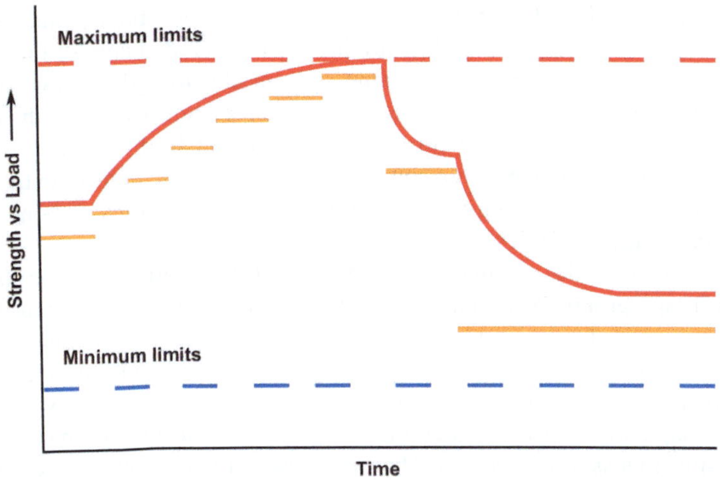

Fig. 6.17 This figure illustrates the adaptation of a muscle to "overload" strength training and to the reduction in muscle loading. Maximum strength is shown by the *red curve*. It is normally greater than the usual maximum load the muscle encounters, shown by the *orange line*. It is periodically increased as the muscle's strength increases; but as muscle becomes stronger the increases become smaller, and adaptation takes longer. After approximately a year maximum strength is reached and the exercise is continued but at a reduced load. Strength falls exponentially and more rapidly than it increased. After adaptation to this level the subject begins bed rest with loads reduced to a minimum. Strength again falls to a much lower level, taking longer to reach this level. The response shown here is characteristic of both muscle and bone, albeit at different rates. Understanding this curve and its exponential response is crucial to understanding the changes that will be seen in bed rest and inflight

mum capacity. If attempts are made to exceed this individual maximum limit, tissue damage and decreased strength can occur.

Adaptation is reciprocal. If weights are suddenly reduced, especially at high levels, the result will be a more rapid exponential decrease until an equilibrium with the reduced load is reached. The final step in normal 1 g adaptation is removal of as many stresses as possible, as in bed rest. The ultimate removal of normal muscle force is spinal cord transection. In this case muscle and bone strength and mass go to their minimum. Both muscle and bone losses in bed rest are complex and even more so in such spinal cord injuries (SCI).

6.6.9 Bone Adaptation

In 1 g bones must be capable of sustaining the large, primarily muscle forces on them as well as much larger transient forces such as can occur in a fall. Bone, like muscle, adapts to the maximum stresses regularly placed on it but much more slowly and with much larger safety margins, typically ten times the usual maximum loads it is subjected to [26]. The medical literature contains increasing numbers of studies that address exercise as a way to increase bone strength. Media now hawk an endless progression of programs and devices to improve "bone health".

Animal studies of bone strength, with directly measured changes in force loads, support a direct correlation between bone and muscle strength. It was found that small changes in bone mass can effect large changes in bone strength. Findings addressing bone mass and strength have been reported from a variety of human studies. These studies ranged from comparing increased bone density in the take off leg to that of the opposite leg in competitive jumpers to studies of bone strength versus force in quadriplegics.

Figure 6.18 provides a remarkable summary of the relation of bone to muscle strength in age groups of normal individuals from 2-year olds to maturity. Lean body mass is an indicator of muscle strength, and increasing bone mineral mass is the indicator of increasing bone strength produced by increasing muscle forces on them. Changes in slope of the curves represent physiological changes with puberty.

Bone adaptation is also a bilateral process. Absence of force on bone from any cause results in loss of its strength. The root causes of bone strength loss can range from changing lifestyle to the total disuse of muscle and bone in paralysis from Polio or spinal cord injury.

Articular cartilage, ligaments and tendons are essential components in muscle and bone action and must carry the loads of bone/muscle force. They often sustain much greater accidental loads and are frequently injured in 1 g, especially in sports. The astronaut population is subject to such injuries. Other than spinal intervertebral discs problems, discussed in Chap. 10, no verifiable in-flight problems in this area have been reported. Little attention has been given to this area by in-flight investigations except small animal studies.

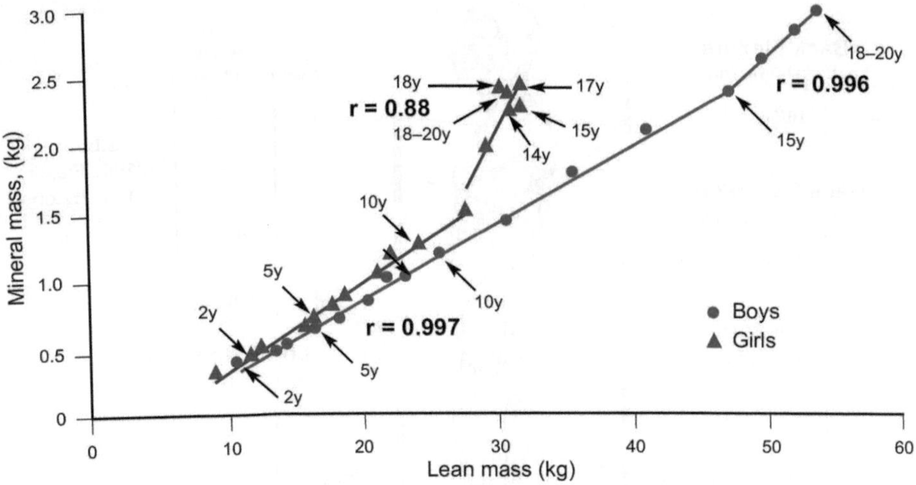

Fig. 6.18 Muscle mass and bone mass were measured by X-ray scanning in a large population of children, teenagers, and mature males and females. Density of bone minerals is plotted as a function of muscle mass. Although muscle force is a primary stimulus for bone strength, it is not the only stimulus. Bone mass increased with puberty and its hormones. The reciprocal of this occurs at menopause and rate of bone lost increases

6.6.10 Muscle in Spinal Cord Injuries

Spinal cord injuries (SCI) that disrupt all nerves at the injury level remove all normal stimuli to muscles below that level. If only the legs are involved it is designated paraplegia or if both arms and legs are paralyzed it is tetraplegia. Tetraplegia creates the ultimate example of disuse atrophy of the musculoskeletal system. It is a worst case model for all other disuse atrophy situations including weightlessness. While a great deal of theoretical and quantitative work remains to be conducted in this field, it has already contributed to our knowledge of bone and muscle loss.

Removing all neurological signals to muscle fibers produces large and rapid changes in muscle that are not completely understood [29]. In the absence of nerve stimuli, muscle fibers shrink rapidly in cross section reducing muscle volume, mass and strength in the weeks that follow. There is a very rapid change in muscle fiber diameters in the months following injury with a 50% reduction being typical. At approximately 24 weeks a transformation of slow, Type 1 to fast Type 2 fibers begins and continues for approximately 2 years. These changes reduce the size, strength and fatigue resistance of muscles [30].

The time course of SCI muscle atrophy is exponential, initially rapid but continuously slowing until a steady state is reached. The rate of loss varies in different muscles and in different subjects. An average time to reach one half of total loss of muscle and strength in this condition is 1–1½ years with large variations in rate and amount of loss between individuals and muscle. Measuring muscle strength is more

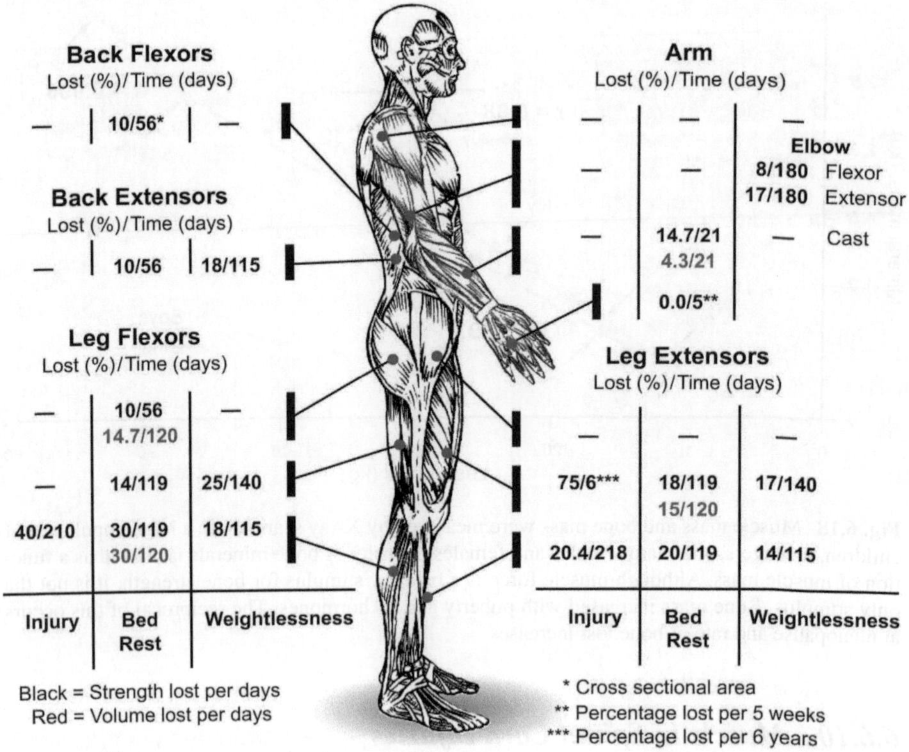

Fig. 6.19 This is a correlation of initial loss of muscle with time from BR and weightlessness with the maximum losses in SCI. Valid comparisons are difficult to make with surrogate measures of strength and with unknown countermeasures and differing regions of interest by the investigators

difficult in muscles without a functional nerve supply and requires direct electrical stimulation of the muscle. Moreover, results cannot be directly compared to normal strength but require matched controls whose muscles were electrically stimulated in the same way. Such complexity probably accounts for finding only one such report.

Instead estimates of strength loss are more often inferred from loss of muscle volume. Even these measurements are scarce and must be used with caution. One such estimate derived from muscle CSA was 18% loss at 6 weeks. Over the next 24 weeks there was an additional mean loss of 15% in leg CSA for a total of 43% in 30 weeks. A single well conducted strength measure in six subjects whose losses were complete some 6 years after injury, showed an average loss of 77% of strength in the thigh muscle that extends the knee (quad. femoris). Qualitative reports are consistent in their stories of large decreases in all muscles below a complete injury level, but in tetraplegia there is relatively less loss in arms than in legs. In paraplegia strength in the arm is usually increased by their increased use. Quantitative results of the few SCI muscle studies that could be found are listed with those of bed rest and weightlessness in Fig. 6.19.

A significant finding is that muscle, after years, even decades of disuse and physiological abuse, can be restored to much of its original function. This will be discussed in the countermeasures section.

6.6.11 *Bone Loss in Spinal Cord Injuries*

When muscle forces are completely removed from bone, the osteoclasts continue or possibly increase removal of bone that is no longer replaced by osteoblasts. Trabecular arches of bone are thinned and often lost completely. In the shafts of long bones mineral is removed from inner surfaces, thinning and weakening them without reducing wall density. Thus to be meaningful, the region of bone studied must be known, for trabecular bone forming the joints loses more mineral mass and strength than cortical bone in the shafts of long bone. Final loss of bone may be approximately 60% of trabecular bone in the joints versus 25% in the shaft of the tibia. The typical time for bone loss to stabilize is 3–8 years. Rate of loss varies with the bone and also varies from person to person. This can be seen in Fig. 6.20, which shows the loss rates and average amounts of bone lost and remaining in a group of patients 6 years after SCI. A follow up 2-year study showed no additional losses and

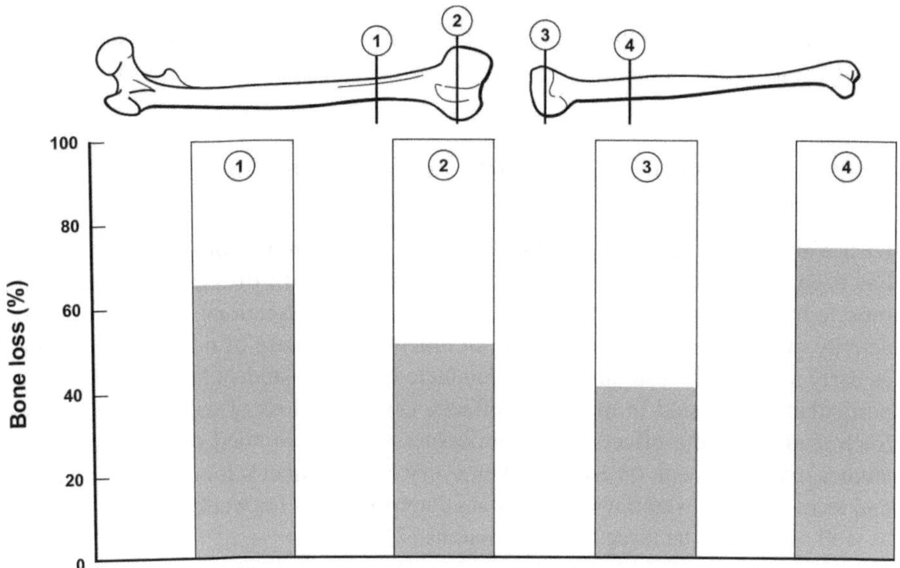

1. Femur, shaft or diaphysis with solid bone.
2. Femur, joint or epiphysis with spongy or trabecular bone.
3. Tibia, joint or epiphysis with spongy or trabecular bone.
4. Tibia, shaft or diaphysis with solid bone.

Fig. 6.20 Bone loss in tibia and femur after spinal cord injury

indicated that the loss process was complete. Most SCI bone studies have focused on the lower limb where loss is greatest.

Bone losses in the lower spine apparently reach 20% at approximately 10 years post injury, after which density may increase to near normal values 30 years or more post injury. This may be produced by forces on the spine during routine care of the patient including sitting and braced.

If the case is paraplegic with sparing of the arms, bone mineral density (BMD) in arm bones may increase through increased usage, but in tetraplegia of long standing where all limbs are affected it appears final arm losses are much smaller than those in the leg, on the order of 20%. It should be noted though that bone losses as well as muscle losses can be altered in SCI by appropriate application of forces [28].

Bone fractures are a constant danger and common problem in SCI. Reported incidence and prevalence vary. Prevalences of 33% and 35% have been reported for subjects 20–30 years post injury. The following is a reported incidence with time post injury.

1 year	1% per year
1–9 years	1.3% per year
10–19 years	3.4% per year
20–29 years	4.6% per year

Knees are the most common fracture site and falls from a wheelchair the most common cause.

A summary of bone losses in SCI has been listed with those from bed rest and weightlessness in Fig. 6.28.

6.6.12 Simulations of Weightlessness and Disuse Atrophy

Disuse atrophy of muscle and loss of strength in the absence of work and exercise has been known since ancient times. Bone atrophy with disuse of its associated muscle has been noted since anatomists began their dissections. Lack of physical activity during illness was studied as an independent cause of numerous problems as early as 1855. Investigators who conducted bed rest studies late in WWII were particularly interested in metabolic effects, especially loss of calcium from bone. Such studies on the effects of weightlessness were continued on Gemini. These studies reached a peak on Skylab when every crewmember's food intake and urine and feces were measured with daily mass determinations for weeks before and after, as well as during, the three Skylab missions [7].

6.7 Changes in Muscle and Bone with Bed Rest

As previously described, a variety of simulations of weightlessness in 1 g were used as early as 1960. These included water immersion, immobilization of a limb and bed rest. Numerous animal studies were also conducted. Continuous bed rest has become the method of choice for spaceflight simulation [13, 31], and results from these studies are used here. Individuals lie flat in bed with legs, trunk and head horizontal. Movement is reduced to a minimum, but some arm and trunk motion is allowed. Necessary shifting of position requires use of hip and back muscle. Six degrees head down tilt in an effort to simulate fluid shifts was standard practice, but it was temporarily discontinued with discovery of postflight ocular problems (see Chap. 5, p. 102). Study populations were male and female selected to simulate flight crews but were usually younger. Durations range from 6 to 370 days[20] but are commonly 30, 60, 70 and 90 days.

It is difficult to find coherent strength data from bed rest studies. Complete functional characterization of the muscle system in such studies would include strength, strength fatigue, endurance and maximum O_2 uptake but such studies have almost never been conducted. Muscle volume, mass and cross sectional areas are increasingly used as a measure of the resulting disuse atrophy. As previously noted, while this measure of muscle loss generally correlates with strength loss, there may be significant discrepancies between it and functional strength loss. Strength directly measured by isokinetic ergometry is used here whenever available. These authors could not find measures of muscle fatigue and endurance. Changes in aerobic function are discussed in Chap. 9.

Keeping normal, active individuals at bed rest for prolonged periods is a formidable, demanding and expensive undertaking [32]. Techniques for conducting this kind of research have evolved over time. It would have been desirable to have used a single group of subjects with characteristics matching those of astronauts for bed rest studies. Such a group could have then been tested for both physiological and functional changes for durations of anticipated flights to establish baselines for the musculoskeletal changes. This should have included evaluation of performance of functional tasks at the end of the study equivalent to those anticipated for flight crews at the end of the planned missions.

As it happened, flight experience, especially flight durations, ran well ahead of bed rest studies that were then left to focus on trying to imitate flight conditions in 1 g simulations that allowed studies that were not or could not be done in weightlessness. A second confounding problem was the demand for countermeasures to prevent adaptation to 1 g. The goal of such countermeasures appears to be maintenance of 1-g adaptations in weightlessness; an impossibility with current technology that will be examined in the next chapter on countermeasures and rehabilitation.

Initially bed rest studies became a cottage industry for various groups and institutions. A broad spectrum of study results has not provided a coherent view of major issues of the problem. Some of the early programs were exceptions and produced data that remains uniquely valuable. As time passed, several groups have established solid programs that avoid many of the problems of early bed rest programs. Several bed rest efforts have evolved, which proceeded to study various problems in an orderly fashion, over increasing times of 30, 60 and 90 days in the U.S. Some groups in Europe have investigated problems in an orderly progression using consistent procedures, for periods of up to 56 days.

[20] The 370-day studies were conducted with 11 subjects in Russia in 1986.

A significant deficiency in all current bed rest studies is their lack of functional relevance. Attention is focused too exclusively on numbers although physical performance can also be evaluated objectively. An increasing trend in measurement is relevant to this. Initially strength, the primary product of adaptation, was once measured, but increasing use of three dimensional scanning devices of muscle volumes and CSA have been substituted. Only one study was found that related CSA to strength and none have related volume to strength.

The authors' original intent was to review many bed rest studies in order to provide orderly arrays of data showing relations between muscle and bone properties and time, and between bed rest and flight. After months of search this effort was abandoned. Only features of bed rest with demonstrable validity are commented upon. Results from SCI, bed rest and spaceflight shown here are inadequate to gain a complete understanding of the problem. Inferences made here are based on any relevant information that could be found from available sources. Available data show that bed rest is a reasonable analog of some aspects of weightlessness and the study of muscle and bone loss. In contrast, data from immobilization of an arm or leg by either casting or by a sling or strap used to simulate weightlessness show these techniques usually to be unsuitable.

6.7.1 *Muscle Changes During Bed Rest*

Regions of the body that would have major muscle loss in bed rest were more or less predictable [17]. Arms and hands are still used but much less than in normal life or in spaceflight. The effects of bed rest on arms are distorted by unknown levels of activity, which are seldom measured during muscle studies. Legs contain the largest, strongest, and in this population, most heavily used muscles of the body. It is here that major losses in strength occur (Fig. 6.21).

There is a chain of locomotor muscles that propel the human body, beginning with the calf muscles that must plantarflex (toes down) the foot. Somewhat less force is required of the thigh to do its share in propelling the body and still less of the hip muscles. This is reflected in both the muscle lost and rate it is lost (Fig. 6.21). If only small forces are required, muscles are correspondingly small. Their rate of disuse atrophy is correspondingly small. Loss of muscle and rate of loss are not proportional to their size but may be proportional to the forces required of them.

As many isokinetic strength measurements of knee extension from bed rest studies as could be found were collected, their origins examined for procedural validity and most discarded. The four remaining points are plotted in red (Fig. 6.22). Even using the best techniques such data will have large, individual variation and the points here are mean values from studies with a number of subjects.

Another set of data points from studies that used changes in muscle volume as a surrogate of strength are plotted in blue. A number of issues relevant to use of bed rest and flight data are illustrated by these curves. There is a large difference in magnitude, and for muscle volume to be used as an estimate its strength, it must be

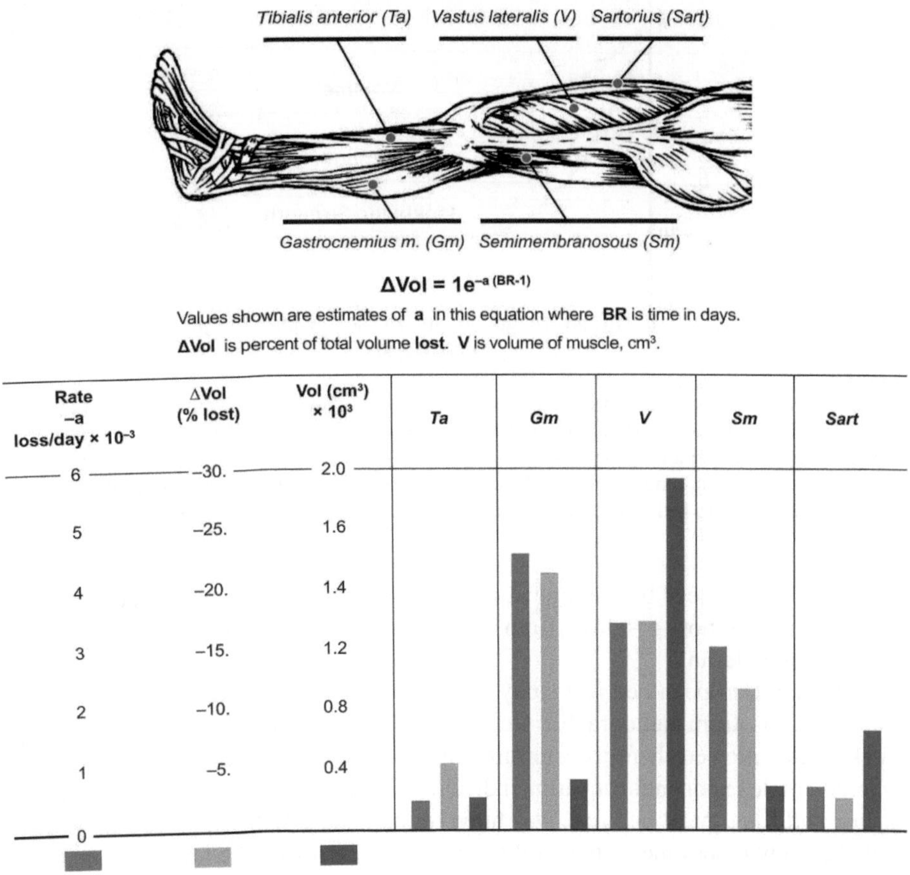

Fig. 6.21 Volume changes of leg muscle in 56 days of bed rest

correlated to measured muscle strength. In this case the curves are congruent in shape, and measured strength loss is simply 2.5 times measured volume loss. In many cases sufficient data is not available to confirm their relationship. Although there are too few points in these two curves to assure great validity, it appears that this muscle group has reached its maximum loss for this condition by about 40 days. The exponential equation in Fig. 6.13 can be fitted to these curves for mathematical expression. As noted earlier, based on excellent European studies and measurements of subjects with SCI injury the time for muscles to reach maximum stable loss is a year or more with a total strength loss of more than 75%. Theoretically the time to adapt to the much smaller losses of bed rest or spaceflight should be much shorter, and is of major theoretical and practical significance as shown in Fig. 6.22.

Muscles that bend the back are large. The lumbar portions also aid in locomotion. Losses in these muscles were expected, but they have varied. A 56-day study measured the CSA of them and found almost equal losses of 10% in those muscles that flex and those that extend the back. Values for the changes in muscle strength

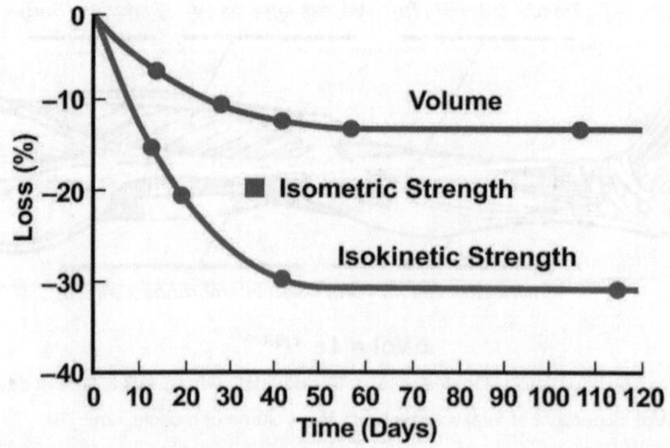

Fig. 6.22 Percentage changes in strength (*red*) and volume of leg extensor muscle from a series of bed rest studies show good correlation for the time course of change. As usual the magnitude of the changes are different and must be correlated for each muscle when volume is used as a surrogate. In this case it is a simple multiplication factor of ~2.7. If the data points are correct this shows a complete adaptation to bed rest for this muscle in a short time (color figure online)

and volumes produced by the time at bed rest are shown in Fig. 6.22. Such changes are smaller than those resulting from loss of nervous connections in SCI shown in the adjacent column.

A 56-day European bed rest study of six normal subjects made muscle volume measurements from radiographic studies of all the hip and leg muscles every 2 weeks. This allowed them to determine the volume changes of each muscle with time. A simple exponential equation indicating fractional loss e^{-at}, where **a** is the rate of loss and **t** is time at bed rest, was fitted to the data, which with loss can be estimated for any other time. A few of these rates are given with the losses in Fig. 6.21. These rates vary with individuals and with the muscles of the individual. They also vary with the situation; e.g., bed rest versus SCI versus spaceflight. Punch biopsies of leg muscles were conducted in a few bed rest studies to examine the fibers. Findings are consistent with those in SCI; i.e., a loss of fiber volume and a shift in fiber types from Type I that is slow to Type II that is fast. A corresponding shift of increased strength in more rapid isokinetic motions was found. An increase in the fatigue factor would be expected, but no reports of this could be found in bed rest.

6.7.2 Bone Changes with Bed Rest

In bed rest the lower limbs are unused except for transient position changes. Hence it could be expected that major muscle and bone losses would occur [17]. Virtually all body weight is removed as are the much larger muscle forces. During bed rest, muscle forces are not entirely absent because unlike SCI, muscle tone is present as are inadvertent voluntary and involuntary motions.[21] Studies of sufficient duration to allow maximum losses of strength in bed rest could not be found.

[21] Only a few motions of appreciable forces can markedly reduce adaptations.

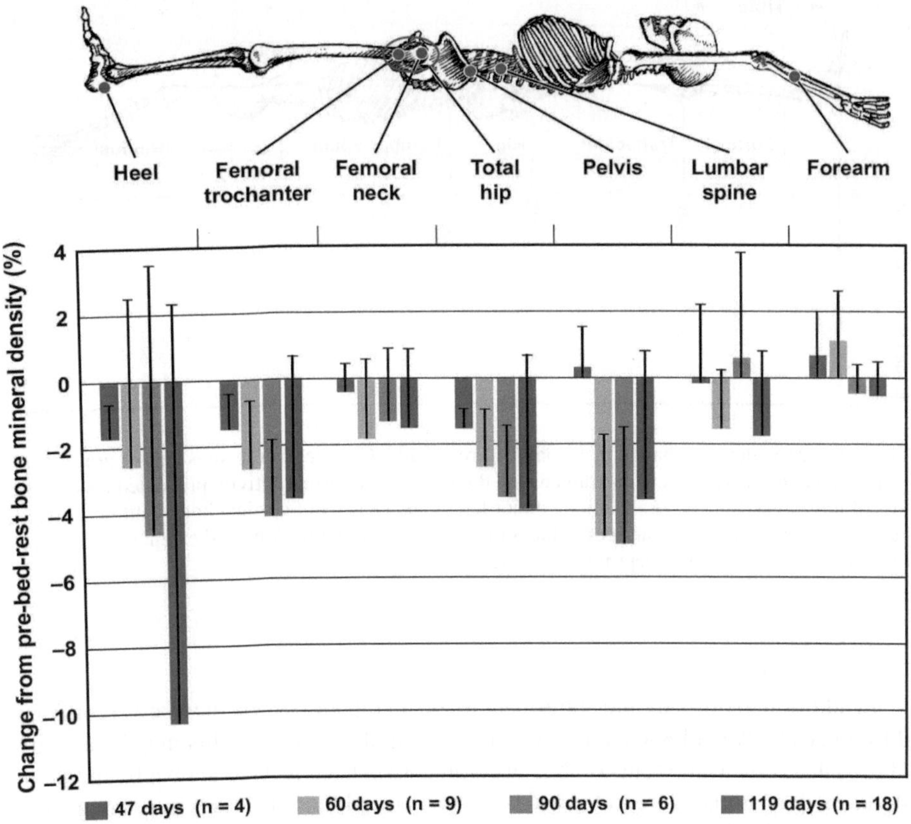

Fig. 6.23 Shown here are changes in bone density measured at the sites shown by DEXA, which were produced by four bed rest studies of increasing duration. The large variation is also indicated by the large standard deviations. Only one side, i.e., one half, of the standard deviations are shown. There are indications that changes have slowed or stopped at three of the sites (trochanter, neck and pelvis) and are slowing at the hip. If true, this would be remarkable because bone changes usually take a matter of years to complete. Conversely the changes are small compared to what occurs with SCI. Changes in lumbar spine, forearm and heel bone (calcaneous) are erratic possibly because of erratic and unplanned loads on these areas

The time required for bone to adapt is several times that of muscle for SCIs. No bed rest studies of sufficient duration to allow bone losses to reach a maximum, probably a year or more, are known. Loads on the spine are less than normal but probably significant from both leg and trunk movements. Since limited arm motion is allowed but not documented, arm bone losses cannot be accurately measured but losses must be less than in legs.

It has proven impossible to collate the variable data from all sources, and they were reduced to a few with the most stable history. Figure 6.23 illustrates a series of NASA planar DEXA studies of bone mineral densities. At the time of the study the hip and pelvis were considered the most likely fracture sites, hence the attention given this region.

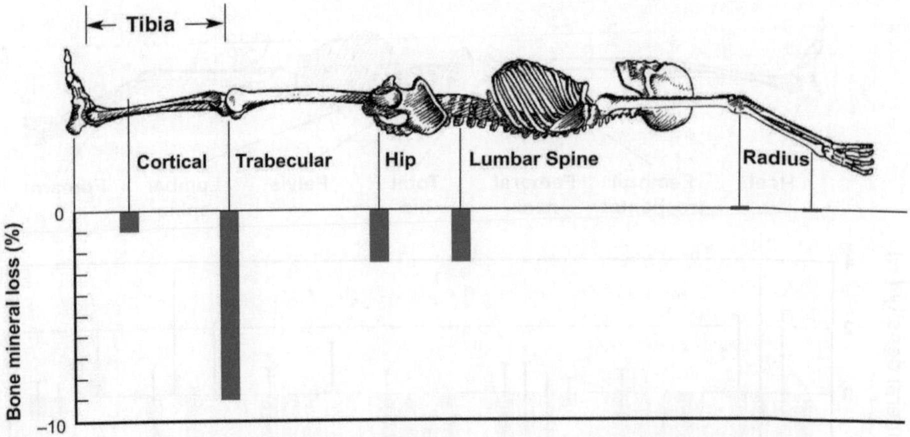

Fig. 6.24 Bone density change in 55-day bed rest study. Changes in the sites shown are typical. Upper body and arms are not constrained, and their relatively normal activity prevented loss. Rapid loss of trabecular bone occurs in all the major leg joints. Density of cortical bone is little changed, but bone lost from its inner surfaces reduces thickness and strength. Time and extent of final losses of bone remain unknown except for a few sites in SCI

In addition to obvious and significant losses in hip and pelvis there are significant differences between losses in the femoral neck and trochanter. This may be a result of the DEXA measurements or the variation between subjects as shown by the large standard deviations. A trend toward or even a completion of losses may be present in the hip and pelvis, but there is too much variation to say this with any assurance. Variation in heel, lumbar spine and forearm are too large to draw any conclusions. Variation in physical activity may cause this. Losses in heel bone (calcaneus) and lumbar spine are variable.

There is surprisingly little data on bone losses in legs during bed rest. A 55-day European study showed a large loss in tibial trabecular bone (Fig. 6.24). Changes in pelvis and spine were consistent with the longer NASA studies. The study also demonstrated an aspect of gains or losses in bone frequently overlooked. Osteoclasts require 2 weeks or more to respond. In this study bone was still being lost for 14 days after the subjects were back on their feet. Timely measurements documented this delay. This is in contrast to the increase in urinary Ca^{2+} that begins almost immediately on exposure to weightlessness.

To provide support for planning and execution of flights, bed rest must provide functional as well as scientific information; e.g., the ability to stand and walk at the end of bed rest should be recorded. Results of inflight bone studies are also necessary to determine the relevance of bed rest to spaceflight. Published results are examined in the next section on flight effects.

6.7.3 *Effects of Spaceflight on Muscle and Bone*

The obvious and primary effect of spaceflight on muscle and bone is the absence of weight, which allows bipedal locomotion and its forces on Earth. In 1 g much of the usual work of the upper body consists of supporting and moving objects against their weight, beginning with the arms themselves. On Earth the arms are adapted to much larger loads by physical work or sports and exercise. In weightlessness virtually all leg loading is gone except during exercise. But in terms of motion, hands and arms are used even more in space than on Earth. Stabilizing one's body position is a major activity, but loads, except during EVAs, are usually restricted to small inertial forces. Muscles rapidly adapt to this new environment [32].

Legs and lower back are more drastically unloaded, for not only are legs relieved of carrying their own and the body's weight, but they are relieved of the much larger muscle forces required to move the body. Figure 6.25 graphically demonstrates the difference in leg activity of an astronaut's day on Earth and in space. Exercise such as jogging or running is a routine daily part of most astronauts' lives on Earth especially before a mission. This locomotor activity determines the strength of leg muscles and bones. In this case it is the 1 g region (blue) between 2.5 and 2.7 BW in the Fig. 6.25. Other exercise may provide some of the FGF in the 0.8–1.6 BW region but the forces in the 1.0–1.3 body BW region are probably generated by walking associated with daily activities.

Without a treadmill inflight FGF would occupy the small area between zero and 0.7 BW (yellow). The total number of steps required to maintain bone is still in

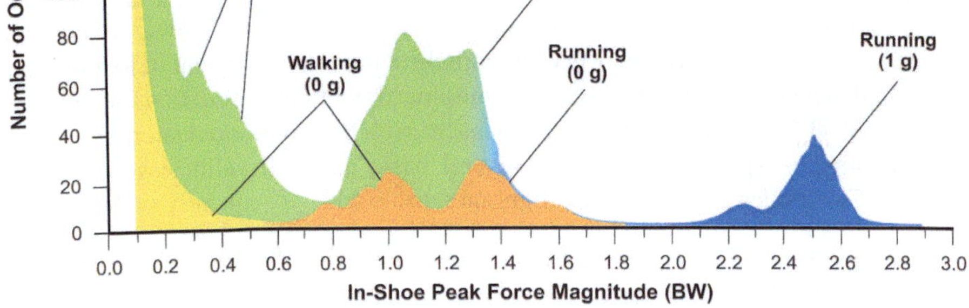

Fig. 6.25 Histogram of foot ground forces of an astronaut on a typical day in space in *yellow* and *orange* and on Earth in *green* and *blue*. The running and walking in space was on a treadmill. Note the great reduction in both number of steps and in magnitude of the force inflight

question; however, even assuming it is small, the difference between 2.4 and 2.6 BWF on Earth and the 1.6 BW on the treadmill inflight is enough to cause a large loss of muscle and bone. Except for being on the treadmill (orange area) the FGF are very small inflight, 0.1–0.3 BW, especially when compared to 1-g forces and number of steps (green).

There are difficulties measuring bone and muscle during spaceflight. Currently the measurement instruments needed are large, heavy and have not been flown so measurements can only be taken pre- and postflight. Also, without the Shuttle program, only two to three American astronauts can fly per year, and the flights are relatively long. This distorts the current monthly average values. Probably more serious is the lack of valid data on actual use of countermeasure equipment.

A considerable amount of muscle and bone information such as pre- and postflight strength and radiometric bone and tissue density has been collected routinely. However, such data has been tightly restricted in the name of medical privacy. The following flight effects on muscle and bone were obtained from published reports.

6.7.4 *Effects of Spaceflight on Muscle*

The first studies of effects of flight on muscle were made on all crewmen on all three Skylab flights [1].

On SL-2 there was no arm exercise device and only a bicycle ergometer for the legs. Thus, the first measurements of strength loss[22] were relatively undistorted by countermeasures, but there were 5 days of recovery before measurement. Even so, there were losses of 10% and 20% in combined leg flexors and extensors (Fig. 6.26). Elbow strength loss was 8% in flexion and 10% in extension. Arm and trunk exercise was added to SL-3. The crew returned with virtually no arm strength losses while leg losses increased. A simulation of walking was added to SL-4. Postflight leg extensor strength was increased (non significantly) and flexor losses were less than half of those measured for SL-3. Loss of body mass was progressively reduced on each Skylab flight as food quantity was successively increased with each mission. This played a role in the reduction in leg muscle volume and strength loss. All Skylab crews were able to walk on recovery but with difficulty. Improvement in locomotion was rapid. The crew of the SL-4 84-day mission was in better functional condition than the SL-3 56-day crew. This was consistent with their measured strength.

The Skylab experience produced several basic and still valid lessons on the effects of weightlessness and spaceflight on muscle and consequently on bone.

- Loss of muscle and strength in legs is large, rapid, and compromises 1 g locomotion postflight.
- Arm losses are significant but have less functional impact postflight.

[22] Isokinetic strength measurements were made pre- and postflight at 140°/s.

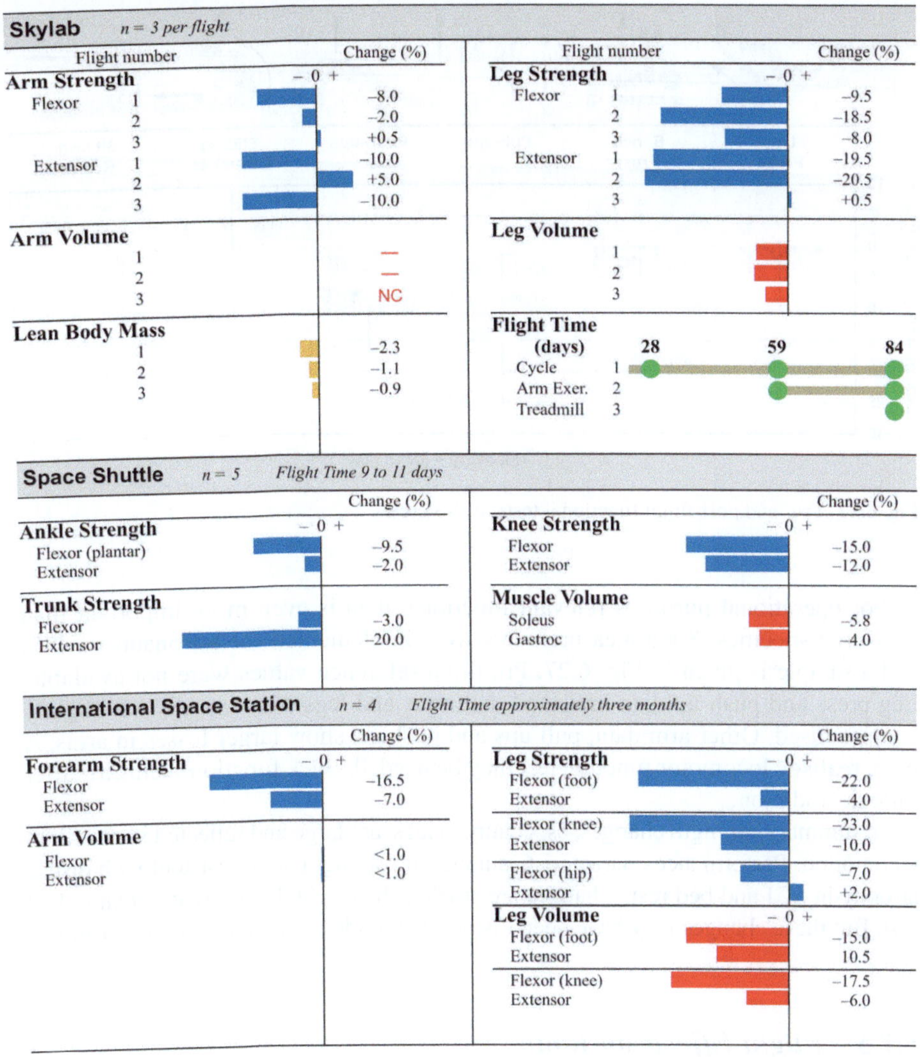

Fig. 6.26 In-flight muscle strength and volume change during Skylab, Space Shuttle, and International Space Station

- Arm strength losses can be controlled with simple countermeasure exercise devices.
- Muscle losses vary considerably for individuals, crews and muscle sites.
- To be effective, countermeasures must be specific in magnitude and site and force characteristics.

The Russian program was very active during this time, rapidly extending flight durations, yet available data are scarce. Leg losses and trunk extension are surprisingly large and rapid on the short Shuttle missions.

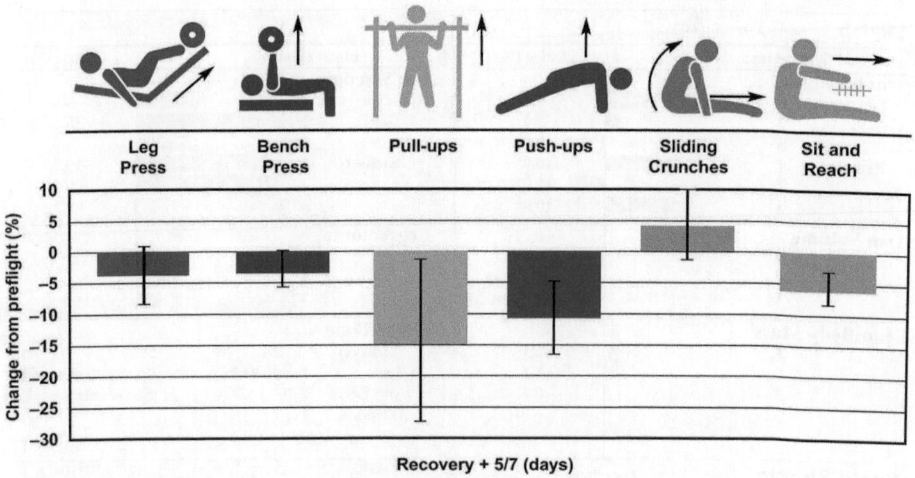

Fig. 6.27 Pre- and post-flight functional tests of ISS crew

For operational purposes relevant functional data is even more important than measured specifics. Such measurements were implemented for astronauts on ISS, and a sample is given in Fig. 6.27. Pre-flight reference values were not available. Leg press and push-up data indicated that arms and legs were equally but slightly compromised. Other arm data, pull ups and pushups show larger losses in arms. A more realistic locomotor function test may be needed. Back function seems to suffer little degradation.

In summary, strength changes, especially in legs, are large and reflected in postflight locomotion. Performance characteristics including strength are consistent with disuse atrophy in SCI and bed rest. Changes are smaller than in SCI and greater than in bed rest. But these changes may have been distorted by undocumented countermeasures.

6.7.5 *Flight Effects on Bone*

X-ray bone absorption studies were conducted on Gemini, Apollo and Skylab [8], although the techniques were relatively crude. Indications were consistent, even at that early stage. Increasing calcium was lost with increasing mission duration through the 84-day Skylab flight, but the rate of loss was slowing by that time. Bone was lost from the heel but much less from the hand bones. Calcium was lost through the Skylab 84-day flight at a decreasing exponential rate.

Chemical markers in blood and urine show that collagen breakdown byproducts from bone are greatly increased in weightlessness as were Ca and PO_4. Other byproducts indicate that osteoclastic activity is increased while osteoblastic activity is unchanged thus supporting the concept of increased bone loss versus its replacement.

Russia used X-ray tomography to record total losses of spinal mineral density of –0.3% to 10.8% that occurred for four cosmonauts on 5–7 month missions. Countermeasures were used but unreported. Two conclusions are possible from these data; appreciable amounts of bone are lost from the spine inflight while individual losses varied from negligible to large.

A joint US/USSR study of 18 cosmonauts on Mir was conducted using DEXA density measurement after flights of 4–14.4 months. Subsequently data were obtained on 60 astronauts and cosmonauts aboard Mir and ISS with flights of 176 ± 45 days. An investigator with access to this data commented on the large individual variability both in terms of sites and extent of bone loss. He tacitly acknowledged the limitations of these data due to variation with time, lack of end points, and lack of information regarding related use of countermeasures. This was followed with the usual implied request for additional studies.

Studies using both DEXA and volumetric QCT measurements were conducted on 14 subjects from 4 to 6 month ISS expeditions. These techniques allowed losses in cortical and trabecular bone in spine and hip to be measured for the first time inflight. This study showed much larger density loss in the spongy trabecular bone than in solid cortical bone in the hip but no difference in the spine. All measures were variable, and heel bone (calcaneus) measurements were so variable that it was recommended that the heel not be used as representative indicator of bone loss. Subsequently these data was used to conduct an engineering strength analysis of the femoral head for upright posture and lateral falls. One of 13 astronauts was found to be in the risk zone of fracture in 1 g after a 6-month flight. Unfortunately, the author could not find a record of exercise for anyone in the group.

At least two ISS bone density studies have attempted to document how inflight exercise relates to bone losses. Gopalakrishnan and Cavanagh [33] made meticulous measurements of the FGF of four ISS astronauts during typical days on Earth and inflight thus keeping records of exercise time and activities. In-flight treadmill FGF were approximately 60% of their maximum in 1 g, well below the stimulus threshold to maintain bone mass. A more recent study of the effects of force loads on the legs from an advanced weight simulator plus correction of caloric and other dietary deficiencies almost eliminated bone loss in two subjects [34].

Thirteen astronauts on ISS expeditions 14–20 (2006–2009) were divided into groups of eight and five depending on their use of IRED or ARED weight simulators (see Chap. 7) during flights of 5.3 ± 2.0 and 4.5 ± 2.0 months. Bone densities of the hip and spine were measured preflight (129 ± 123 days) and postflight (12 ± 11 days). Blood and urine were collected 180 and 45 days preflight, up to five times inflight, on landing, and 30 days thereafter. No pre- or postflight strength or exercise data are given. Both treadmill and cycle ergometer were scheduled, but time and intensity are not available. Squat loads and repetitions on IRED and ARED were reported as equal, but values are not given.[23]

[23] The detailed data on prescribed load and adherence in referenced Smith S, et al. [35]. Benefits for bone from resistance exercise and nutrition in long-duration spaceflight was not available to the author at the time of this writing.

The resulting changes of weight and mass are discussed in Chap. 10, p. 284. There were significant reductions in loss of measured bone density in the ARED group. There were also indications of an increase in muscle mass in the ARED group, including an increase in lean body mass and decreased repetition, which make interpretation of the cause of this reduction in bone mass problematic.

Loss of body mass from decreased caloric intake is known to reduce muscle mass and consequently strength and maximum force on bone, which would indicate that maintenance of nutrition in the ARED group would aid in the maintenance of bone. However, findings of increased lean body mass and decreased body fat in the ARED group indicate a slight increase in strength produced by exercise in force from exercise. This was in contrast to a trend toward increased fat and reduced lean body mass in the IRED group.

Force produced by the representative exercise in both machines was stated to be equal. Presumably, this is the static force of an equivalent weight. In the case of IRED this is the maximum force during the movement, but the ARED has an inertial component that, depending on acceleration during the lift, will add a large additional force to the static component, possibly explaining some of the difference seen in the ARES and IRED groups (see Chap. 7, p. 208).

The other factor that must be considered is the treadmill exercise used by both groups, which can produce both strength and aerobic stimuli. Changes in leg strength magnitude of foot to ground forces are unknown, but treadmill force loading may have been increased for the ARED group. Effects of aerobic exercise also become significant with locomotor exercise. While weight lifting must produce an increase in respiration, but the increase is typically small and brief compared to that produced by even 30 min of jogging.

While bone and muscle strength may be maintained by weight lifting on Earth and in space, their endurance is another matter yet to be addressed. The traditional 1-g use of strength exercise for upper body and jogging or treadmill for locomotor exercise also seems appropriate here. An in-flight bone study using the drug Alendronate as an antiresorptive countermeasure has been completed. Early results indicated that when added to a more rigorous exercise regimen losses were sharply reduced [35].

Effects from SCI, BR and weightlessness on bone loss is shown in Fig. 6.28. The Investigators obviously had differing areas of interest so only the hips and lumbar spine can be compared, and that must be done with caution. The relatively short times and unknown countermeasures of BR and flight do not provide data for quantitative comparisons. Much longer times without countermeasures and with multiple loss rates are needed for a comprehensive view of the three situations and base lines for BR and flight effects.

6.8 Recent Data

It is interesting to note that when referring to the human body and weightlessness, impaired health, detrimental effects and medical risks, are often cited [36, 37]. In this book we have taken a different approach. With one possible exception (MOS),

Bone loss (%)			Site
Spinal cord injury	Bed rest (90 days)	Spaceflight (180 days)	
		9.1*	Radius cortical
		1.2*	Radius trabecular
14.8*	+0.5	4.9	Spine
29.0*	4.8	7.7	Pelvis
35.0*	1.3	6.8	Femur neck
44.0*		7.8	Femur trochanter
58.0			Femur cortical
48.0			Femur trabecular
64.0*			Knee
34.0			Tibial trabecular
25.0			Tibia cortical
	4.4	2.9	Heel (calcaneous)

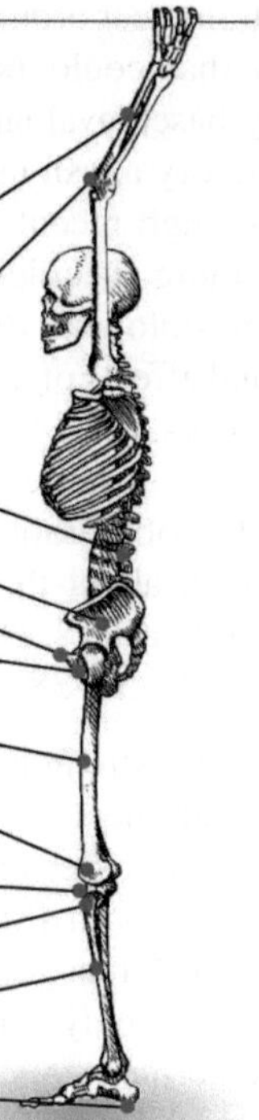

Fig. 6.28 The initial hope for this figure was to show a comparison of the maximum bone loss in spinal cord injury, bed rest, and spaceflight. Such hopes soon vanished for a variety of reasons. Different procedures and measurements made most of the existing bed rest and flight data incompatible. In spite of these severe limits, it is believed that significant value is present in the limited data. Studies of almost 100 spinal cord injury patients by Eser and colleagues [15] set a standard both in theory and execution. The four sites shown are state of the art bone mineral and dimensional losses measured by pQCT and correctly analyzed. Traditional bi-dimensional DEXA studies may approximate pQCT and several are used. *Unmarked* data is pQCT, which is now being employed in bed rest and pre- and postflight studies. Some factors are obvious. Losses are larger where muscle forces are greatest. Losses are more or less prevented in arms during bed rest and spaceflight. More frequent measurements are needed in bed rest and spaceflight

the changes produced by all exposures to weightlessness seem to be appropriate, reversible somatic adaptations to a new environment. A different region of the spectrum of somatic adaptations in bone and muscle is regularly seen and experienced on Earth, but usually not recognized as such. Upper limits of the spectrum of muscle and bone adaptation may be seen in professional athletes. The DNA and mechanosensors of myocytes and osteocytes function on Earth, ISS, or Mars to maintain a physiological state consistent with the usual environment. This level of usual of physical condition may not be adequate to manage mechanical jams or hours of continuous hours of activity in a stiff space suit.

This immediately raises a still to be answered question: What are the criteria for normal health or acceptable physical condition in space or on the Moon? Few athletes train year around. Yet most astronauts have launched in the best physical condition that could be managed under the circumstances. There has not been a reality-based evaluation of required strength of muscle and bone in spaceflight or exploratory missions. This question is addressed briefly at the end of this book.

Although recent ISS reports on bone loss have added to the literature on this topic, more specificity is needed. Given the great and obvious individual physical and physiological differences, meaningful interpretation of effects of weightlessness and effects of countermeasures depend on knowing quantitative details of these differences. NASA policies that require any identifiable individual difference on a mission be obscured for the reasons of individual personal privacy are however limiting. The obscuration of details make generalizations difficult. Much more needs to be known about the effects of weightlessness as a function of gender, time and counter measures. However, a few general conclusions could be drawn from recent ISS reports [36–38]:

- Female strength losses were greater than male on the initial post flight measurements
- Strength losses increased with flight duration
- Exercise with a simulation of weights in 1 g, i.e. constant forces plus unspecified inertial forces as used in ARED is a more effective countermeasure to strength loss than only elastic forces used in IRED. See Chap. 7 for further comment on countermeasures.
- Strength losses are regional and correspond to usage of that region in weightlessness rather than to their prior use in 1 g

Mineral losses in bone during the first decade of ISS was recently reported on [37]. This report compared results from pre/post flight two-dimensional DEXA scans in selected populations of 33 ISS astronauts and 28 cosmonauts and 7 astronauts who flew on MIR. Changes in of BMD at five sites were compared on ISS and MIR flights. ISS flights produced a mean loss of approximately −3% compared to −6% on MIR but flight durations are unknown as are countermeasures and other details. BMD T-Scores for 33 ISS astronauts for four sites were derived from pre/post flight DEXA scans. The forearms showed no change. Lumbar spine and two femoral sites showed a remarkable range of individual preflight values and a large range of losses. No post flight value reached −2.5, JSC's minimum value for bone health. It should be noted that BMD T-Scores, which are supposed to relate T-Scores to probability of a fracture in the site scanned, were based on very different older populations, often in clinical situations. BMD changes in the lower body and legs for 26 ISS astronauts whose exercise counter measures were not specified were compared to seven who used the ARED. Mean values for the 26 was −5.6% and −5.0% for the seven who used ARED. Approximately half of the section was discussion of the DEXA BMD T-Scores and use of QCT scans. A diversity of opinion among experts on this issue was documented by references. The author's summary of this section is that "the ISS astronaut cohort represents an at risk population" (for bone damage).

6.9 Summary of Flight Effects

There are many potential effects of space flight on muscle and bone but the two predominant and demonstrated effects are nutritional and weightlessness. Nutritional effects are common to Earth and space. Humans have well developed nutritional instincts developed over hundreds of millions of years and eons of practical experience plus some 200 years of scientific experience with this subject. In space it is usually a logistics problem that has been recognized and is being worked on aboard ISS (see Chap. 10). Conversely, loss of all weight forces is still a relatively new and unique human experience. Moreover, its major effects are usually secondary to reduction in usual muscle activities, especially bipedal locomotion, and not the loss of weight.

Results from studies of effects of spaceflight are consistent with the concept that this loss of muscle and bone is a rather extreme example of disuse atrophy. Changes seen are consistent with the patterns seen in SCI and bed rest. In days, muscle quickly begins to adapt by losing actin and myosin protein from muscle fibers with a consequential reduction in size and strength of muscles. The extent and time course of these changes vary by both individual and muscle. The course is exponential with time and requires several years to reach a stable condition. In SCI losses of muscle and strength can approach 70%. Changes are much smaller in bed rest and spaceflight but their full extent have yet to be measured. Removal of muscle forces causes a prompt but much slower response in bone where normal replacement of bone stops and removal continues or is increased. The result is that 3–8 years after a SCI up to 60% of bone is lost in some sites. Both muscle and bone losses can be fitted to exponential equations from which the total loss, time to reach total loss, and loss at any point in time can be calculated. To date this has been done only for four bone sites in SCI. Extended time periods and or a large number of subjects and measurements are required to fit such curves. A much more limited measure of the changes seen is in common use; i.e., the percentage rate of loss at the start of bed rest or spaceflight. This is easily confused with long-term changes and has the potential for other errors, some of which are illustrated in Fig. 6.14.

Prevention of such adaptations in muscle and bone can be prevented by replacing the motions and forces usually experienced on Earth by some means, usually called countermeasures, in weightlessness. This may be difficult for the required forces are complex and must specifically and individually match those of the 1 g forces they simulate. This was crudely demonstrated by IRED and ARAD. The principles involved are discussed further in the following chapter on countermeasures.

Characterization of human muscle is still incomplete and that of bone is less known. This was demonstrated on a much larger and longer scale by the now generally recognized failure of DEXA T-values as an effective 1 g clinical tool. Characterization of a living bone's ability to carry the force loads imposed on it, often miscalled strength, is complex and difficult. Factors involved include size, shape, nature of its materials and their structural arrangements. Except for a few flexibility measurements there are no practical measurements of a subject's bone functional capacities. Many recognized experts on bone and virtually all clinicians have taken a measurable aspect of bone, DEXA's two dimensional density scans, and these are often as a measure of functional strength. The use of QCT to produce a more realistic model of bone should be a step closer to measure of functional reality.

6.9.1 Muscle and Bone in Space; Past, Present and Future

One of the first studies of the effects of space flight was the erroneous finding of remarkable losses of bone after the first short flights—such as during the Gemini project. This was based on two-dimensional X ray density changes which are now measured using two X ray wavelengths and a half century's technical advances instead of film. However, many inherent errors in the use of DEXA still remain. Some may be removed by QCT and the improved analyses it allows, but its use as a measure of bone integrity, remain to be proven.

In 1973 the first isokinetic pre/post flight strength measurements were made by an Ad Hoc addition to SL-1. From the results of this, and subsequent studies from Skylab 2 and Skylab 3 plus crew experience, a few pounds of arm exercise devices were added to SL-2 and 3½ lb of locomotor exercise gear to SL-3. After the then long duration flights of up to 86 days, all crewmen were able to walk onto the recovery ships unaided. It has been impossible to find any published, validated post flight function data to compare with this. There are many stories of recent postflight crew conditions, and they range widely from incapacitation to getting up and walking away from the vehicle on landing after many months inflight. Based on historic precedence, this is probably heavily dependent on the individual but postflight muculoskeletal performance probably comes under the NASA personal privacy agreements.

As for the future; at this time the human body is often presented as the limiting factor in human space exploration. In support of this a picture of broken bones from osteoporosis is often painted. While a more complete knowledge of muscle and bone would benefit both space exploration and many humans on Earth, when a vehicle capable of say a Martian visit and return is ready, and if the right individuals are chosen, the current knowledge of bone and muscle may be adequate if properly applied. A key part of correct application are countermeasures, the subject of the next chapter. In conclusion, the problems of muscle and bone loss have been partially characterized, but more complete characterization is needed for realistic mission planning and evaluation of countermeasures—especially for long-term missions. Control of adaptation is obviously incomplete and addressed in the next chapter.

References

1. Thornton WE, Rummel JA. Muscular deconditioning and its prevention in spaceflight. In: Biomedical results from Skylab. NASA SP-377. Volume 21. U.S. Washington, DC: GPO; 1977. pp. 191–7.
2. Section 3 Functional performance evaluation. In: Extended duration orbiter medical project. NASA SP-1999-534. NASA Johnson Space Center, Houston, TX; 1999.

3. Linenger J. Off the planet: surviving five perilous months aboard the space station Mir. New York, NY: McGraw Hill; 2000.
4. Wolff J. Das gesetz der transformation der knochen. Berlin: A. Hirchwald; 1892.Published with support from the Royal Academy of Sciences in Berlin. English translation by Maquet P and Furlong R. Springer-Verlag, 1986
5. Mullender MG, Huiskes R. Proposal for the regulatory mechanism of Wolff's law. J Orthop Res. 1995;13(4):503–12.
6. Mack PB, Vose GP, Vogt FB, LaChance PA. Experiment M-6, bone demineralization. In: Gemini midprogram conference. NASA SP–121; 1966, p. 407.
7. Weedon G, et al. Mineral and nitrogen metabolic studies. In: Biomedical results from Skylab. NASA SP-377. Volume 18. Washington, DC: U.S. GPO; 1977, pp. 167–74.
8. Vogel J, et al. Bone mineral measurement. In: Biomedical results from Skylab. NASA SP-377. Volume 20. Washington, DC: U.S. GPO; 1977, pp. 183–90.
9. Thornton W, Ord J. Physiological mass measurements in Skylab. In: Biomedical results from Skylab. NASA SP-377. Volume 19. Washington, DC: U.S. GPO; 1977, pp. 175–82.
10. Thornton W, Hoffler G, Rummel J. Anthropometric changes and fluid shifts. In: Biomedical results from Skylab. NASA SP-377. Volume 32. Washington, DC: U.S. GPO; 1977, pp. 330–8.
11. Lang T, Keyak J, Le Blanc A. Background on bone loss. In: Cavanagh PR, Rice AJ, editors. Bone loss during spaceflight: etiology, countermeasures, and implications for bone health on Earth, vol. 6. Cleveland, OH: Cleveland Clinic Press; 2007. p. 64.
12. Cavanagh PR, et al. Foot forces during typical days on the international space station. J Biomech. 2010;43:2182–8.
13. Greenleaf J, editor. Deconditioning and reconditioning. Boca Raton, FL: CRC Press; 2004.
14. Spalding KL, Bhardwaj RD, Buchholz BA, Druid H, Frisén J. Retrospective birth dating of cells in humans. Cell. 2005;122(1):133–43.
15. Eser P, Frotzler A, Zehnder Y, Wick L, Knecht H, Denoth J, Schiessl H. Relationship between the duration of paralysis and bone structure: a pQCT study of spinal cord injured individuals. Bone. 2004;34(5):869–80.
16. Frotzler A, Berger M, Knecht H, Eser P. Bone steady-state is established at reduced bone strength after spinal cord injury: a longitudinal study using peripheral quantitative computed tomography (pQCT). Bone. 2008;43(3):549–55.
17. Belavý DL, Miokovic T, Armbrecht G, Richardson CA, Rittweger J, Felsenberg D. Differential atrophy of the lower-limb musculature during prolonged bed-rest. Eur J Appl Physiol. 2009;107(4):489–99.
18. Risk of impaired performance due to reduced muscle mass, strength, and endurance. In: Human research program, human health countermeasures element, evidence book. NASA Johnson Space Center, Houston, TX; 2008.
19. Risk of accelerated osteoporosis. In: Human research program, human health countermeasures element, evidence book. NASA Johnson Space Center, Houston, TX; 2008.
20. Shubin N. Your inner fish: a journey into the 3.5-billion-year history of the human body. New York, NY: Vintage; 2008.
21. Guyton AC, Hall JE. Contraction of skeletal muscle. In: Textbook of medical physiology, vol. 6. 11th ed. Amsterdam: Elsevier; 2006. p. 61–7.
22. McArdle WD, Katch FI, Katch VL. Skeletal muscle, structure and function. In: Exercise physiology, vol. 18. 5th ed. Philadelphia, PA: Lippincot, Williams & Wilkins; 2001. p. 358.
23. Netter FH, Woodbine RT, Crelin ED, Kaplan F. Musculoskeletal system. In: The Netter collection of medical illustrations, vol. 8(1). Dover Township, NJ: Ciba-Geigy Corp; 1987. p. 169–72.
24. Klein-Nulend J, Bacabac R, Bakker A. Mechanical loading and how it affects bone cells. Eur Cell Mater. 2012;24:278–91.
25. Netter FH, Woodbine RT, Crelin ED, Kaplan F. Musculoskeletal system. In: The Netter collection of medical illustrations, vol. 8(1). Dover Township, NJ: Ciba-Geigy Corp; 1987. p. 94–7.
26. Bassey EJ, Littlewood JJ, Taylor SJ. Relations between compressive axial forces in an instrumented massive femoral implant, ground reaction forces, and integrated electromyographs from vastus lateralis during various 'osteogenic'exercises. J Biomech. 1997;30(3):213–23.

27. Bergmann G, Deuretzbacher G, Heller M, Graichen F, Rohlmann A, Strauss J, Duda GN. Hip contact forces and gait patterns from routine activities. J Biomech. 2001;34(7):859–71.
28 De Lorme TL, Watkins AL. Resistance exercise: technic and medical application, vol. 2. Prentice Hall, NJ: Appleton-Century-Crofts; 1951. p. 2–3.
29 Giangregorio L, McCartney N. Bone loss and muscle atrophy in spinal cord injury: epidemiology, fracture prediction, and rehabilitation strategies. J Spinal Cord Med. 2006;29(5):489.
30 Rittweger J, Gerrits KH, Altenburg TM, Reeves ND, Maganaris CN, Haan AD. Bone adaptation to altered loading after spinal cord injury: a study of bone and muscle strength. J Musculoskelet Neuronal Interact. 2006;6(3):269–76.
31 Stuempfle KJ, Drury DG. The physiological consequences of bed rest. J Exerc Physiol. 2007;10(3)
32 Narici MV, De Boer MD. Disuse of the musculo-skeletal system in space and on earth. Eur J Appl Physiol. 2011;111(3):403–20.
33 Gopalakrishnan R, Genc KO, Rice AJ, Lee S, Evans HJ, Maender CC, Ilaslan H, Cavanagh PR. Muscle volume, strength, endurance, and exercise loads during 6-month missions in space. Aviat Space Environ Med. 2010;81(2):91–104.
34 Smith SM, Heer MA, Shackelford LC, Sibonga JD, Ploutz-Snyder L, Zwart SR. Benefits for bone from resistance exercise and nutrition in long-duration spaceflight: evidence from biochemistry and densitometry. J Bone Miner Res. 2012;27(9):1896–906.
35 Leblanc A, Matsumoto T, Jones J, Shapiro J, Lang T, Shackelford L, Smith SM, Evans H, Spector E, Ploutz-Snyder R, Sibonga J. Bisphosphonates as a supplement to exercise to protect bone during long-duration spaceflight. Osteoporos Int. 2013;24(7):2105–14.
36 Hayes J. The first decade of ISS exercise: lessons learned on expeditions 1–25. Aerosp Med Hum Perform. 2015;86(12):A1–6.
37 Sibonga JD, Spector ER, Johnston SL, Tarver WJ. Evaluating bone loss in ISS astronauts. Aerosp Med Hum Perform. 2015;86(12):A38–44.
38 English KL, Lee S, Loehr JA, Ploutz-Snyder RJ, Ploutz-Snyder LL. Isokinetic strength changes following long-duration spaceflight on the ISS. Aerosp Med Hum Perform. 2015;86(12):A68–77.

Chapter 7
Countermeasures to Loss of Muscle and Bone During Spaceflight

7.1 Introduction

Bone and muscle loss is currently one of the major concerns for long duration spaceflights and will become especially important in exploratory missions [1, 2]. Extended stays on planetary bodies such as asteroids, the Moon and Mars will also encounter this problem for their gravity is only a fraction of that on Earth (1/6 and 1/3 respectively). Prevention of this loss of bone and muscle, especially bone, has been a major effort of NASA's life science. Currently it is being given special attention as an essential part of future exploratory missions. However, a comment by a leading investigator in this area made some years ago is even more pertinent with the release of a summary of exercise countermeasures during the first decade of ISS. "It is remarkable that so little progress has been made in understanding and alleviating this important problem [3]." It took less than half that time from Goddard's first successful rockets for man to reach the Moon. In addition to time and effort, hundreds of millions of dollars have been spent on this problem both on Earth and in space to date (Fig. 7.1).

Loss of muscle and bone is the musculoskeletal system's adaptation to a new gravitoinertial environment. Weightlessness usually plays a smaller direct role in loss of muscle and bone in spaceflight than its indirect result. Weightlessness precludes much of the normal physical activities that determine one's usual bone and muscle capacity. This is a very familiar problem on Earth. Disuse Atrophy and its effects on animals and humans have been known for millennia, as have ways to prevent it. Weightlessness makes the major activities that stimulate strength and mass in the major bones and muscle on Earth impossible in weightlessness without mechanical aid. If mechanical aids, countermeasures, can be designed to replace the forces and motions to which muscles and bones are adapted on Earth, weightless adaptation with loss of muscle and bone can be controlled. Under these conditions

W. Thornton, F. Bonato, *The Human Body and Weightlessness*,
DOI 10.1007/978-3-319-32829-4_7

Fig. 7.1 After four plus decades in space maintenance of 1-g adaptation is still a significant problem

functional capacities can be modulated in space by control of countermeasures. For example, increases in functional capacity by exercise in space, known as training, can be made in some situations. This was first seen on SL-3 when elbow flexor strength in one subject was increased in flight using simple elastic and isokinetic devices added to that mission.

Absence of bipedal locomotion produces the largest adaptation to weightlessness. Early spacecraft were too small to accommodate countermeasures to this and flights were too short to require them. Once volume and mass were available, the countermeasure problem became one of producing devices to properly load the Musculoskeletal System (MSkS) in space for extended periods of time. Some of the current exercise devices are questionably adequate but may be too large, heavy and power hungry to be considered for extra orbital flight.

Arguably the most serious problem has been, and to some extent still is, the need for a better understanding of the biophysics and biomechanics involved and a resulting coordinated pursuit of countermeasures. Just as a concept can have a significant consequence, so can a lack of one. An example of this was the Interim Resistive Exercise Device (IRED) that was designed for strength training. IRED provided only elastic force, to which human muscle is poorly species adapted and thus its failure as a countermeasure was guaranteed before the first drawing of the device was ever made.

Although progress in exercise and in-flight studies during Shuttle, Mir/Shuttle and early International Space Station (ISS) expeditions has been made, additional advances will require scientific rigor and discipline. Meaningful quantitative results in the physical sciences requires specificity. All too often essential details of countermeasures are either missing or buried in statistics. An exemplar of this non-specificity is the term "resistance exercise," which is applied to a wide range of exercise that may or may not contain mechanical resistance, a precise physical term. Meaningful results from countermeasures used during Shuttle and much of ISS would have benefited from more scientific rigor and discipline. Usage of devices was sometimes uncontrolled and unreported or recorded.

An understanding of biomechanics and mechanics in general, especially the concept of mechanical impedance, would help progress the development of effective countermeasures. Concepts of mechanical impedance and impedance of human force generators and their coupling, if known, should be employed either by users or designers of exercise device countermeasures. Some of these problems have been examined and some corrected, but an adequate countermeasure to the absence of locomotion for all subjects remains to be proven. Conversely, adequate strength exercise now appears to be available on ISS, as is cardiorespiratory exercise. A protocol with 2 ½ h per day of exercise is currently in effect on ISS. It has been said that loss of muscle and bone in legs, pelvis, and lower back has been successfully countered in a few astronauts. Whether this is preservation only of measured isokinetic strength and density of DEXA recordings, or is in fact preservation of all necessary muscle and bone function is difficult to determine.

A critical issue in planning missions is the extent and level of 1-g adaptations that must be retained. Only exercise under 1-g gravitoinertial forces will maintain all aspects of 1-g physical condition. In the following sections the major emphasis is on the so far unsolved problem of losses in locomotor areas. Each of the problems identified in this will be discussed. A brief history illustrates some of the problems of the past and provides a point of departure for the present and future.

7.1.1 *History*[1]

Most reviews of use of inflight exercise countermeasures begin with the rubber tubing with handles flown on Mercury and Gemini [4], but this was only a stressor for a brief EKG stress test. To one author's knowledge (WT) the first exercise countermeasure was designed, built and tested at Brooks Air Force Base in 1966–1967 for the USAF Manned Orbiting Laboratory (MOL). It was a continuation of a failed contract, designated the Total Body Exerciser (TBE). Because of the weight, space and power limitations of spacecraft at that time, a concept and space for the TBE had been designed into the vehicle and its planned missions. Its transverse bar could be translated from floor to maximum extended standing reach of a subject. Loading of the bar could be varied to accommodate a range of strength and aerobic exercise. One author's (WT) challenge was to determine the nature and magnitude of the forces on the bar and develop a generator for them within stringent weight limits.

No theoretical or experimental guidelines were available at that time. Major Ken Cooper was developing his aerobics program and was also part of the MOL medical team and aerobic rather than strength exercise became the primary concern. Based on electromechanical equivalence and generator to load matching theories, one author (WT) began with design of variable loads of constant, inertial and resistance forces to allow estimation of the mechanical impedance of muscles for optimum coupling. Not surprisingly, the resulting choice of load was a simple combination of constant force springs and a geared fly wheel for inertia equivalent to a 1 g weight. A simulator for weightless performance verification was a more difficult and complex task. After verification and life cycle testing, its performance was further verified in JSC's Zero g aircraft. A flight qualified prototype was available when the MOL project was cancelled. It was still available, and advocated as an adjunct exercise for Skylab. When this was refused for Skylab, the Astronaut Office (AO) supported my pre/post flight isokinetic strength measurements on the three missions.

Adding a cycle ergometer to Apollo missions was initially seriously considered by JSC Life Sciences. Size and weight limits could not be met by contractors until the author (WT) collaborated with the owner of Del Mar Avionics in a no cost government agreement to develop a miniaturized prototype cycle ergometer with load capacity of 250-W. A passive flight prototype was developed and demonstrated to JSC Life Sciences managers for Apollo, and three decades later to ISS managers, as a potential design for exploration flights. All Apollo missions were flown with only a commercial ExerGenie® but it was largely unused.[2]

NASA's first countermeasure device was its bicycle ergometer on Skylab [4]. These cycles were beautifully produced to JSC Life Science's performance specifications by Marshall Space Flight Center, except for a hidden fabrication flaw. This

[1] There are some discrepancies between the following text and some published reports including the first decade summary of exercise on ISS. Statements made here are based on large archives of original records at JSC and the University of Texas Medical Branch.

[2] It was a variable capstan with a rope and handles, a true resistance device.

was discovered in the Skylab Medical Experiments Altitude Test (SMEAT) when it mechanically failed repeatedly during attempts to ride it for extended periods at its maximum rated load of 300 W. It performed normally in SMEAT and on Skylab with a limit of 260 W.

When crew experience, and results from SL-2 pre/post isokinetic strength measurements, demonstrated the cycle ergometer to be inadequate as a musculoskeletal exercise strength countermeasure, the Astronaut office supported the author's (WT) urgent search for upper body strength exercise devices. Flight requirements limited its weight to a few pounds and required flight qualifications and readiness for launch on SL-3.

A set of extension springs with handles[3] and a commercial passive isokinetic device[4] [5] were modified and flown as an Astronaut Office DSO. Their regular use in suggested exercises proved effective for arms and back with virtually no measured strength loss, while losses of locomotor strength continued unabated. On SL-4 a Teflon® foot plate and 90 kg bungees[5] were flown and added to a discarded onboard bicycle restraint harness allowing a simulation of climbing a slippery hill with a 20 kg load. This preserved leg extensor strength and enabled the crew to walk on recovery after 84 days in flight [6].

Russia filled the gap in human spaceflight between Skylab and Shuttle with vehicles and flights of increasing duration. They developed a wide variety of musculoskeletal (MSK) studies and countermeasures, unknown to us at the time. Chief of these was a treadmill with both active and passive modes. A Penguin suit provides elastic across joint loads and elastic compressive loads across segments of the body. Their cycle ergometer was and is called a "Veloergometer". They also produced a resistive exercise device called a force loader, and copied, and for many years, used the elastic extension device flown on SL-3 and SL-4.

As the reality of Space Shuttle operations approached no apparent consideration was being given to MSK or neuromuscular effects of these flights. There had been no development of countermeasures flight hardware since Skylab. Flight operations provided a very small fund to explore this deficit. Henry Whitmore had been remarkably productive in research, development, design and fabrication in the USAF. We resumed a collaboration that will be subsequently identified as TW. A treadmill compatible with Shuttle operations was the first product of this collaboration. The prototype was a small passive unit with demonstrated performance equal to the better commercial units. Size was dictated by Shuttle flight specifications that it would be disassembled and stowed in a single locker for launch and assembled on orbit without tools.

Reports from STS 1 landing (Chap. 9) and egress aroused enough interest in the Astronaut Office to fly bungees and an upper torso harness that allowed jogging in

[3] These devices could be found in popular department stores at that time. The flown assembly was afterwards adopted by the Russians and flown for decades on their vehicles.

[4] This device was known as the Exergym®

[5] This was based on the average body weight of crewmember that was 70 kg. This 90 kg unalterable overload was deliberate.

place under one body weight axial loads. STS 2 returned with striking pictures of its commander jogging in place. The prototype treadmill easily passed performance verification and flight qualification testing and it was flown assembled on the mid-deck of STS 3. In the absence of other exercise gear this preproduction prototype became a part of STS flight hardware. This attracted the attention of a Russian support contractor who subcontracted Whitmore to design, build and install what may still be the largest human treadmill in the space program. It accommodated up to three EVA suited cosmonauts in a large vacuum chamber in Russia. Whitmore designed, built and tested it in his shop, took it to Russia and installed it without incident, performance deficiencies, time or cost overruns. This was in striking contrast to implementation of exercise countermeasures in NASA programs [7].

TW had been working on replacement of the Shuttle Treadmill (STM) and the addition of other exercise items. A flight simulator was constructed at Brooks Air Force Base with a vertical TM and horizontal subject for verification of constant force subject loads and recording of foot/ground forces. Del Mar Avionics added strain gauges to their standard TM and accelerometers to their Holter recorders per specification. This allowed recording and analysis of each and all steps of every user of the TM. The Holter with accelerometers allowed ambulatory recordings of locomotor activity and indications of upper body activity in an extended JSC bed rest study.[6]

The TW Shuttle tread mill was flown from STS 3 until its replacement by a JSC Life Sciences cycle ergometer a decade later. A rowing machine was also developed, tested and flown by TW with good results. Use of the Shuttle treadmill as a countermeasure exercise on Extended Duration Orbiter (EDO) missions was explored by JSC life sciences [8]. They also began development and testing of exercise countermeasures (ECM) for ISS. This included a series of cycle ergometers and a treadmill. TW designed and life tested an ISS prototype. Design, test and production of a modified treadmill was awarded to a contractor who did not test the tread. Another contractor's vibration isolation was added and the assembly designated Treadmill Vibration Isolation System (TVIS) [7].

A joint effort with Russia studied the effects of exercise with astronauts on Mir and cosmonauts on ISS. Mir had its treadmill and some strength devices. According to the ISS exercise supplement [9], the JSC contractor TM was mounted on a vibration isolation device (TEVIS) and an elastic strength device, Interim Resistance Exercise Device (IRED), were added to ISS Expedition 1 for joint US and Russian use. The treadmill failed and a JSC cycle ergometer mounted on a vibration isolation device (CEVIS) was added to Expedition 2. At this point it becomes difficult to translate, comprehend and reconcile the subsequent chain of reported events. NASA shows TVIS and CEVIS was "off nominal operation" for several periods and that TVIS was not "operational" several times while the Russian report states simply that during Expeditions 1–6 the TEVIS "did not operate for 89 days due to failure of the vibration isolation system and for 79 days due to the treadmill" [10].

Exercise protocol for cosmonauts was extensive and elaborate [11]. After a short time the treads failed on the US treadmill [11], which left ISS without locomotor

[6] Mechanical design and construction was done by Henry Whitmore.

exercise for four expeditions,[7] wrecking exercise schedules. A commercial 1-g unit was modified as a replacement. A NASA Interim Resistive Exercise Device (IRED) was also added and in 2008 was replaced by an Advanced Resistive Exercise Device (ARED). The European Space Agency (ESA) module has an inertial Flywheel Exercise Device (FWED) [12]. Both NASA and Russian treadmills were eventually replaced with upgraded units, and the TVIS now became the COLBERT. Recently, a large series of exercise protocols were developed and reported on for ISS and some are apparently being followed with regular onboard evaluations [13].

Numerous bed rest and inflight exercise studies have been conducted and several will be discussed. The current opinion is that arm muscle and bone losses can be prevented but that muscle or bone loading by treadmill locomotion has been inadequate for most. At least one study reports virtually eliminating mineral loss in leg bones by a combination of heavy strength and treadmill exercise and adequate diet[8] [14]. It should be noted that although food should not be considered a countermeasure, absence of an adequate diet will insure loss of muscle and bone. (See Chap. 10.) A follow up study done on ISS that added a bone conserving pharmaceutical shows essentially complete conservation of bone [15].

7.2 Background

An approach to the bone/muscle loss problem begins with its adaptation to a new environment with reduced forces on the body and arms and virtually no forces on the legs. The body, and especially the legs, are adapted to large forces in 1 g. Muscles and bones of the lower legs are at the bottom of a column of force bearing bones. They carry the heaviest loads of the body in 1 g, and suffer the largest losses in weightlessness. Current research demonstrates that the stimulus for the myocytes and osteocytes to increase or maintain muscle and bone strength is force, as described in the last chapter.

Figure 7.2 illustrates the magnitude and distribution of some of the forces in the lower legs. Standing quietly with body weight evenly distributed produces a single foot ground force of 0.5 body weight (BW), which is carried directly by the bone. Muscle forces are small.[9] This is all changed when the muscle contracts and lifts the heel off its ground support. The bone is now supporting both body weight plus almost two BW in muscle force.

Running in 1 g requires two BW or more foot ground force (FGF), making weight an even smaller fraction of the much larger forces on bone and muscle. Also note that the inertial component, shown in orange, of FGF is generated through acceleration

[7] Expanders appear to be elastic cords used in a variety of arm exercises.

[8] Although food should not be considered a countermeasure, absence of an adequate diet will insure loss of muscle or bone. See Chap. 10.

[9] Intermittent muscle forces of a few percent of forces used in walking keep the body in balance.

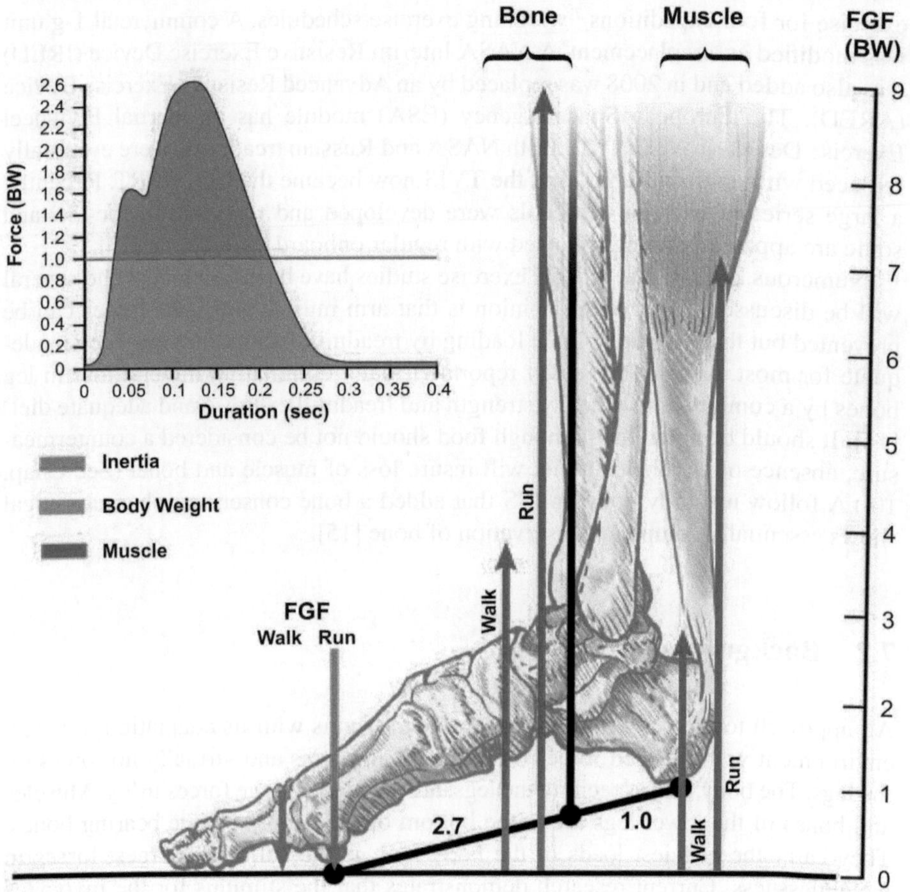

Fig. 7.2 Diagram of biomechanics and forces in foot and lower leg with toe rise and when running slowly. Weight forces are in *blue* and muscles forces in *red*. Relative magnitude of weight and muscle forces are easily seen. When standing quietly (not shown) except for negligible "trimming" forces to balance the body, muscles are inactive, and each lower leg directly supports body weight producing 0.5 BW FGF by each foot. When the heel is lifted, as in a slow walk, the lever arrangement comes into play completely changing forces on muscle and bone. The muscle must now generate approximately 2.7 times BW, and the bone must carry both weight and muscle forces multiplied by the lever arm and produced by the muscle or a total force of 3.7 BW. To run requires considerably more than 1.0 BW FGF to project the body clear of the ground. A typical FGF of 3 BW in a slow run is shown with the increase in bone and muscle forces. Weight is now almost trivial compared to the muscle forces. Strength of bone and muscle is determined by the level of force in 1 g. Note: the FGF waveform shown as a rapid jog is not correlated with forces in the diagram

of body mass by muscle action. This typical example of the magnitude and distribution of force in 1 g illustrates issues critical to effective countermeasures:

- Foot ground forces that must be replaced consist of 1 BW of constant force and much larger inertial forces.

- Muscle force, not weight, is the primary load on bone in this population, most of whom are joggers and runners.
- Locomotor forces are large, especially the internal forces.
- Countermeasures must begin with 1-g force stimulation of muscle to maintain 1-g levels of adaptation.

To prevent the large losses that will otherwise occur, the forces must be replaced in magnitude and force characteristics. The number of repetitions to maintain endurance properties of muscle and bone on Earth or in space are still unknown. To determine the effectiveness of any strength counter measure the forces must be accurately known or measured and the muscle and bone characteristics measured before and after application of the countermeasure. Bone poses a problem because its strength cannot be measured without removal from the body. Measurement of the bones' mineral content and geometry is used to calculate an estimate of their strength.

A second essential characteristic of muscle is its ability to repeat force cycles and, of bone, to sustain them. This varies with muscle and its adaptation. Arm muscles normally are limited in their number of heavily loaded repetitions. An increasing decrement in force occurs with repeated motion, and the decrement for a stated number of cycles is known as a fatigue factor. Legs make thousands of cycles a day, and this ability is termed endurance. Bone under repeated stress develops microscopic disruptions, and this can go on to stress fracture. Ability to resist such failures is called fatigue resistance.

These characteristics of muscle and bone are developed through repeated activity. In the legs, one of the current unknowns is the minimum number of repetitions required to maintain endurance and fatigue resistance. Probably hundreds, and possibly thousands of cycles are required. Classic strength (resistance) training does little to enhance fatigue resistance. There is no way to measure bone fatigue with bone in situ.

A relatively few cycles of force on an approximately daily basis at the level to which the muscle or bone is usually subjected appear adequate to maintain its strength. However, if the level is less than usual, then muscle, followed by bone, will reduce to the strength needed for the reduced level. This is illustrated in Fig. 7.3. In this figure, C_0 is the fixed minimum level of strength. This fixed reserve is apparently maintained under control of DNA. This lowest level varies from muscle to muscle and bone to bone and with the site in the bone, for example as in Table 6.1 in the previous chapter.

This is the fixed portion of strength, the lower limit labeled C_0. There is also an upper absolute maximum limit, which is set by the individual's DNA not shown here. These maximum and minimum limits are individually variable. Our concern is the usual level shown by A_1, which varies with an individual's usual maximum force load. If all loads on muscle, and subsequently bone, are removed, the time course of change is that described by X in the given equation. In the cases of spinal cord injury (SCI) discussed in Chap. 6, without normal muscle activity virtually all bone strength is lost except that maintained by DNA, as shown by X_1.

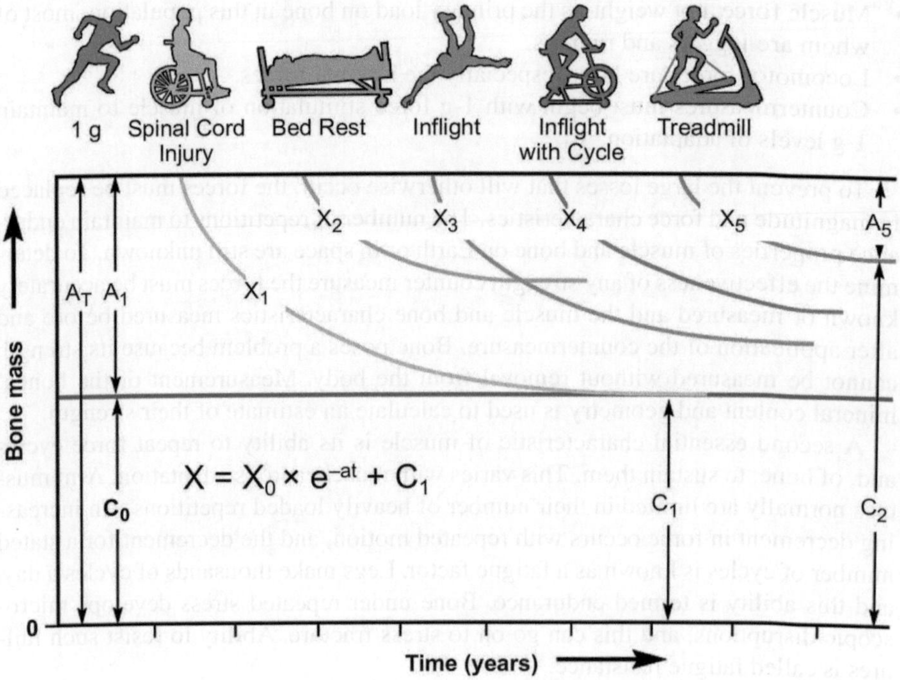

Fig. 7.3 Profiles of bone mineral loss in various force environments. X, loss as an exponential function of time, t, is estimated by the equation whose constants are derived from measurement. C is the amount of bone remaining at the end of the process, which can last six-plus years as in this case. The rate of loss, a, varies with the bone and the maximum stresses on it. A_T is the total amount of bone at the beginning of the process and in 1 g is determined by the usual maximum forces on the bone. C_1 is bone remaining, the limit of maximum loss, when all forces are removed as in SCI. This limit depends on minimum forces in the environment. Note: Curves X_1–X_5 have been distorted, time has been compressed, to show the estimated changes in loss in leg bones in weightlessness. Although cycle forces are small they will reduce the total mass lost. Current treadmill force loads are inadequate and still allow large losses. Unfortunately these must be estimates for loss data were inaccessible and constants could not be derived. Muscle losses follow this profile, but the time constant is much shorter, approximately one-third of that bone loss

If instead the person is placed at bed rest, leg muscles are used very much less than usual but still are inadvertently used. Only a few such motions per day will maintain the minimum level shown by X_2. In weightlessness legs are moved with less force than in bed rest, but there is a steady small level of nervous impulses that maintain a reasonably constant weak contraction, or muscle tone, that stimulates bone. Add the forces of a cycle ergometer on a regular basis and a new strength level is present, X_4. If a treadmill with simulated g-forces is added such that maximum foot ground forces (FGF) are one half of those in 1 g, then less muscle strength is lost, X_5.

The goal of inflight countermeasures to date appears to be to add loads such that no muscle or bone is lost. However, complete and exact replacement will require 1

g. Conversely the losses can be reduced to acceptable values. Strength is only one aspect of countermeasures. A different but also necessary countermeasure is the repetition of force as indicated by the number of steps taken per day.

7.2.1 Countermeasure Forces and Force/Motion Relations

An understanding of forces in a countermeasure is needed to understand their effects. It is a dictum of physics that all forces (e.g., electrical, mechanical) have the same effect. Conversely, the profile and magnitude of these forces vary with their motion. This is illustrated by the force and displacement curves in Fig. 7.4.

Gravity is an intrusive force in every particle of mass that attracts every other particle (see Chap. 1), and any unit of mass near the Earth is mutually attracted to it by a constant force of 1 g. This is the primal constant force to which our ancestral fishes' bodies had to adapt on moving to dry land. Weight is not the only or largest force that determines the strength of our muscles and bones because a second intrinsic force is generated by every particle of mass whenever it is moved. This inertial force is indistinguishable from gravity,[10] but its variation with motion produces very different effects in the human body than those caused by the constant force of gravity. Weight on Earth is always combined with inertia, and muscle and bone are adapted to this combination.

Elastic forces are not unusual in the environment and are frequently met in and essential to a human body, as in the muscle, joints, tendons and intervertebral discs. As a spring or elastic element is compressed or extended it absorbs energy, and an increased force is exerted by it. Figure 7.5 illustrates the need for this. A small force spike is present on heel strike.[11] It would be very much larger, and large enough to stress bone, if it was not partially absorbed by the elastic tissue in the leg. Another example of elasticity is seen in the modern leg prostheses in where an S-shaped steel spring replaces the formerly rigid peg legs, with functional benefits.

Mechanical resistance is the remaining member of this quartet of passive forces (Fig. 7.4d). Its characteristics are very different from the other three, which store (conserve) the energy produced when work is produced by motion. In gravity when a weight is lifted from a lower to higher level, Energy = Force × Distance, the energy is stored and will be released on impact if the weight is suddenly dropped to its previous level. When a mass is accelerated it acquires inertial energy, which may be suddenly released, as too often happens in vehicle crashes. When an elastic element is compressed or stretched, X, energy is stored, as in the case of a drawn bow. But when work is done moving against a resistance, the energy is simultaneously trans-

[10] This is Einstein's "equivalency" of gravity and inertia, an integral and essential part of his general theory of relativity.

[11] Heels are usually, but not always, the first foot contact on falling back to ground at the end of a step when foot velocity is greatest. Other portions of the foot may make first contact in some people.

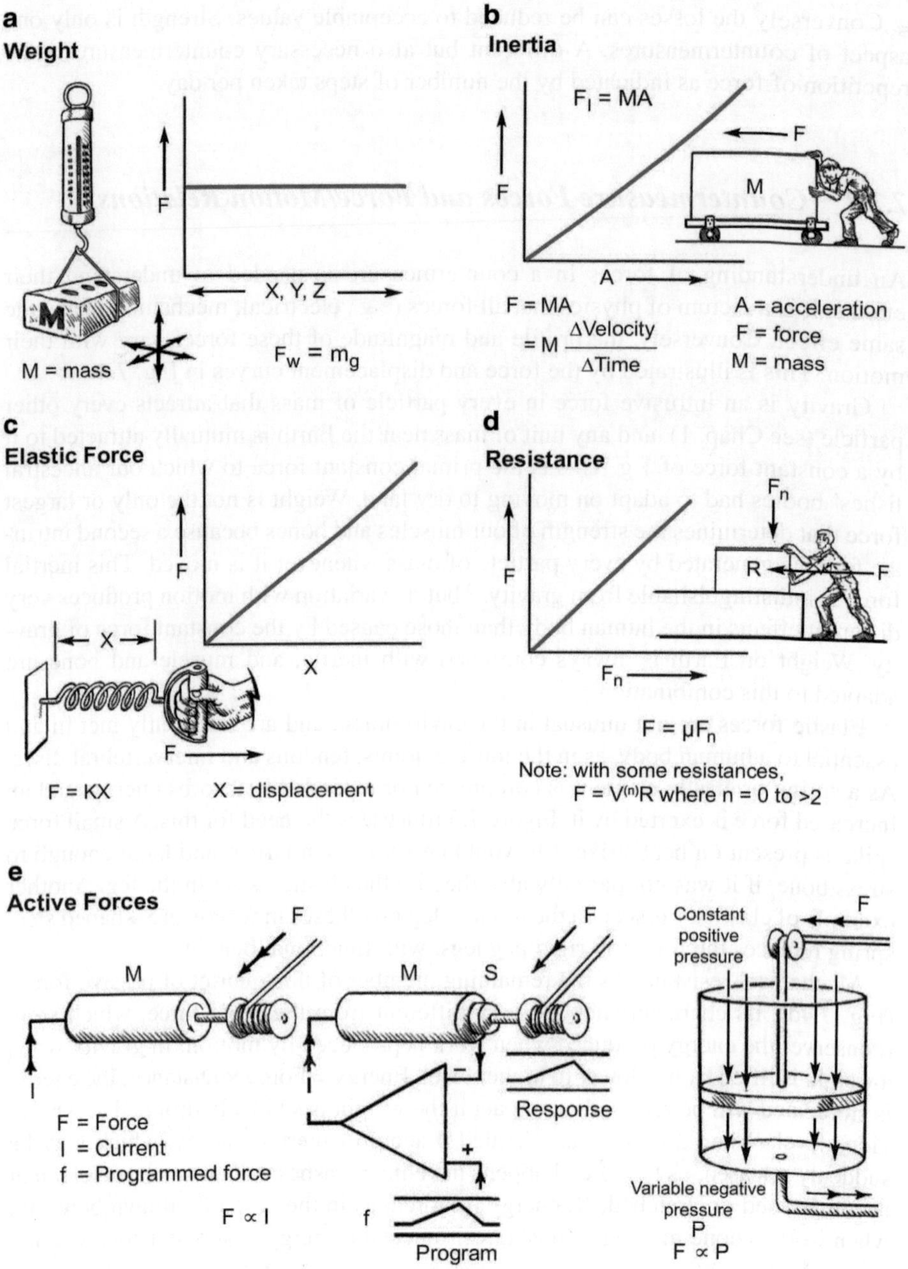

Fig. 7.4 Forces commonly encountered and their variation with displacement. (**a**) Weight is constant for purposes here and unaffected by change of position in any direction. (**b**) While the forces of weight and inertia are equivalent in the general theory of relativity, their behavior with motion is different. Since inertial force is a function of acceleration, to obtain a constant inertial force, velocity must increase as the square of time. (**c**) Elastic force varies directly with displacement.

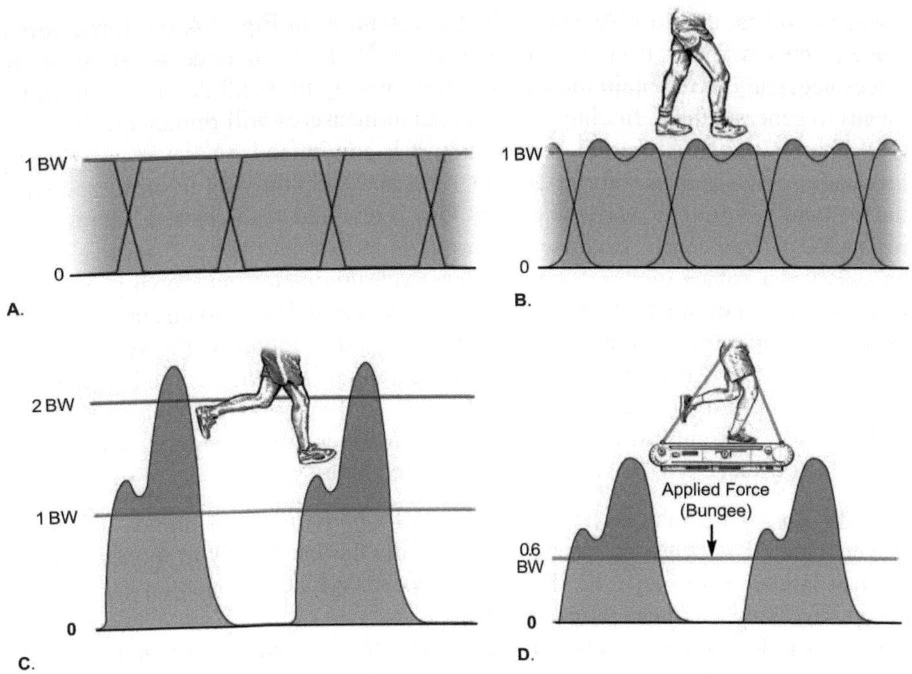

Fig. 7.5 Waveforms of foot to ground forces. *Red* is the level of subjects' body weight. *Blue* is weight component, and orange is inertial force. (**a**) Theoretical waveform of a human walking in 1 g without inertia. (**b**) Idealized waveforms of subject in 1 g walking very slowly show each leg's FGF. (**c**) Recorded waveforms of a man jogging in 1 g. (**d**) Recorded waveforms of man jogging on treadmill inflight with simulated body weight of 0.6 g

Fig. 7.4 (continued) This can be significant when attempting to generate a constant force, such as weight on a treadmill. A common way to minimize this effect is to lengthen the elastic element relative to the amount it is stretched. (**d**) Friction is shown as an example of mechanical resistance. It is often used in exercise devices for expediency. The starting transient or stiction is not shown. There are a great range of mechanical resistances available and dashpots, cylinders with a piston in fluid, or Faraday disc, a conducting disc in a magnetic field, produce reliable, variable forces, which vary with velocity. Mechanical resistance is also an intrinsic component of muscle. (**e**) Active Forces can be produced in an endless variety to generate a force of any desired characteristic. ARED uses a simple but large and heavy cylinder with a piston under a constant differential pressure to produce a constant force to simulate weight. The negative pressure is easily varied to produce a range of simulated weights up to several hundred pounds. A much more efficient alternative is a motor, M, whose force and velocity are controlled by current and voltage, as depicted by the central sketch. By adding electronic control, such as a microprocessor as conceptually depicted, with sensors, S, in appropriate feedback loops, virtually any force level and characteristic can be generated. The large cylinders and their pumps plus heavy inertial fly wheels needed to simulate weight forces in ARED can be replaced by a single small motor under electronic control

formed into heat and lost. As shown by the equations in Fig. 7.4, the force versus velocity curves varies from system to system. Without an understanding of the forces necessary to maintain musculoskeletal capacity in weightlessness and of the means to generate them, finding efficient countermeasures will remain elusive.

While resistance is present in the body it is minimized wherever possible by lubricating fluids and smooth surfaces in joints and other moving tissues, especially the tendons, to minimize waste of energy. One is tempted to say there is no place for resistance in "resistance" exercise to strengthen muscle and bone. Conversely, it is necessary in aerobic exercise devices such as cycle ergometers and rowing machines to stimulate and dissipate the body's energy production. It is also cheaper to reproduce than say inertial force, hence resistance is often found as the sole force in exercise devices. "Resistance" is almost universally misunderstood and misused as a description of exercise and devices.

Resistance is complex, often erratic and sometimes misunderstood. Three categories of resistance will be very briefly described here;

Fluid or drag—this may be from movement of a body in gas or liquid—the force generated, F, depends on characteristics of the fluid and body in motion, a constant labeled R in Fig. 7.4d. Force also depends on the exponential value, n, of velocity, V, where n can vary from 0 to 2+.

Friction—F, is a force produced by movement of one surface over another when they are pressed together by a normal force labeled Fn. Figure 7.4d F is dependent Fn and nature of the surfaces A large force transient called stiction, unlabelled here, must be overcome to start the movement.

Electromagnetic force—is generated on a sheet or disc when it is moved through a magnetic field. Such forces are simple to generate with permanent magnets.

Active forces, provided by electric motors or other power sources, can have almost any characteristic desired. When placed in a feedback loop under computer control they can exactly reproduce any combination of the four passive forces described (see Fig. 7.4e). There are few pure forces in nature because natural forces are usually combinations of the basic forces. Some illustrations of combined forces and their biophysical effects follow.

We have seen how the body increases its locomotion speed by leg forces that exceed body weight (BW). The legs can accelerate the body's mass until the vertical component of the inertial force exceeds body weight raising it clear of the ground for short repetitive weightless parabolic flights. This occurs with each step in jogging and running. This phenomenon can be appreciated from foot ground force diagrams (Fig. 7.5). Imagine locomotion without inertia. Energy can no longer be stored and released to add to or subtract from the constant weight of the body, and the result will be a walk with FGF that cannot rise above nor fall below body weight (Fig. 7.5a). The result will be a series of trapezoids as the feet alternately support the body, which now cannot be thrown clear of FGF contact. The only way speed can be increased is by limited increases in stride length and rate, limited since ground contact must be maintained, and by increased step rates. However, body weight must always be supported and there is a constant total FGF of one BW.

In reality FGFs are very different. In walking BW is always supported by contact with the ground, but FGF is increased as the body's center of mass is accelerated upward, and the inertial force is added (Fig. 7.5b). FGF falls below BW when the center of mass falls downward. However, the major component of FGF in walking is weight.

But Jogging in 1 g is usually the activity that controls leg strength levels in astronauts, and this is clearly determined by inertial forces as shown in orange in Figs. 7.2 and 7.5c.

Forces on other parts of the body are also affected by the combination of forces they meet. Since all mass has inertia, it also plays a role in the lifting and moving of weights in 1 g, for example, the interplay of forces in an arm curl. All muscle forces vary with length of contraction (Fig. 5.6, previous chapter). Geometry of a simple hinge joint, as in the elbow, would accentuate this by further reducing forces near extremes of muscle contraction. This is partially overcome by arrangement of the several arm muscles such that in full arm extension there is considerable force available, but above the midpoint of motion available forces rapidly decrease. To make a complete curl, before available force is decreased, velocity of motion is increased storing energy in the moving weight which "carries it through" the region of decreasing force. Absence of inertial force in IRED insured its failure as a strength exercise.

This involves a subject that has been avoided, the nervous system, which controls all normal muscle activity. To coordinate and synchronize muscle activity such as locomotion requires large areas of the brain and spinal cord with response from force sensors. Repeated activities, especially complex actions such as locomotion are retained and stored in unconscious memory programs. One of the first such programs we acquire and that lasts a lifetime is learning to walk and later to cycle. Even something as simple as increasing strength in exercise elicits reprogramming. The rapid strength gain seen at the beginning of a new exercise program is produced by learning processes in the nervous system. Changes in the muscles themselves come later. The importance of this is that countermeasures that use differing forces and motions will cause reprogramming and replace usual patterns, with postflight consequences. A common example of this can be seen in the gait of sailors after long voyages.

A glimpse of these forces in action and on Earth has been given. The magnitude of their losses in weightlessness can be seen in Fig. 6.23 in the previous chapter. To counter such losses a variety of means and devices or countermeasures have been devised and used. Significant examples of flown devices follow with brief descriptions of them. In the following sections where possible, the use and effects of their usage in bed rest and spaceflight are given.

7.2.2 Countermeasure Hardware

These are the means by which forces encountered on in 1 g are simulated in weightlessness. A full description of all the proposed exercise hardware for spaceflight would be a large catalog, and that of flown hardware would be a book. The following list is a severe distillation of flown hardware plus an elemental biophysical description of each.

Hardware had evolved into two functional categories

- Arms, head, neck and upper body
- Legs and lower body

Although the principles of forces and their applications are the same for arms and legs, upper body and arm motions are complex and generate low to moderate force with relatively few repetitions. In contrast leg, motions are typically simple, powerful and repetitive. Exercise devices accommodate these differences; e.g., treadmills, but they frequently are used in both regions; e.g., the ARED. Their division here is arbitrary. The devices are shown approximately in order of their evolution and use in the U.S. space program.

7.2.3 Upper Body Devices

Figure 7.6 illustrates NASA Devices.

(a) A commercial capstan resistance device is shown. It can be adjusted to provide extensive range of forces and velocities within a variety of exercises. Resistance forces produce rapid fatigue and are unpleasant to use but can be partially effective with dedicated use.
(b) Spring device with a moderate range of force that increases with extension. It is easy to use but ineffective without skills in its use. It is useful for arms only.
(c) A resistance only device with a maximum velocity, which could be set to a range of fixed velocities thus mimicking isokinetic motion [5], was an ad hoc addition to SL-3 and SL-4 by an Astronaut Office DSO. It was capable of producing large forces.
(d) The Interim Resistance Exercise Device (IRED) [16–19] was a contractor project that was unfortunate in concept, design, execution and use on ISS. It was designed to be a strength exercise using elastic force only. This was a fundamental flaw because simulation of weight requires both a constant and inertial force.

    IRED consisted of a vertical series of rims with elastic spokes connected to hubs connected by a splined shaft to a pulley and cable to provide increments of force equivalent up to 300 lb using a handle or bar. Forces were set by adjusting the starting position of wheels to apply initial tension on the spokes and hubs, but forces varied with extension of cable. There were numerous failures and

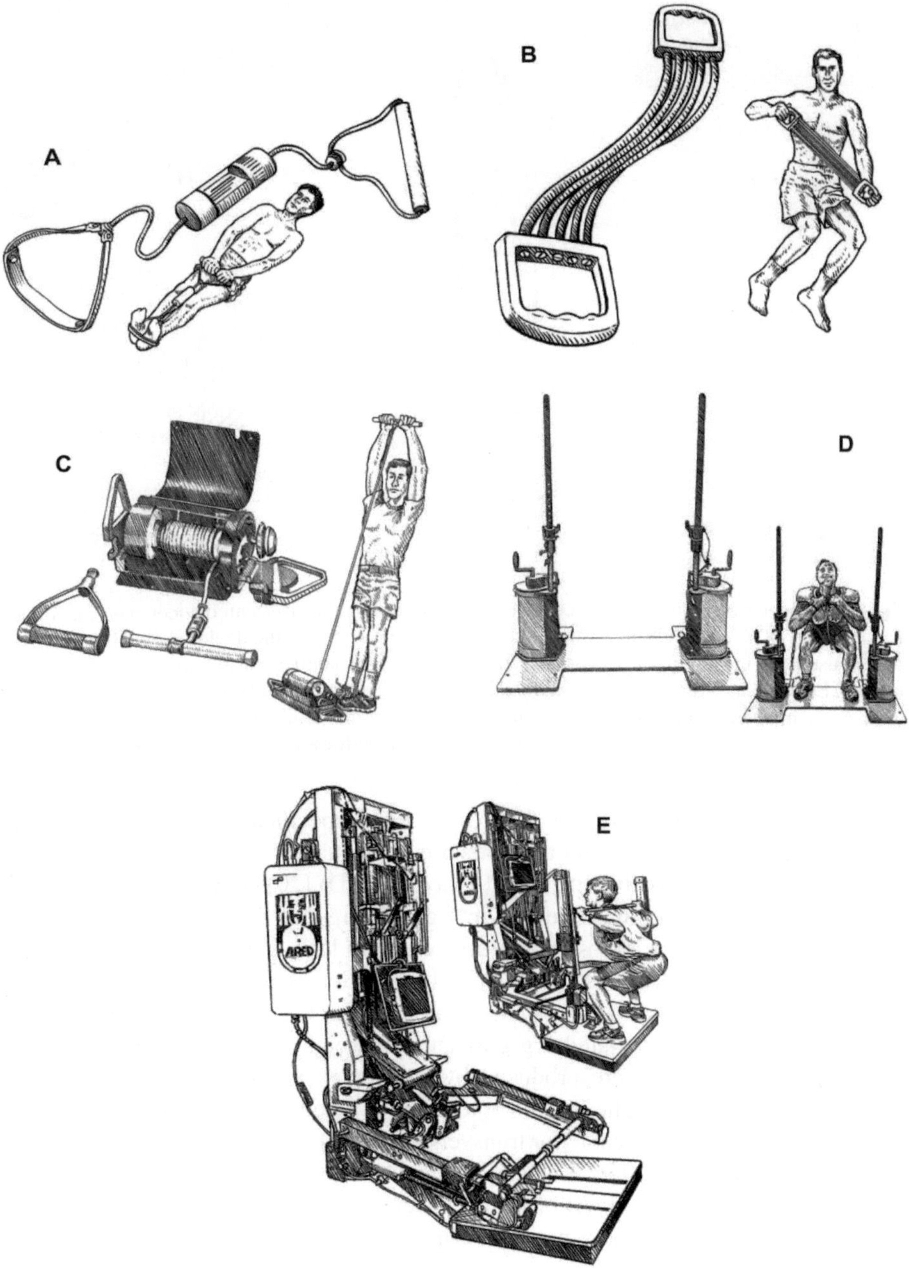

Fig. 7.6 Upper body NASA exercise devices. (**a**) Exergenie® from Apollo era. (**b**) Modified commercial device used on SL-3 and SL-4. (**c**) Exergym, a resistance device with centrifugal brake to approximate isokinetic exercise. Used effectively on SL-3 and SL-4. (**d**) IRED, a resistance only strength exercise with a variety of faults in idea and practice. Used on early ISS expeditions until it was replaced by ARED. (**e**) The Advanced Resistive Exercise Device (ARED)

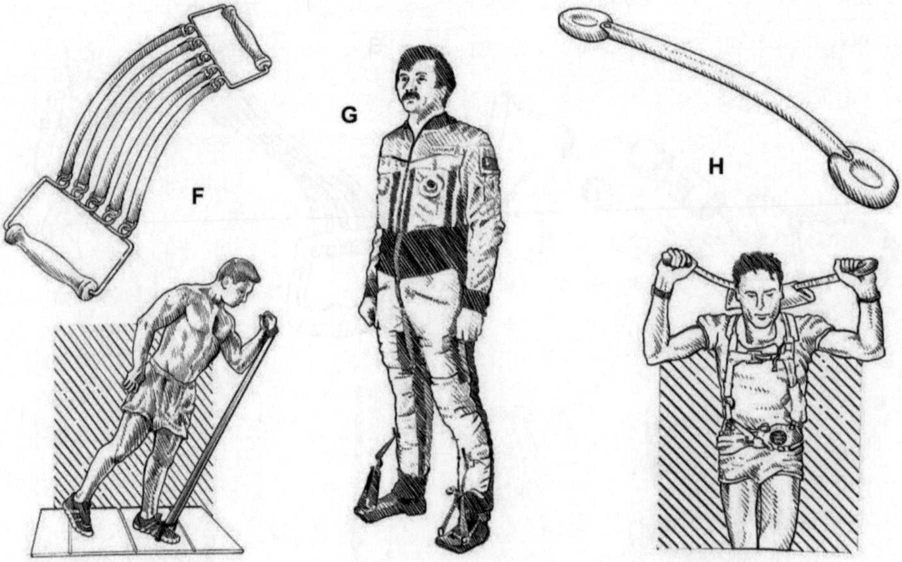

Fig. 7.7 Upper Body Russian Exercise Devices. There are other Russian devices mentioned but not shown. (**f**) This set of elastic springs with a handle was a copy of the device shown in Fig. 7.6a. It was used for more than a decade. (**g**) The Penguin Suit applies elastic stresses that oppose motion. Elastic bands that oppose plantar flexion are attached to toes of shoes. This "suit" has been worn by cosmonauts from early days of the Russian program and is in routine current use today. (**h**) It appears that elastic rods are used in the bending mode for a variety of exercises

performance problems, and failures to meet specifications in 1-g test and inflight. Elastic forces without inertia are inadequate.

(e) The Advanced Resistive Exercise Device (ARED) [20] combines constant and inertial forces to simulate weights in 1 g. See Fig. 7.4. It uses controlled negative pressure versus atmosphere pressure in two large cylinders with pistons to maintain a relatively constant reciprocal force. See Fig. 7.4e. This force is coupled by a gear train to a rotating moment of inertia adding an inertial mass equivalent to the simulated 1 g gravitational weight produced by the pistons. Variations in weight are produced by a complex arrangement of arms with adjustable leverage. The levered forces are coupled to a vertical bar for axial subject loads and to a cable for transverse loads. Maximum bar motion is 30 in. Maximum load is 600 lb. A cable and handle allows 76 in extension maximum and up to 250 lb load. Allowed displacement and loads are interdependent Details of this device have not been available.

Figure 7.7 illustrates Russian upper body devices.

(f) Russia used versions of item (b) in Fig. 7.6 until recently.

(g) Their Penguin suit contains elastic elements that exert reasonably constant longitudinal loads on the body and across the joints when they are moved away from their neutral position. This is an effort to simulate loads in 1 g, but the forces are limited to 50% body weight, too small to be effective. In spite of this, the suits have been worn from early days of the program until the present. In 2007 this was combined with a complex measurement system with 20 "tension dynamometers" to record, store, process and transmit data. This combination is called Penguin 3 but is not shown.

(h) Russia has some described exercise programs such as this one, which uses some form of elastic cords or bungees in an exercise mode. It also has an electric muscle stimulator "Tonus 3" on ISS that is not shown.

Figure 7.8 illustrates NASA Lower Body Exercise countermeasures.

(a) Although the leg forces generated by bicycle pedaling are negligible, compared to FGFs, it is still regularly used for cardiorespiratory conditioning. As an endurance exercise it can produce near maximum O_2 uptakes thus providing a means of stress testing. It was the only planned exercise on Skylab. The unit flown provided a range of constant loads from 25 to 300 W over pedal speeds of 50 to 100+ RPM. Its moment of inertia was adequate, and its loading was electromagnetic and smooth. It was recently discovered that its use was limited to 260 W on Skylab after two failures at its design rating of 300 W in SMEAT.

(b) The first locomotor exercise was flown on Skylab 4 [6]. It simulated subject weight with bungees attached between the harness and deck. Unfortunately, this obviously inadequate loading arrangement has been only modestly improved and a major cause of inadequate leg force loading on subsequent treadmills. It is claimed that recent improvements allow 1 BW loading for many users. The ground reaction force is from the resistance of feet slipping over a sheet of Teflon. The resultant is a force vector aligned with the body's long axis tilted to the treadmill's normal. Skylab bungee loads were made for a minimum of 80 kg BW. FGFs are unknown for the 70 kg crew on SL-4. Two of the crew also jumped or ran in place under the bungee loads adding the missing inertial force. This arrangement was investigated as a contingency for ISS [21].

(c) A treadmill that allowed normal walking and jogging was developed by TW using the same bungee harness arrangement as on SL-4.[12] Track inertia was provided by gearing to a small high speed flywheel. An adjustable brake controlled tread speeds. In a passive, subject driven treadmill there must be a force parallel to the tread to overcome the friction of the mechanism. In 1 g this is done by tilting the treadmill some four to five degrees relative to the vertical in 1 g. In weightlessness it amounted to "leaning into" the bungees (Fig. 7.8c).

(d) NASA replaced the STS treadmill with a series of cycle ergometers. A European built unit is now part of the US equipment ensemble [22]. This Cycle Ergometer with Vibration Isolation System (CEVIS) consists of a box with pedals, a control box and in weightlessness, some form of restraint system (Fig. 7.8e). Published specifications are: constant workload over range of pedal speeds of 50–120 RPM from 25 to 350 W adjustable in 1 W increments.

---

[12] Henry Whitmore of Whitmore Enterprise, San Antonio, Texas.

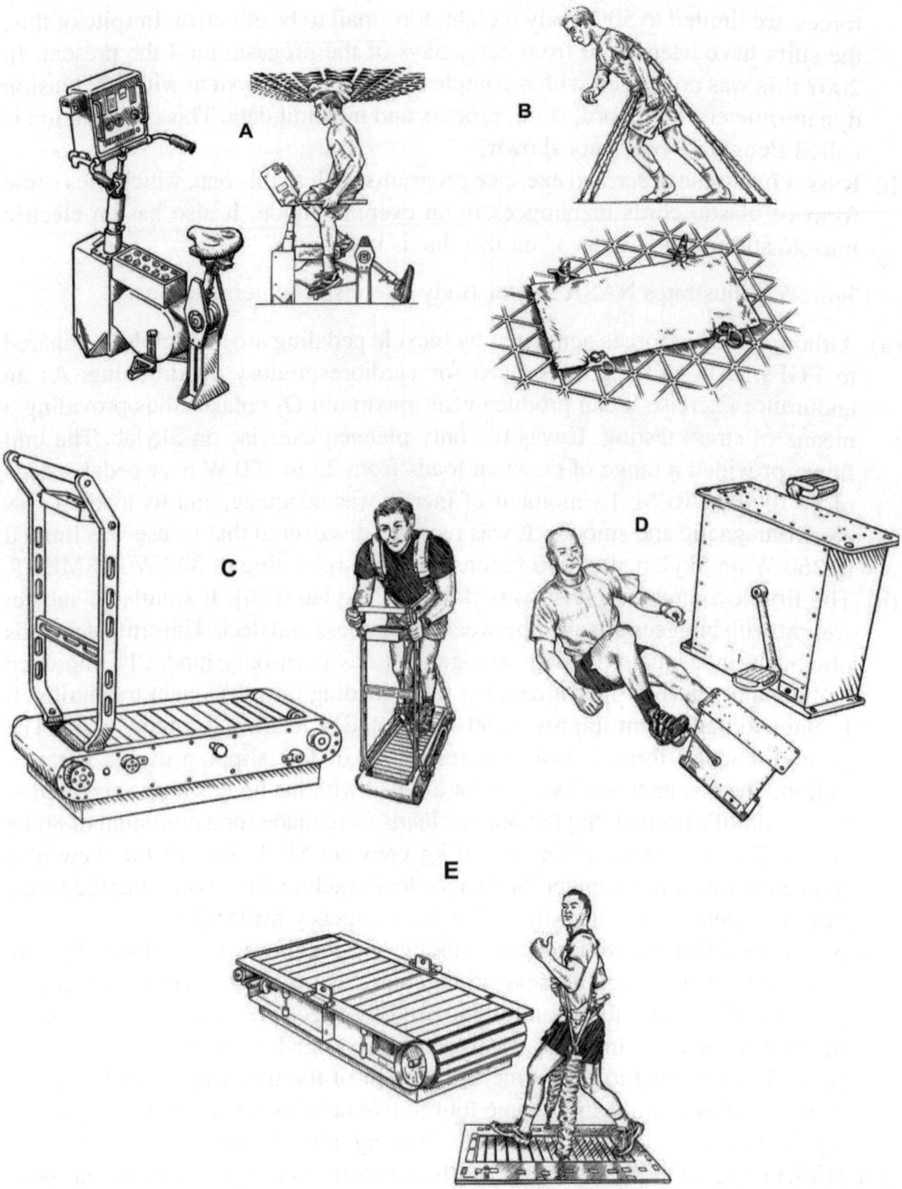

Fig. 7.8 NASA Lower Body Exercise Devices. (**a**) Skylab's bicycle ergometer was a simulation of conventional 1 g bicycles with adjustable load to a rated 300 W. Inflight the harness and seat were abandoned, and the small pedal forces were countered by holding the handle bars and sometimes with aid of head and neck against the overhead deck as shown. (**b**) An approximation of locomotion was provided for SL-4 by adding bungees to the discarded bicycle constraint harness and a Teflon foot plate. Like all friction devices it was so unpleasent to use that it was dubbed "Thornton's revenge." (**c**) Whitmore's all aluminum Shuttle treadmill was a remarkable mechanical achievement that could be assembled without tools and disassembled for locker stowage.

Fig. 7.9 Russian lower body exercise devices. (**f**) Russia added a passive treadmill to Salyut, and its use has continued with upgrades including a motor drive, which is currently used on ISS. (**g**) Russia's cycle ergometer on ISS is designated Veloergometer (VB-3). (**h**) In addition the Russians have a motorized treadmill mounted on the vibration isolation system

(e) During the Shuttle program JSC developed a series of treadmills, but modified commercial units were used in TVIS and COLBERT.

Figure 7.9 shows Russian lower body devices.

(f) Russia's treadmill began life as a passive unit. Its maximum BW loading of 60 kg is provided by bungees through pulleys and cables to a single point on each side of a harness. This treadmill has been upgraded over the decades into a passive or motor driven unit.

Fig. 7.8 (continued) Speeds of 8 miles per hour at a minimum equivalent grade of 5–6% were possible. (**d**) A variety of NASA cycle ergometers were flown. The current unit is a European unit designed for ESA. A cycle's small counter forces allow minimum constraint. (**e**) NASA treadmills evolved through Shuttle to a currently used commercial unit mounted on a vibration isolation device (not shown) and designated TVIS. Harness and loads (bungees shown) to simulate body weight remain inadequate. This was replaced by a larger modified device device mounted on friction damped springs and designated COLBERT and is in current use

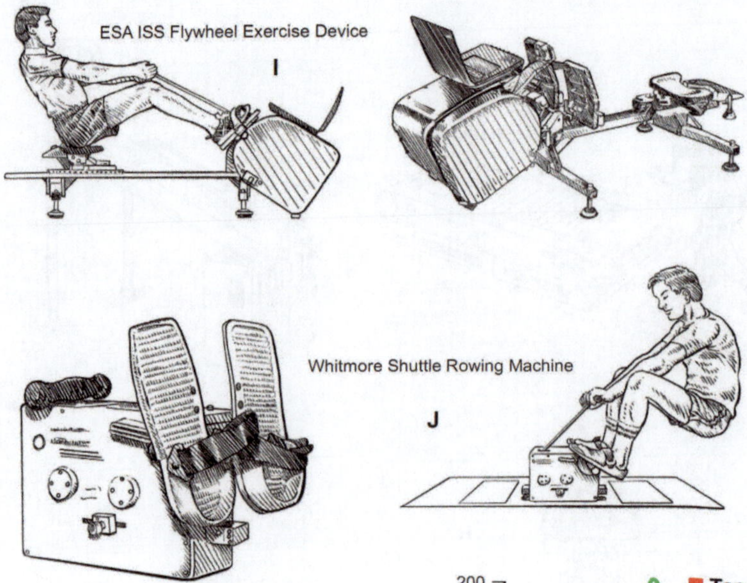

Size	10 × 12 × 24″ with foot supports attached
Weight	26.5 lbs (Shuttle locker stowable)
Load inertia (slugs)	>10
Stroke	10 ft maximum
Stroke rate	0 to 30+ min^{-1}
Operational envelope	Approx. 3.5 × 4 × 5 ft
Power, mean	300 watts continuous 400 watts – 15 minutes
Maximum load force	450 lbs
Noise level	<65 dBA at 3 ft

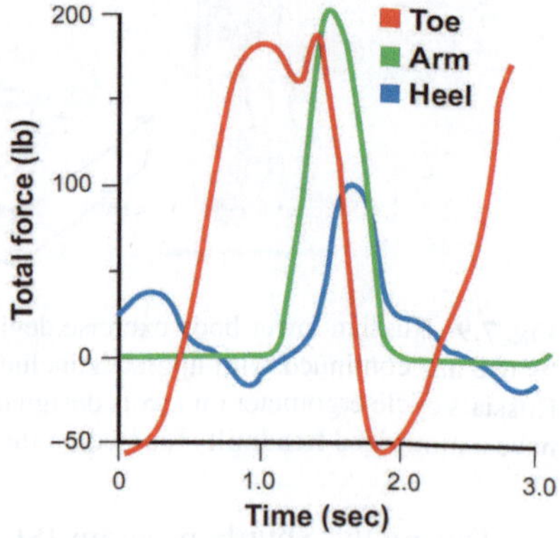

Fig. 7.10 (**i**) ESA ISS Flywheel Exercise Device (FWED) and (**j**) Whitmore Shuttle Rowing Machine

(g) Russia has a cycle ergometer (Veloergometer) with specification of a 100–250 W load adjustable in 25 W increments and constant over 40–120 RPM. It was designed to be compatible and comparable to the ESA cycle ergometer and results are used interchangeably [18].

(h) In 2013 the Russian treadmill was replaced with an upgraded unit apparently equivalent to the US COLBERT mounted on a much simpler vibration isolation unit. At the time of this writing, technical information on these units was not available to the author [23]. It is reported to have foot to ground force reaction measurement [24].

Two additional countermeasure devices that can be used for whole body exercise are shown in Fig. 7.10.

(i) European Space Agency (ESA) ISS Flywheel Exercise Device (FWED) [25, 26]. A simple large flywheel with an axle and cord around it can provide a large force according to the rate at which the cord is pulled. To maintain a constant force, the cord handle's displacement must be increased exponentially. Large amounts of energy can be stored during displacement of the cord. When it reaches its limit, it will immediately reverse direction and rewind at the maximum rate imparted unless restraining force is applied. In a simple elbow flexion each repetition of necessity requires both concentric and eccentric force. Depending on the arrangement of the apparatus, including safety features, such devices can cause injury. A dead man foot brake is incorporated in this model. The illustration is with the device set up for operation in 1 g. It has been on ISS since 2000.

(j) TW Shuttle Rowing Machine and force profiles. Whitmore's rowing machine was flown a number of times on Shuttle, but objective records of its use were not kept. Good simulation of actual rowing with its large viscous resistance $F_R = R_X^2$ plus inertia $F_I = M \cdot A$ is relatively simple to simulate using a Faraday disc and flywheel. In weightlessness support for BW is not required, and a small efficient package is possible.

Rowing provides large back and arm forces, plus leg forces, greater than a cycle ergometer. Its aerobic demands are large. It is a more efficient exercise than cycling that offers aerobic benefit only.

7.2.4 *Countermeasures in Bed Rest and 1 g*

Bed rest validation of equipment and procedures to be used as countermeasures in space would be an expected part of their development cycle. However, several problems are to be expected in this phase of development. Use of hardware in bed rest, such as a treadmill, is difficult to do without violating the basic integrity of bed rest. An uncounted number of bed rest studies with countermeasures have been done and a large number of them have been published. There was such diversity in protocols and counter measures that collation of results had doubtful validity. Three references are listed to illustrate the diversity [27–29]. There is also still a question of how well bed rest replicates the musculoskeletal effects of weightlessness.

7.2.5 *Countermeasures in Bed Rest*

After weeks of searching, one paper only [30] could be found that described a countermeasure in bed rest that was subsequently used in spaceflight.[13] Nine men and five women used a countermeasure in a 17-week bed rest study while 18 control

[13] Several bed rest studies used the FWED but always in combination with other exercise such that

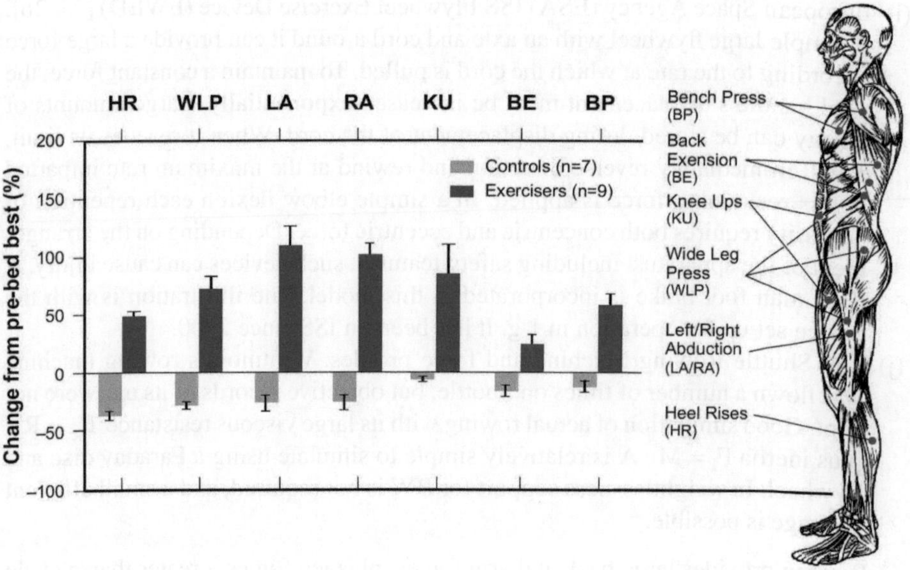

Fig. 7.11 Change in muscle strength after 17 weeks of bed rest in groups without exercise and with exercises indicated

subjects undergoing bed rest performed with no exercise. A horizontal mechanism allowed the experimental subjects to perform prescribed 1 g equivalent strength exercise with weights, while prone or supine, 6 days per week. Three days were allocated to upper body and 3 days to lower body exercise. The protocol used progressive increases in load, equivalent to the De Lorme protocol, and the results were similar to the illustration in Fig. 6.16 (previous chapter) with an untrained subject beginning weight exercise [31].

Typical weight loads were 72% of 1 RM except for 20 heel raises, which were 58% of 1 RM.[14] After 17 weeks bed rest the difference in isotonic strength (IRM) between the exercisers and controls for percentage change for various exercises from before bed rest are shown in Fig. 7.11. These results are remarkable in that large gains of strength were made while minimally exposed to BW forces. Also note the contrast to the losses of non-exercisers. They will be compared to the effects of countermeasures in spaceflight in the following section.

Changes in bone density are shown in Fig. 7.12. Although the strength-only exercises produced a remarkable reduction in loss of bone density, it was not complete in certain critical areas. This indicates that the stresses imposed on these bones in 1 g, presumably locomotor stresses, are not completely replaced by the exercise coun-

its effects were contaminated.

[14] One RM max is a single maximum effort action.

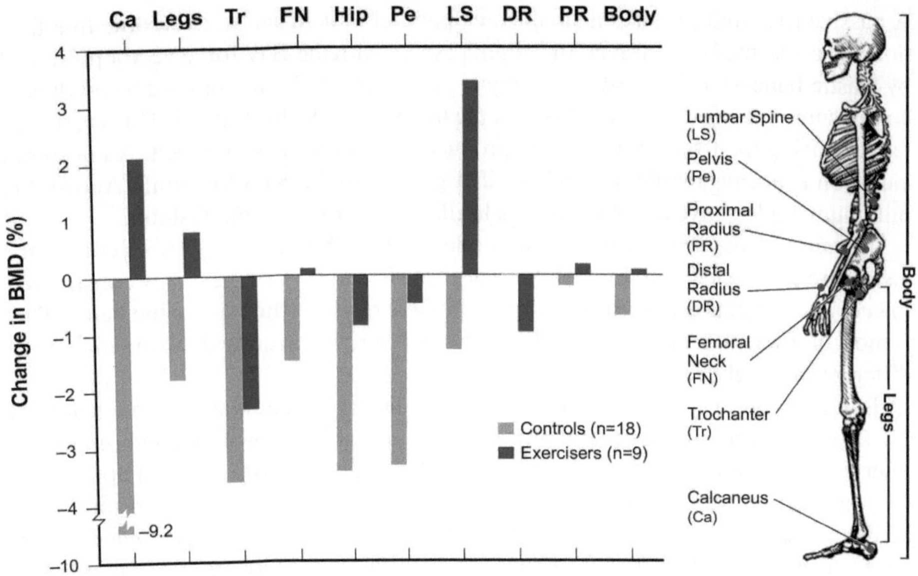

Fig. 7.12 Percentage change in bone mineral density after 17 weeks of bed rest

termeasures used here. Conversely it indicates that while a countermeasure may not totally prevent loss it can still reduce the loss that would have otherwise occurred.

7.2.6 Countermeasure Device Evaluation in 1 g

Usually there is some urgency in acquiring an operational countermeasure, and often it will be flown before bed rest studies can be conducted. An example of actual events in preflight testing of countermeasures is seen in the history of IRED [16]. An extensive paper on ARED preflight evaluation [20] states that the decision to deploy IRED, antecedent of ARED, was based on bed rest, probably on the study just described. There was an extensive 1-g testing of IRED including bed rest [16–18], and several deficiencies and fatigue failure were found, and modifications made. Its ARED replacement was also flown after extensive 1-g testing but without bed rest studies. The initial ISS treadmill copied features of a prototype by Whitmore Enterprises, which had been exhaustively tested, including life testing by human subjects. A contractor's emulation of it made material modification, and it was flown without life testing the modified treads. The treads of the first JSC treadmill quickly failed inflight, leaving ISS without locomotor exercise for four expeditions [11].

An example of the need for a sound understanding of the forces in weight versus those in weightlessness was seen in an erroneous evaluation of effects of an ISS treadmill on a subject in subject driven (passive) versus motor driven (active) modes [32]. In an active treadmill internal mechanical resistance is overcome by the motor.

A passive treadmill requires a small force parallel to its tread to overcome this friction and cause the belt to travel. In weightlessness with the BW force vector produced by elastic bungees (Fig. 7.8d), the subject automatically leans forward to produce a component of the force needed to drive the tread at the desired speed. This is equivalent to raising the front of it in 1 g to produce a vector to drive the belt. A minimum elevation is always required to do this in 1 g, 5–6° in the STS treadmill. Above this minimum angle active and passive treadmills behave in the same fashion.

To compare the active and passive modes in the ISS treadmill, NASA contractor tests were conducted with the treadmill at zero elevation. A harness was constructed to keep the subjects on the treadmill in its passive mode as they drove the belt.[15] This is most inefficient, and consequently the modes were incorrectly declared to be very different in weightlessness.

Before flying the TW STS treadmill it was tested for equivalence to a standard 1-g active treadmill using several physically trained subjects walking and running at a series of measured speeds and grades. Treadmill speeds, EKG, oxygen uptake and CO_2 production were measured. The STS treadmill was then placed on top of the active treadmill and the series repeated. There were only a few percentage differences in any parameter. There is essentially no difference in locomotion on properly designed and used active versus passive treadmills.

7.3 Counter Measures in Spaceflight

7.3.1 Procedures and Problems

Most of the bed rest studies were conducted in accord with accepted scientific practice. Two matched populations are put in identical conditions of bed rest. One group (experimental) uses procedures or devices to be tested and results are compared to the base line group that did not use the procedure (control). There are too few people in space at a given time to conduct such controlled studies. Alternatively, a baseline group without the countermeasure can be compared with similar times and conditions inflight but over multiple flights. This was attempted in the Extended Duration Orbiter study of treadmill effects on muscle performance over a 4-year period [8].

But to be meaningful the countermeasure parameters must be known. In the case of a treadmill this is FGF profile and number of repetitions. Less rigorous surrogates for this would be treadmill speed, duration, body mass and in-flight force loads used to simulate weight, in this case bungee loads. Such data from in-flight usage of the treadmill was not obtained nor was their usage controlled during EDO, Shuttle/Mir and early ISS.

[15] This was equivalent to dragging a chain on a track. This had to produce great differences in the two modes. Conclusions were that effects of the two modes were very different and use of the passive mode questionable.

Although the pre- and postflight studies of strength fatigue and aerobic capacity were conducted, little could be said about countermeasure effects under the circumstances. It was observed that leg muscles were most affected by weightlessness. Early Russian countermeasure studies appear to have had similar problems because their investigators could only conclude that losses appeared to be inversely related to amount and "intensity" of exercise.

Once flight durations were extended to months and beyond, some form of countermeasure was always employed thus making it impossible to establish a baseline for comparison. In addition, difficulties in accurate documentation of countermeasure use continued. A series of studies of bone loss on 3-month flights on Shuttle/Mir, Mir, and early ISS was conducted without real documentation of the countermeasures used. Failure of NASA's IRED and then the treadmills added to the confusion during the first year of ISS. One consistent conclusion was that bone and presumably muscle losses in the limbs, pelvis and spine were large and not prevented by exercise. The exercise was not described in detail or quantitatively but only in qualitative terms such as intense, extensive, etc.

The result has been a transition from a scientific approach to an empirical one. An obvious exception to this has been Cavanagh's extended series of studies, which produced the measurement methodology required for use in space with his small but unique, reasonably complete documentation of use of countermeasures available on ISS [33, 34]. Their measurements documented the low equivalent body weights used on the treadmills. With the availability of a simulated (constant + acceleration force) weight machine provided by ARED and the imposition of its more regular use, the problem of muscle and bone losses in weightlessness is sometimes claimed to have been solved.

To illustrate the problems in prevention of musculoskeletal (MSK) losses, results of some inflight studies follow, beginning with the first documented use of MSK countermeasures in Skylab. Two studies of bone loss in early ISS are still typical. A small properly conducted study of MSK treadmills in weightlessness follows. Two reports of combined countermeasures and a recent summary of the first decade of strength countermeasures on ISS conclude these examples.

7.3.2 Countermeasures in Flight

Skylab with its three crews of three astronauts each produced basic insight into the loss of muscle and bone [6]. On Skylab the only planned exercise device was a bicycle ergometer (Fig. 7.8a). Pedal forces from this are far below the threshold of normal leg muscle strength for 1 g locomotion; i.e., cycling did not significantly affect loss of leg strength muscle.

Losses of strength on SL-2 (Table 6.3 in the previous chapter) were significant in both arms and especially legs, and the crews were calling for increased inflight exercise. An elastic spring extension device and the "isokinetic" friction Exergym (Fig. 7.6b, c) were added to SL-3 and used by the crew in unscheduled time without a record of use. After 8 weeks in flight, postflight arm strength was maintained. Leg extensor strength loss increased only slightly more than on SL-2 possibly indicating that it had reached equilibrium with existing forces, but flexor losses almost doubled.

A form of locomotor exercise was added to SL-4 (Fig. 7.8b) in which elastic bungee forces simulated BW in 1 g.[16] To "walk" on this device the foot has to slip over the Teflon sheet against a large frictional resistance.

The ground reaction forces were never measured, but inflight photos show Jerry Carr bent to an estimated 45° angle to generate the necessary force to slide across the surfaces. Also the bungee appears to be stretched, which would sharply increase the equivalent BW. A reasonable range of FGF in this simulated walking would be 1.5 to 2+ BW. The other two crewmembers used more forceful forms of leg exercise by simply running and jumping in place under the equivalent of 1.1 BW. Depending on the vigor with which this is done it can produce up to three BW per leg. Although details are unknown the results were remarkable—no loss of leg extensor strength in 12 weeks of weightlessness and leg flexor strength reduced to −6%, less than its loss on the initial 4 week mission.

In 1 g as a leg is swung forward clear of the ground for the next step, its weight must be supported by bone and muscle of the opposite leg, but in weightlessness the only forces on that leg are inertial, which may account for at least part of the flexor losses. Given that the same force-loading scheme is used in current treadmills, the same lack of leg weight may account for part of the inadequate FGF and loss of flexor strength seen in current treadmills in weightlessness.

Other relevant products of this study include:

- The losses in arm and leg are rapid and significant;
- Leg losses are much larger than in arms;
- These losses can be prevented by exercise, but it must be specific and of adequate magnitude;
- In arms this is relatively simple and can be done with conventional 1-g strength exercise;
- Exercise devices can be simple but effective;
- Adequate locomotor exercise is more complex and a single force plus friction to replace BW is inadequate;
- Diet may also play a significant role in muscle and bone losses;
- Scheduled times and mandatory, documented versus elective application of countermeasures are necessary.

Although Russia was rapidly extending flights for beyond Skylab's 12 weeks and a variety of countermeasure devices were developed and used, adequate information to objectively determine the essential countermeasure limits and their effects is not available. A series of combined US-Russian studies began with the Mir/Shuttle program when nine astronauts each spent 3 months on Mir. The same lack of objective data on countermeasures and their effects continued and extended into

[16] Bungees were made such that a minimum of 80 kg force versus crew BW 70 kg was applied.

ISS [11]. Russian and NASA, have continued but have produced little objective data on the effectiveness of countermeasures. These studies document the continued large losses of bone in spite of application of exercise countermeasures. However, measurement of input forces, FGF, and bungee loads and speeds on the treadmill remain unreported and apparently unknown. There seems to be general agreement that upper body losses are less than lower and can be prevented by now available strength exercise.

Final magnitudes of the losses with countermeasures are still to be determined. Force magnitude, type, and repetition are the stimuli that determine loss or gain of function in muscle and bone but these remain unmeasured and unknown. Muscle function is production of force and displacement (motion). Bone's mechanical function is simple: to safely carry the combined repetitive forces of muscle and weight. Even if one's interest is solely in research aspects of muscle and bone, the stimulus must be known for results to be meaningful. Measuring only such quantities as time on a treadmill in weightlessness without its speed or body loading is of no quantitative value. It was difficult to measure FGF or arm forces in 1 g and impossible in flight under Skylab's conditions, but it is now simple to directly measure FGF on a treadmill or ARED.

One often-referenced report [35] is on calf muscles in ten subjects in space for 161–192 days. Individual preflight exercise activity was recorded as days per week of resistance exercise plus days per week of swimming, cycling and running. In-flight cycle ergometry is reported as minutes per week with half of the users reporting load in watts. Treadmill use was reported in minutes per week with seven subjects reporting speeds of 2–3 miles per hour and one jogging 5.5 miles per hour. There is no report of equivalent BW loading. Strength exercise did not include a single force. Estimates of repetitions were made but without force.

Mean strength changes in calf plantar flexion a week[17] after landing was a loss of 17% isometric and 24% isokinetic at 180°/s. There was a weak correlation between time on the treadmill and muscle volume but no correlation between muscle strength and volume. If leg bone density had been measured the inadequate loading would also have reduced it. Without knowledge of type and magnitude of leg strength exercise and of equivalent BW forces used on the treadmill, one can only conclude that the combined exercises were inadequate, while it is known from other sources that treadmill loads are inadequate in both NASA and Russian programs.

Another frequently referenced study of bone mineral loss in space entailed 11 subjects on ISS for 4–6 months. There is no mention of exercise beyond "an array of MSK conditioning exercises being present and possibly effective against heel [calcaneal] bone loss." Changes in bone density during this study are listed in Table 7.1. Magnitude of the changes is significant, especially in this critical region. However, without a baseline of change without countermeasures there is no way to estimate effects of the countermeasures used especially because they were not listed. To have any meaning the counter forces must be known and quantified. This

[17] There is a very rapid gain in strength after landing. In this case actual loss was almost certainly larger.

Table 7.1 Total percentage change in bone mineral density in 11 subjects during 4–6 months on ISS

	Femoral		Proximal
	Trochanter	Neck	Femur
Cortical	−2.5	−1.86	−2.0
Trabecular	−12.1	−15.6	−12.1

Fig. 7.13 Foot ground forces running on Earth and in flight

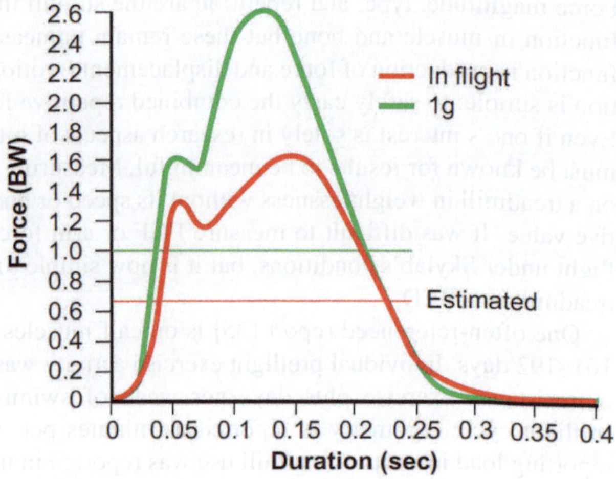

and the previous strength study may be measuring the maximum loss that would occur under the circumstance.

Development of an instrument that records the FGF individual profiles for a day's time was undertaken, tested, and reported [33]. Four astronauts on ISS missions of 181 ± 15 days were recruited to wear the recorder for a sampling of a day's activity preflight and 4 days inflight, plus some endurance diagnostic activities inflight. Strength and volume of arm and leg muscles and bone densities were measured pre- and postflight. Daily records of all exercise were kept [34].

Findings were unique, often definitive, and numerous. For example a quantitative survey of daily physical activities on Earth versus space was recorded for the first time (Fig. 6.23 in previous chapter). This report also produced records that provide a definitive answer why treadmill exercise in weightlessness does not prevent loss of muscle and bone (Fig. 7.13). This figure compares waveforms of running on Earth versus on ISS on the TEVIS treadmill. Comparison of scheduled versus actual exercise times showed average times spent in exercise in Fig. 7.14. Both amplitude and number of FGF cycles on Earth are significantly greater than in flight. Postflight measurements documented the predictable effects of such force deficits on the locomotor system are shown in Figs. 7.15 and 7.16.

One study to date, which shows a striking reduction in the loss of bone mineral in the hip, pelvis, and spine region, deserves particular attention [14]. Its major fea-

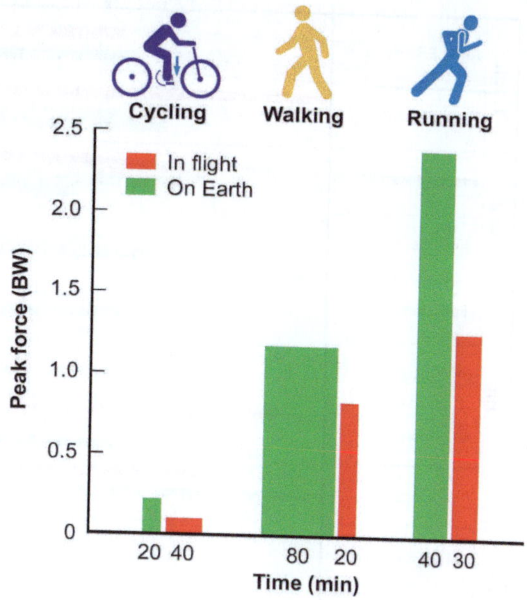

Fig. 7.14 Peak forces (measured) and time per day (estimated) for three leg exercises

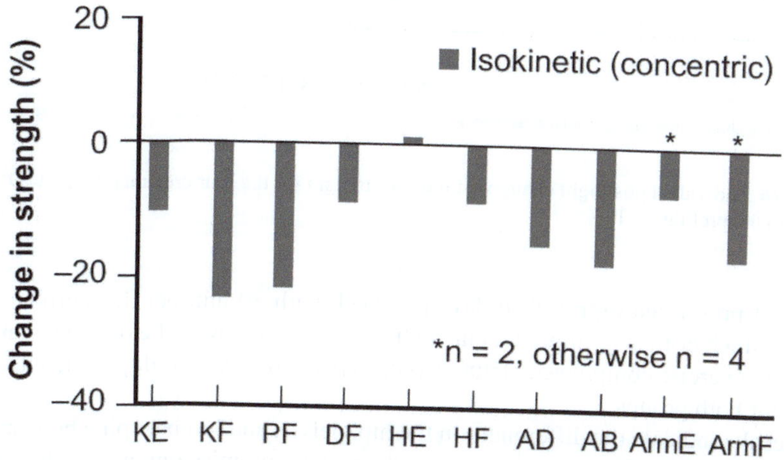

Fig. 7.15 Average postflight changes in the isokinetic (concentric) strength of knee extensors (KE), knee flexors (KF), plantar flexors (PF), dorsal flexors (DF), hip extensors (HE), hip flexors (HF), adductors (AD), abductors (AB), elbow extensor (ArmE), and elbow flexors (ArmF)

tures were an adequate diet and the addition of the ARED strength training with lower body exercises. A comparison group used the IRED device. The study entailed 13 astronauts on ISS expeditions. All were well matched in all aspects except for the strength and treadmill exercise they performed. IRED was available and used by eight subjects (six males, two females) and ARED was used by five (three males and two females) for 5.3 and 4.5 months respectively. Both groups were scheduled for

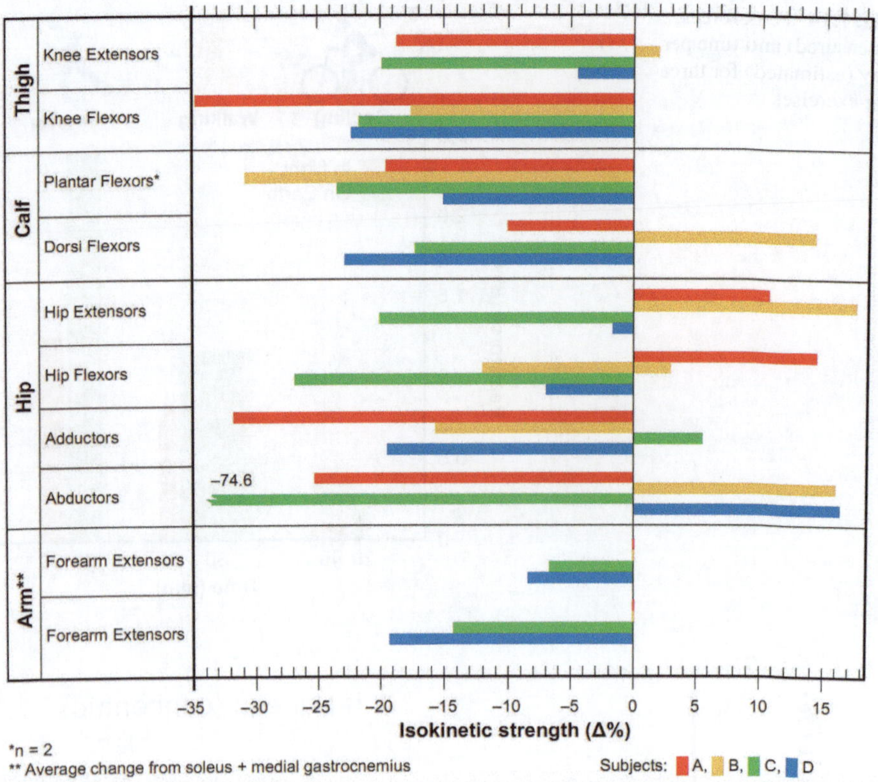

Fig. 7.16 Individual postflight changes in muscle strength for the four crewmembers (*A–D*) compared to the preflight values

2.5 h of prescribed exercise[18] 6 days per week with 30 min per day aerobic exercise.[19] Much critical preflight and in-flight data are unknown. The only data on total in-flight exercise compliance, defined as days performed versus days scheduled was 75% for both groups.

Results indicate the differences in the hip, pelvis, and lumbar spine bone minerals when the ARED group is compared to previous missions and to the IRED group (Fig. 7.17). Changes in muscle strength are not given. The author correctly emphasizes the squat strength exercise because of the large force it imposes on the key muscles of the locomotor chain. Loading of this squat was 0.9 BW for IRED and 0.87 for ARED. The report notes differences in the forces produced by IRED and ARED.

---

[18] Includes setup and clean up time.

[19] The report's author is likely accurate in listing the treadmill as aerobic for loading was too small to affect strength.

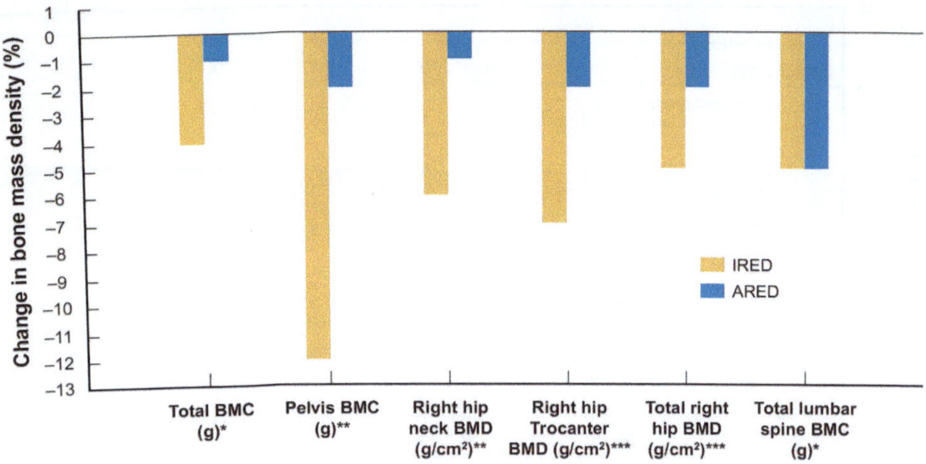

IRED = interim resistive exercise device; ARED = advanced resistive exercise device; BMC = bone mineral content; BMD = bone mineral density.
*Significant effect of spaceflight, $p < 0.05$.
**Significant interactive effect of spaceflight and exercise type, $p < 0.01$.
***Significant effect of spaceflight and exercise type, $p < 0.05$.

Fig. 7.17 Loss of bone density using IRED and ARED

IRED's elastic loading varies with displacement of the bar but has no inertia thus no force from acceleration. Acceleration forces from the inertia of the ARED would further the discrepancy between the machines. However, a bothersome feature of the squat as done on these devices is that the total squat load in weightlessness is reduced by absence of BW. This could be 70% or more leaving an effective BW of only 20% in the squat, which appears to be much too small to produce the effects given. With the data given it is impossible to do a biomechanical analysis of the exercise loading. If losses of muscle and bone in the legs can be prevented by strength exercise then the other characteristics of muscle and bone adaptation need examination. The authors of this study were contacted apropos input forces, and apparently there was a failure in the FGF plate in the study. Input forces are being reevaluated.

A second important and unequivocal aspect of this study is its emphasis on adequate in-flight nutrition, especially protein and caloric intake. After Skylab it had been assumed that diets would always be adequate, but lessons painfully learned [36] were apparently forgotten. Caloric deficits and weight loss became a part of Shuttle, Mir, and Shuttle-Mir missions. Effects of such deficits on loss of muscle and muscle strength are generally accepted, but direct effects on bones are not well documented, if indeed known.

Metabolic records from this study allowed the individual loss in pelvic bone to be measured as a function of caloric and protein intakes. Whether these losses are a reflection of muscle loss or a direct effect is unknown, but they are significant. However, caloric intake in this group was adequate because there was no weight loss, in contrast to the matched IRED group.

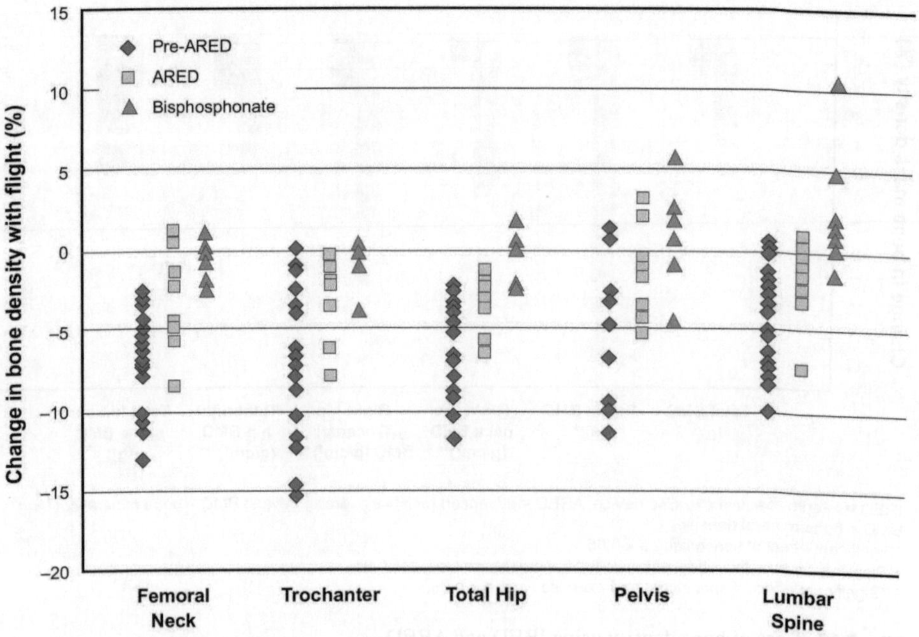

Fig. 7.18 Change in DXA BMD after long-duration spaceflight: 1 $p < 0.05$, pre versus postflight; 2 $p < 0.05$ (bisphosphonate group significantly different from pre-ARED); 3 $p < 0.05$ (bisphosphonate group significantly different for ARED). Pre-ARED (n = 18); ARED (n = 11); bisphosphonate (n = 7)

A follow up study that added a drug[20] to block osteoblastic activity reports that bone loss and urinary Ca++ increase was insignificant [37]. The initial population of ten astronauts was first reduced to eight because of stomach upsets and another dropped out. The remaining seven astronauts participated in the study with an average time on orbit of 5.5 months. Exercise was scheduled for 2.5 h per day that included set up and hygiene time with 1 h strength (resistance) and 1.5 h week aerobic exercise on cycle or treadmill. There is no quantitative information on actual exercise in any of the three programs that are reported in this series of studies. This study has now been completed and published. Apparently the study was partially successful in prevention of loss of bone density [15].

Figure 7.18 compares bone loss in the three groups, which are sequential in time and labeled pre-ARED, ARED and drug. Emphasis is placed on the addition of strength exercise to the three groups and on the use of ARED in the two groups. There were statistically significant ($p < 0.05$) pre and postflight mineral density changes in the pre-ARED group and statistically significant differences in mineral losses between pre-ARED and the drug groups in all bone regions measured. On inspection of the data there are obvious reductions in bone mass loss when an ade-

[20] Alendronate, 70 mg, an antiresorptive drug taken orally each day.

Table 7.2 Mean values obtained from isokinetic studies of nine ISS astronauts

Strength (ft/lb)	*60°/s*		*ft/lb Max*
	Pre	*Post*	*% Change*
Bench press	169	176	+4.0
Leg press	494	474	−4.0
Knee extension	127	121	−4.7
Knee flexion	76	67	−11.8
Endurance	*180°/s*		*ft/lb Total (n = 10)*
	Pre	*Post*	*%*
Knee extension	1703	1560	−12.9
Knee flexion	1062	923	−13.1

quate strength exercise facility (ARED) and adequate diet were added. There is a further improvement when a drug is added. Moreover the calculated strength and fracture resistance of bone preserved by the drug was normal as was urinary Ca++. Other biomarkers of bone loss were improved.

Also for the first time musculoskeletal injuries and exercise postflight are commented upon. Apparently a follow up study of the role of exercise in this series is in progress.

Table 7.2 shows pre- and postflight means of isokinetic measurements of nine astronauts after approximately 6 month ISS expeditions.

This group all had ARED, treadmill, cycle ergometer and a near isocaloric diet. Bone changes could be expected to parallel strength changes. While there were large individual variations, a familiar and consistent pattern, first seen on Skylab, remains. Muscle and consequently bone losses in the upper limbs can be controlled by strength counter measures.

Some authors recognize several of the crippling limitations of the section on strength changes [38] citing a lack of a mandatory standardized in flight exercise program. The problem could be addressed with valid recordings of force loads on the muscles that must be known to evaluate either effects of weightlessness, or effects of exercise. Measurement and valid records, not new rules, will be required to do this. Cavanagh recognized this and made the only pre and inflight measurements to date using instrumented shoes (see Fig. 6.22). Currently the only significant force generators for the lower body and legs are the treadmills and ARED. For treadmills with a flexible belt tread it is a simple matter to insert four strain gages under the tread support plate to measure foot ground forces normal to the tread. The same arrangement can made with ARED. Until such measurements are made only qualitative guesstimates are possible.

Large, time dependent variations are bound to occur when the sampling periods are so widely separated as in this report. Base line measurements of 6 and 2 months preflight, no inflight measurements and 1, 2, and 6 weeks post flight assures large variations. ISS was and is a unique opportunity to significantly increase our knowledge of the means to control effects of weightlessness on muscle with a large mixed

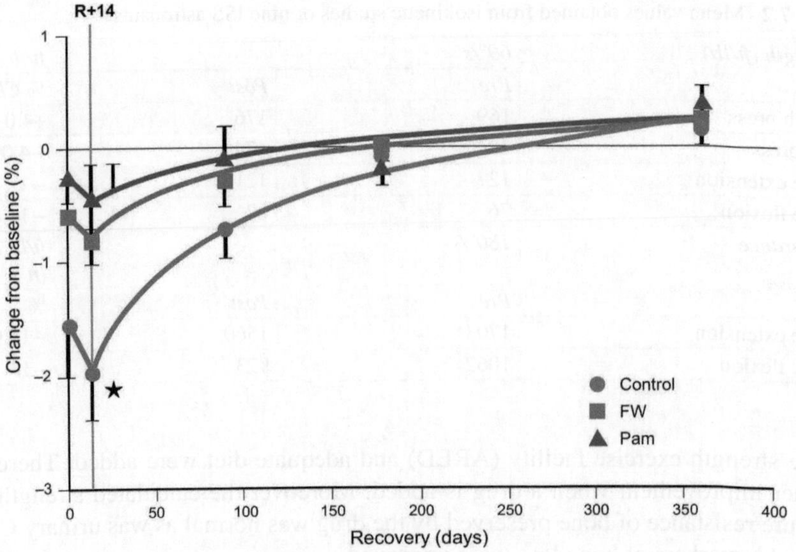

Fig. 7.19 Mean loss and recovery curves of trabecular bone after 90 days bed rest with three groups (*FW* flywheel exercise, *PAM* pamidromates), followed by 1 year of normal activity. Note continued loss post BR–(0)

population exposed for the longest periods to date. Predictably the IRED strength exercise with only elastic forces was less effective than the ARED with constant and inertial forces. The key questions of the characteristics of loss of strength and effective ways to control this loss remain unanswered. It is hoped that the crucial role of force in the exercise counter measures will be recognized and are being measured in the second decade of ISS.

7.4 Recovery of Muscle and Bone Loss

7.4.1 Bedrest (BR)

A well controlled 90 day bed rest study of loss and recovery of leg muscle and bone used 25 men in three bed rest regimes; nine with BR; (C) 9 injected with pamidronates prior to BR (PAM); seven with flywheel exercise (FW) during BR [39]. Figure 7.19 shows the fitting of exponential curves to the three test groups and the small standard deviations in contrast to the flight data in Fig. 7.20. Table 7.3 summarizes the changes.

Post BR the subjects followed for a year with pQCT scans to measure bone density and muscle cross sectional area. Some of the results are shown in Fig. 7.19 [40] and Table 7.3. Although CSA was used as a surrogate measure of muscle

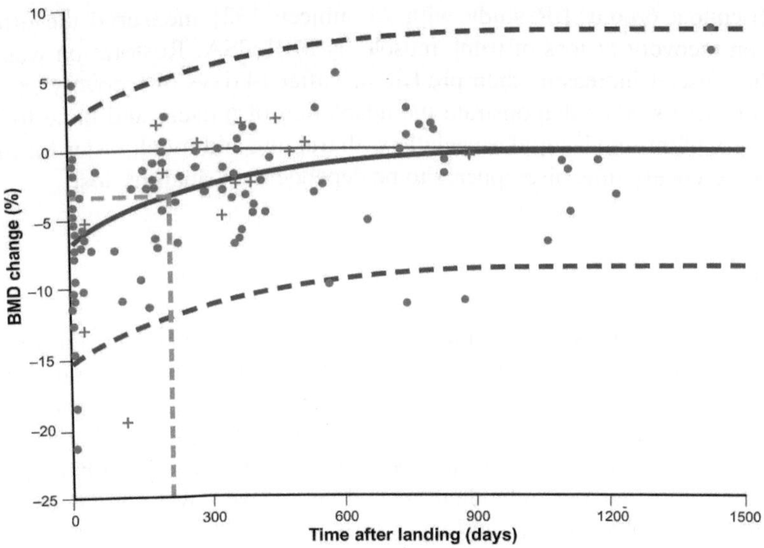

Fig. 7.20 Recovery of bone lost inflight from femoral neck on return to 1 g. Solid curve is mean of data points and dotted curves are 95% confidence limits. Mean loss at return was 6.5%. Time to recover half of that was estimated to be 211 days with total recovery at 900 days

Table 7.3 Bone and muscle lost and regained

	Muscle CSA[a]			Bone mineral	
	Loss (%)	Time[b] (days)		Loss (%)	Time[b] (days)
Control	25.4	102	Tibial-trabecular	6.0	632
			Cortical	2.0	256
Exercise	17.5	82	Tibial-trabecular	2.8	610
			Cortical	0.8	187
Drug	25.2	55	Tibial-trabecular	3.6	368
			Cortical	0.5	148

Loss is in percentage at end of BR and time is days after BR for full recovery of loss
[a]Cross sectional area
[b]Time to regain loss

strength, it used the same muscle group that allows a valid comparison. The drug had no effect on muscle, but it reduced bone loss. Exercise also reduced muscle and bone loss. Cortical bone loss had recovered by one year's usual activities, but there was still a slight loss of trabecular bone, which was predicted to recover in 2 years. The observations of the investigator are pertinent to all normal bone loss, "The recovery of bone mass appeared to be tightly regulated, and generally followed neuromuscular recovery (and) underline the capacity of the adult skeleton to adapt to mechanical stimuli."

A subsequent 60-day BR study with 24 subjects [32] measured the effect of exercise on recovery of loss of trunk muscle by MRI CSA. Restoration was rapid with some muscles increasing their pre BR size after 14 days of recovery training.

Both of these studies demonstrate the adaptation of muscle and bone to force. The recovery of muscle is rapid especially with training and much faster for muscle than bone. Recovery time also appears to be dependent on quantity lost.

Postflight

The 84-day SL-4 crew with its limited countermeasures walked off the recovery helicopter and were walking and standing on arrival for extensive evaluations the following day. At least one crewmember was jogging 4 days post flight. In those days readaptation was left to the crew, and was typically rapid. Flights were relatively short in the Shuttle program and other than acute effects of vestibular readaptation or fluid shifts in a few individuals, post flight recovery was complete in a day. When missions extended 18 days or longer during Shuttle, recovery sometimes took longer in some individuals.

When the Shuttle-Mir program began, a new generation of managers and life scientists were in control. A clinically based group treated the healthy, space adapted astronauts in some ways like patients. Astronaut Jerry Linenger described his long, medically based rehabilitation in some detail. It has been replaced by a more rational readaptation program. Currently it is informally reported that after 3–6-month missions the crew are now able to resume limited ambulation in 3–4 days. Re-adaptation with strength training, buoyancy supported ambulation, and increasing locomotion continued as rapidly as tolerated has the typical crew returning to normal duty in 3 weeks. Although this is an individual matter, with a mandatory fitness protocol that begins a year preflight and continues through flight and readaptation, there has been great improvement in postflight recovery. A recent returnee said that he was walking 4 h after recovery.

Great concern has been expressed, especially to the media, over two aspects of bone loss in space. When trabecular bone is lost, some of the arches or trusses are destroyed and not replaced in re-adaptation. It is proposed that when ageing and normal loss of bone occurs, these individuals may be at increased risk for fracture. It appears that the amount of trabecular bone lost is completely replaced over time. Because this is done in response to maximum forces, it would be remarkable if this were not arranged in a way to meet these maximum stresses, whether it were exactly as before or not. A second concern is traumatic fracture in bones weakened by loss of mineral. Two such events have followed flight to date. While traumatic fracture is a real concern in pathologic osteoporosis, especially in extreme cases such as SCI and in the osteoporosis of ageing, it can be questioned whether or not this the same situation as in normal healthy individuals with reduced bone mass.

Relevant to this problem are the large temporal differences in adaptive gain and loss of muscle. The response time of bones to changes in stress is several times that of muscle. On return to 1-g bone continues its loss of mineral for a week or more

Table 7.4 Estimated recovery times for bone loss inflight from various body regions

Region	50%	100%
Femoral trochanter	255 days	1150 days
Femoral neck	211 days	1300 days
Pelvis	97 days	600 days
Lumbar spine	151 days	750 days
Calcaneous	163 days	640 days

(Fig. 7.19). In contrast strength increases very rapidly, initially through neurological mechanisms and in a few days time by anatomical replacement and significant gains in recovery strength. During rehabilitation of maximum losses of muscle and bone in SCI of long duration, this difference in recovery time is a real concern because the rapid gain of muscle strength can produce fractures in bone with its slower response. It should be noted that this is a different realm from the disuse atrophy of space, which will not approach that in SCI. However, this delay in reaching maximum bone loss post flight should be considered in measurement of it.

There are numerous studies of the postflight re-adaptation of bone mineral. One study of recovery of muscle volume and cross sectional areas following 2 months of bed rest was found but was not included because of its short duration and undemonstrated relation of muscle geometry to strength. There have been spinal disc injuries post flight and these are discussed in Chap. 11 on accommodations.

Sibonga and colleagues used data from 42 astronauts and cosmonauts on 54 missions lasting 126–208 days, plus two flights of 311 and 489 days [41]. Seven astronauts and 22 cosmonauts flew on Mir and 12 astronauts and five cosmonauts on ISS. The investigations are concerned only with magnitude of losses inflight, not their causes, and the time of recovery from them. One such curve is shown in Fig. 7.20 and estimated recovery times from five bone loss sites are shown in Table 7.4. These data are based on a series of exponential curves fitted to the data points as shown in Fig. 7.20. The recovery times are presumably based on normal activity and would be shortened by increased muscle forces in a recovery program.

Summary of Countermeasures to the Effects of Weightlessness on the MSkS

When weight and usual physical activity loads are "removed" by weightlessness there is a rapid response by muscle with loss of mass, strength and endurance. The first measurements of, and countermeasures to, effects of weightlessness on Skylab demonstrated that the smaller losses in the arms could be prevented by relative simple measures, and that the larger losses of the legs could also be countered. However, this required larger repetitive forces, including forces on the body equivalent to its weight in 1-g locomotion. This has yet to be done routinely in all subjects, and results in significant losses in leg muscle and bone loss for many.

An alternative approach to countering leg losses was weight lifting in which a relatively few repetitions of heavy body loading would preserve bone. The in-flight studies that indicate it can be done left a number of unanswered questions. It is to be hoped that provision of an adequate subject load on a treadmill equivalent to that in 1 g will be pursued because locomotor exercise has a number of advantages.

Proper evaluation of countermeasure effects could be found in only one bed rest and one in-flight study. Measurement has regressed in practice. Bone strength is directly dependent on muscle strength. Hence muscle strength is an essential measurement in evaluation of a countermeasure to bone loss.

Given that it has become relatively simple to measure muscle size during the measurement of bone density, muscle volume now has become an accepted surrogate for the measurement of strength, yet it is well established that the correlation between muscle volume and strength is poor or often nonexistent.

Anyone reading this must ask why are centrifuges and drugs not discussed? To be effective a centrifuge needs space, power and physical support, much more than spacecraft of the forseeable future can provide. Short radius centrifuges (SRC) produce a new array of physiological problems, but there is no point in trying to delay the inevitable. The fixed idea of gravitoinertial forces as a cure all for effects of weightlessness assures the continuation of major expenditures on SRC's. One day there may be Star Wars spacecraft or reduced-g outposts capable of supporting a properly conceived and designed centrifugal countermeasure.

As to the question of medical countermeasures for bone loss, it has been demonstrated that medications can preserve bone mineral in ageing or medically compromised populations and that this retention is beneficial in 1 g. It has also been demonstrated that there are significant side effects from these medications. It has now been demonstrated that one of these medications can prevent loss of mineral in the disuse atrophy in space. However, preservation of bone in an ageing population is a very different situation from that of physically trained, healthy individuals on a trip to Mars. Side effects in space are a very different situation from on Earth with a nearby doctor's office. The population is also very different. Will this medically preserved bone in a population scrambling around on Mars have adequate bone strength in practice? The calculated strength and fracture probabilities are based on very different populations and activities from that of astronauts.

If bone strength is preserved, then what about muscle strength? Medication based clinical practice will have a ready answer—hormones and more medications. This is a slippery slope. The physiological solution to the problem is finding the will and skill to apply adequate force loads to legs deprived of locomotor stress. This will be addressed further in the final summary of all aspects of MSK losses in weightlessness.

7.5 Summary of Loss of Muscle and Bone in Spaceflight

By the end of Skylab (1974) the nature and essentials of the problem had been demonstrated. When physical activity is unrestrained in weightlessness locomotor activity is still precluded and arm forces reduced. The last four plus decades have seen uncounted investigations by simulations in 1 g, in humans and other animals, studies often seeking other causes of the losses. Inflight studies usually focused on the end result, changes in bone, with few studies on the proximate cause, loss of muscle force. The variety and application of countermeasures has been limited only by the imagination of the investigators. There has been no evidence that adequate replacement of 1-g equivalent force is not the primary solution to this problem.

If removal of the stimuli of locomotor activity is the cause then the best replacement to date is the treadmill with the subject loaded by a body weight equivalent constant force. The majority of FGF measurements that have been made show replacement loads of only 50–60% which is consistent with the resulting loss of muscle and bone seen, in spite of significant amounts of time spent on the treadmill. Recently it is claimed that magnitude of the body force loads approach body weight with losses virtually eliminated in some subjects but not in others. This is unpublished at the time of writing. Other attempts at reducing bone loss include drugs, which alter mineral metabolism, and heavy strength exercises which stress legs and back. Recently heavy cycle or other exercise that increases oxidative metabolism at the molecular level has been advocated. A significant finding was an inadequate diet and inadequate documentation of food intake.

An inflight drug study showed preservation of mineral in bone. Another inflight study with an adequate diet and heavy strength exercise plus treadmill reported a virtual elimination of loss.

In spite of all the above, effective and efficient countermeasures have not been demonstrated. Arguably the simplest, most efficient and most physiological is a combination of a versatile strength machine, a technically refined ARED, and a treadmill properly coupled to the subject. Properly loaded locomotion provides the repetitive forces needed to maintain leg/back strength and fatigue resistance in muscle and bone and in the process maintains aerobic capacity. Adequate upper body strength can be maintained using ARED.

A major problem has been the lack of definition and measurement of forces on Earth and especially in flight. For any reasonable analysis of the combination of human locomotion on a treadmill, the magnitude of the forces must be known. Also, knowledge of their time profiles and relations can often be equally important.

A basic reliable treadmill has been in place for more than a decade however it does not have the necessary FGF or body load measurements that are obligatory for proper use in flight. Such measurements are dual purpose: first, they monitor the adequacy of performance of both machine and its individual user; second and equally important, they provide the necessary information for an analysis of the mechanical coupling of body to machine: mechanical impedance matching. This is a design tool that directed the coupling of the human body to the oscillating body mass measurement device to produce unmatched accuracies on Skylab. The extraordinary performance of Whitmore's exercise devices had the same basis. Unless such analytical tools are used, more years of cut and try as exemplified by IRED and ARAD can be expected. The use of short radius centrifuges, will not maintain all the many adaptations to 1 g by the body's complex system. Realistic compromises will be necessary and should be defined.

Given that the treadmill is such a key element in maintenance of the bone and muscle for bipedal locomotion and since it has been so difficult to employ this effectively in weightlessness, a brief commentary follows. The rather obvious facts that 1-g locomotor exercise with its large forces is used as a potent stimulator of bone maintenance [42, 43] is increasingly accepted. Knowledge of the magnitude of forces and their temporal profile is used as a countermeasure to evaluate their effectiveness. This was recognized by an increasing number of investigators and resulted in studies by Cavanagh and others [44–46] who found major discrepancies in foot-ground forces in weightlessness when compared to those in 1 g [47]. There are several obvious sources of these discrepancies including inadequate force magnitude, distribution of forces on the body and differences in compliance of applied force. (Earth's gravitational force has virtually infinite compliance). Both magnitude and character of forces have been studied in simulation and flight. One aspect of loading that is yet to be commented upon is the effect of absence of simulated weight loads of arms and legs on treadmill exercise inflight. Such effects may have been seen in results from SL-4.

Although significant compliance was found in gravity simulation loads, measurement of this and related elements were insufficient to allow a meaningful analysis of effects [48]. A significant factor in effective simulation of g loading is the simulated ground reaction produced by the treadmill itself. Ideally this would be a rigid or non-compliant counterforce in the Z and Y axes with an X-axis surface moving at selected, constant rates. So far as can be determined the impedance [49] the displacement of these surfaces with applied force, has not been measured. Addition of "vibration isolation" to TVIS and COLBERT may well have resulted in distortion and reduced effectiveness of the force necessary for maintenance of muscle and bone. TVIS used a combination of passive and active elements while COLBERT uses passive resistance and compliance (springs) to distort transmission of FGF to the mass of the spacecraft. The effects of this distortion of the foot forces on the subject remain unknown.

Until an integrated program of required measurements plus analysis using solid physical and engineering concepts and practices, such as mechanical impedance [49, 50] and impedance matching [51], the process of physical countermeasures will remain a matter of speculation, conflicting opinions, empiricism and unnecessary expense in time and money. Such analysis should begin with the human mechanisms that make bipedal locomotion possible on earth, should include treadmills in 1 g as well as in weightlessness and proceed through each physical element, the substitute g loads, their application and all relevant aspects of the treadmills and their use in space. Such an approach is familiar to physicists and engineers but generally foreign to physiologists and especially physicians whose skills are equally foreign the physical scientists, but integration of differing knowledge is what biophysics and bioengineering are about.

7.5.1 *The Future*

Until a new coordinated approach is made using evidence based current theory of the of these problems and rigorous state of the art design of countermeasure hardware, only the slow sometimes regressive efforts seen to date can be expected. It is hoped that it will be in time when the fundamental role of spaceflight, human exploration, is resumed. It is also hoped that the allure of quick fixes such as drugs, cycle ergometers, and short radius centrifuges will not disrupt the development of fundamental practical countermeasures. There is no doubt that a cycle exercise will always be present as will some form of bone drug. The cycle ergometer can provide a harmless diversion if time is not critical. But it should be at least as small and efficient as the Del Mar Apollo prototype. A short radius centrifuge is another matter. The current strength device and treadmill can be redesigned into items of usable size and weight by use of appropriate biomedical engineering, such as active force generators alluded to in Fig. 7.4.

On reaching the Moon or Mars, maintenance of physical capacity should become more easily managed given that some gravitation will be present. There will be ample mass for weight and more opportunities for exploration of the human body in a new environment.

References

1. Risk of impaired performance due to reduced muscle, strength, and endurance. In: NASA's human research program evidence book. Houston, TX: NASA JSC; 2008.
2. Risk of accelerated osteoporosis. In: NASA's human research program evidence book. Houston, TX: NASA JSC; 2008.
3. Cavanagh P. Bone loss during spaceflight. In: Cavanagh J, Rice A, editors. Cleveland, OH: Cleveland Clinic Press; 2007. p. 1–2.
4. Lem J. A.I.F Ergometer. In: Biomedical results from Skylab. Washington, DC: U.S. Government Printing Office. NASA/SP–377; 1977. p. 441.
5. Lem J. A.II.E Exerciser. In: Biomedical results from Skylab. Washington, DC: U.S. Government Printing Office. NASA/SP–377; 1977. p. 485–6
6. Thornton W, Rummel J. Prevention of muscle deconditioning in spaceflight. In: Biomedical results from Skylab. Washington, DC: U.S. Government Printing Office. NASA/SP–377, 18; 1977. p. 191–7.
7. Korth DW. Exercise countermeasure hardware evolution on ISS: the first decade. Aerosp Med Hum Perform. 2015;86(12):A7–13.
8. Greenisen M, et al. Section 3: Functional performance evaluation. In: Extended duration orbiter medical project. Houston, TX: NASA JSC. NASA/SP-1999-534; 1999. p 3-1–3-17.
9. Hayes J. The first decade of ISS exercise: lessons learned on expeditions 1–25. Aerosp Med Hum Perform. 2015;86(12):A1–6.
10. Kozlovskaya IB, Yarmanova EN, Yegorov AD, Stepantsov VI, Fomina EV, Tomilovaskaya ES. Russian countermeasure systems for adverse effects of microgravity on long-duration ISS flights. Aerosp Med Hum Perform. 2015;86(12):A24–31.

11. Kozlovskaya I, Grigoriev A. Russian system of countermeasures on board of the International Space Station. Acta Astronaut. 2004;55:233–7.
12. Norrbrand L, Tous-Fajardo J, Vargas R, Tesch PA. Quadriceps muscle use in the flywheel and barbell squat. Aviat Space Environ Med. 2011;82(1):13–9.
13. Loehr JA, Guilliams ME, Petersen N, Hirsch N, Kawashima S, Ohshima H. Physical training for long-duration spaceflight. Aerosp Med Hum Perform. 2015;86(12):A14–23.
14. Smith SM, Heer MA, Shackelford LC, Sibonga JD, Ploutz-Snyder L, Zwart SR. Benefits for bone from resistance exercise and nutrition in long-duration spaceflight: evidence from biochemistry and densitometry. J Bone Miner Res. 2012;27(9):1896–906.
15. LeBlanc A, et al. Bisphosphonate ISS flight experiment. Houston TX; NASA JSC. JSC-CN-30935; 2014.
16. Moore A, et al. Results of the International Space Station interim resistance exercise device man-in-the-loop test. Houston, TX: NASA JSC. NASA/TP-2004–212062; 2004.
17. Amonette W, et al. Ground reaction force and mechanical differences between the interim exercise device (iRED) and Smith machine. Houston, TX: NASA JSC. NASA/TP-2004-212063; 2004.
18. Amonette W, et al. Evaluation of the horizontal exercise fixture in conjunction with the interim resistive exercise device (iRED). Houston, TX: NASA JSC. NASA/TP-2009-214798; 2009.
19. Sibonga JD, Spector ER, Johnston SL, Tarver WJ. Evaluating bone loss in ISS astronauts. Aerosp Med Hum Perform. 2015;86(12):A38–44.
20. Bentley J, et al. Advanced resistive exercise device (ARED). Houston, TX: NASA JSC. NASA/TP-2006-213717; 2006.
21. Loehr J, et al. Use of a slick-plate as a contingency exercise. Houston, TX: NASA JSC. NASA/TM-2003-210789; 2003.
22. Jason N, et al. Comparison of the U.S. and Russian cycle ergometers. Houston, TX: NASA JSC. NASA/TP-2007-214760; 2007.
23. Schoon R. ISS trashing first treadmill, COLBERT and Russian model to pick up the pace in its place. Latinos Post. 2013. www.latinospost.com.
24. Yarmanova EN, Kozlovskaya IB, Khimoroda NN, Fomina EV. Evolution of Russian microgravity countermeasures. Aerosp Med Hum Perform. 2015;86(12):A32–7.
25. Flywheel exercise device (for Space Station). ESA Doc N EUC-ESA-FSH-044.
26. Norrbrand L, et al. Quadriceps muscle use in the flywheel and barbell squat. ASEM. 2011;82(1):13–9.
27. Belavy D, Miokovic T, Armbrecht G, Rittweger J, Felsenberg D. Resistive vibration exercise reduces lower limb muscle atrophy. J Musculoskelet Neuronal Interact. 2009;9(4):225–35.
28. Akima H, Kubo K, Kanehisa H, Suzuki Y, Gunji A, Fukunaga T. Leg-press resistance training during 20 days of 6 head-down-tilt bed rest prevents muscle deconditioning. Eur J Appl Physiol. 2000;82(1–2):30–8.
29. Smith SM, Zwart SR, Heer M, Lee SM, Baecker N, Meuche S, Macias BR, Shackelford LC, Schneider S, Hargens AR. WISE-2005: supine treadmill exercise within lower body negative pressure and flywheel resistive exercise as a countermeasure to bed rest-induced bone loss in women during 60-day simulated microgravity. Bone. 2008;42(3):572–81.
30. Shackelford LC, LeBlanc AD, Driscoll TB, Evans HJ, Rianon NJ, Smith SM, Spector E, Feeback DL, Lai D. Resistance exercise as a countermeasure to disuse-induced bone loss. J Appl Physiol. 2004;97(1):119–29.
31. De Lorme T, Watkins A. Resistance exercise: technic and medical application, vol. 2. New York: Appleton-Century-Crofts; 1951. p. 2–3.
32. Smith C, et al. Physiologic responses to motorized and non-motorized locomotion. Unpublished report by JES Tech to NASA JSC.
33. Cavanagh PR, Gopalakrishnan R, Rice AJ, Genc KO, Maender CC, Nystrom PG, Johnson MJ, Kuklis MM, Humphreys BT. An ambulatory biomechanical data collection system for use in space: design and validation. Aviat Space Environ Med. 2009;80(10):870–81.
34. Gopalakrishnan R, Genc KO, Rice AJ, Lee S, Evans HJ, Maender CC, Ilaslan H, Cavanagh PR. Muscle volume, strength, endurance, and exercise loads during 6-month missions in space. Aviat Space Environ Med. 2010;81(2):91–104.

35. Trappe S, Costill D, Gallagher P, Creer A, Peters JR, Evans H, Riley DA, Fitts RH. Exercise in space: human skeletal muscle after 6 months aboard the International Space Station. J Appl Physiol. 2009;106(4):1159–68.
36. Thornton W, Ord J. Physiological mass measurements. In: Skylab in Biomedical results from Skylab. Washington, DC: U.S. Government Printing Office. NASA SP-377, 19; 1977. p. 175–82.
37. LeBlanc A, et al. Antiresorptive counter measure for spaceflight bone loss: preliminary results presented NASA Investigators' Workshop; 2012.
38. English KL, Lee S, Loehr JA, Ploutz-Snyder RJ, Ploutz-Snyder LL. Isokinetic strength changes following long-duration spaceflight on the ISS. Aerosp Med Hum Perform. 2015;86(12):A68–77.
39. Rittweger J, Felsenberg D. Recovery of muscle atrophy and bone loss from 90 days bed rest. Bone. 2009;44:214–24.
40. Hides JA, Lambrecht G, Richardson CA, Stanton WR, Armbrecht G, Pruett C, Damann V, Felsenberg D, Belavý DL. The effects of rehabilitation on the muscles of the trunk following prolonged bed rest. Eur Spine J. 2011;20(5):808–18.
41. Sibonga JD, Evans HJ, Sung HG, Spector ER, Lang TF, Oganov VS, Bakulin AV, Shackelford LC,LeBlanc AD. Recovery of spaceflight-induced bone loss: bone mineral density after long-duration missions as fitted with an exponential function. Bone. 2007;41(6):973–8.
42. Morseth B, Emaus N, Jørgensen L. Physical activity and bone: the importance of the various mechanical stimuli for bone mineral density. A review. Norsk Epidemiologi. 2011;20(2):173–8.
43. Lanyon LE. Using functional loading to influence bone mass and architecture: objectives, mechanisms, and relationship with estrogen of the mechanically adaptive process in bone. Bone. 1996;1891(Suppl):37S–43S.
44. McCrory JL, Baron HA, Balkin S, Cavanagh PR. Locomotion in simulated microgravity: gravity replacement loads. Aviat Space Environ Med. 2002;73(7):625–31.
45. McCrory JL, Derr J, Cavanagh PR. Locomotion in simulated zero gravity: ground reaction forces. Aviat Space Environ Med. 2004;75(3):203–10.
46. Genc KO, Mandes VE, Cavanagh PR. Gravity replacement during running in simulated microgravity. Aviat Space Environ Med. 2006;77(11):1117–24.
47. Genc KO, Gopalakrishnan R, Kuklis MM, Maender CC, Rice AJ, Bowersox KD, Cavanagh PR. Foot forces during exercise on the International Space Station. J Biomech. 2010;43(15):3020–7.
48. DeWitt JK, et al. Kinematic and electromyographic evaluation of locomotion on the enhances zero-gravity locomotion simulator: a comparison of external loading mechanisms, TP–2007–214764, Jan 1, 2008.
49. Neubert V. Mechanical impedance. State College: Jostens Printing and Publishing; 1987.
50. Hogan N. Adaptive control of mechanical impedance by control of antagonistic muscles. IEEE Trans. on automatic control, (AC-29)8; 1984.
51. Horowitz H, Hill W. The art of electronics. New York: Cambridge Univ. Press; 1989. (1) p. 29 and (13) p. 879.

Chapter 8
Loss of Aerobic Capacity During Weightlessness

8.1 Introduction

Physical activity depends on the cardiorespiratory (CR) system's ability to deliver energy to the muscles to sustain their action. The g loads of high-performance aircraft also put large stresses on the CR system. But in a matter of days to weeks, cardiorespiratory capacity, usually called aerobic fitness,[1] is sharply reduced by lack of physical activity, whether on Earth, in bed rest or in weightlessness (Fig. 8.1).

Unlike fluid shifts and vestibular conflicts, reduced aerobic fitness was a familiar phenomenon on Earth, and simple countermeasures to it were available. Advances in technology and instrumentation had also allowed cardiology to make significant contributions to theory and practice in this area. All of these factors allow this problem to be dealt with more successfully than other adaptations to weightlessness such as bone and muscle loss. However, many questions about the CR system in space remain to be answered.

When it was decided to fly humans in space, special attention was given to the CR system in the selection and evaluation of the first seven astronauts. At the Lovelace Clinic, Dr. Arthur Masters' "two-step" clinical cardiac stress test was used in the search for any cardiac disease or anomalies. A Swedish bicycle ergometer was used to evaluate aerobic fitness [1]. At Wright Air Development Center a Harvard Stress Laboratory step test was done, and performance on the treadmill using Balke's protocol of increasing elevation with time became the standard of comparison for physical fitness [2]. For many years, time on the treadmill test was the figure of merit for physical performance in the Astronaut Office.

Mercury astronauts' CR systems subsequently handled the stresses of high-g loads on launches and entry without difficulty. Reduction in postflight aerobic fitness

[1] Activities that require large amounts of air.

W. Thornton, F. Bonato, *The Human Body and Weightlessness*,
DOI 10.1007/978-3-319-32829-4_8

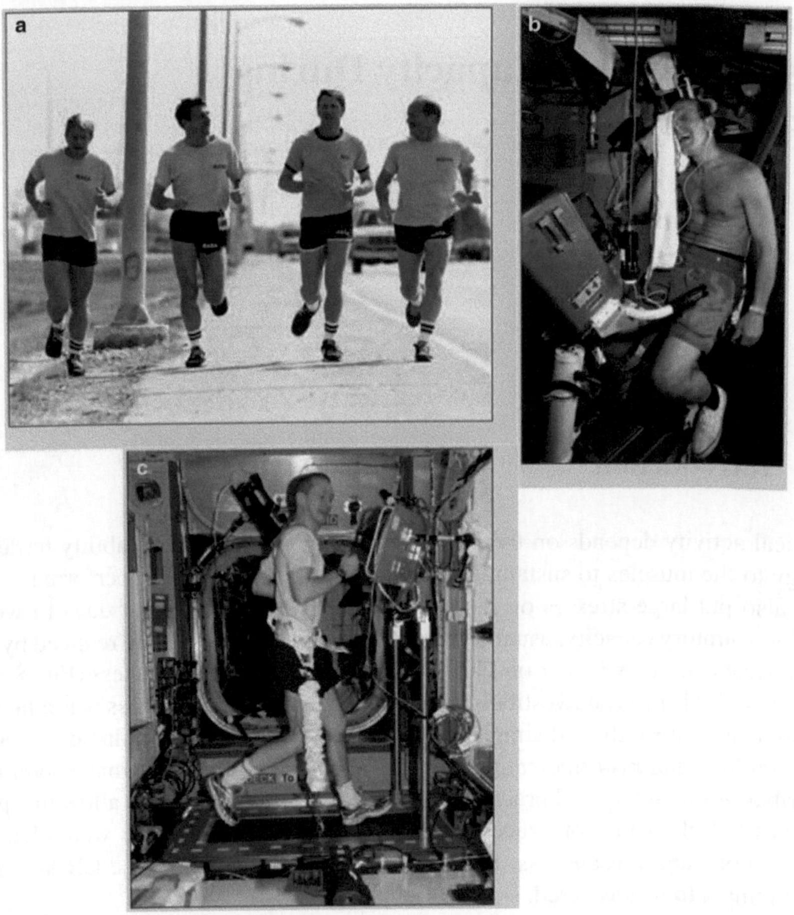

Fig. 8.1 Jogging was the primary aerobic exercise for the Astronaut Office. This is not surprising since approximately one third of the body is dedicated to bipedal locomotion, and jogging provides necessary aerobic, strength and endurance exercise to maintain both the cardiorespiratory and musculoskeletal systems. In the absence of adequate equivalent body weight loads on the treadmill in weightlessness, cycle ergometry has become increasingly used to maintain aerobic condition. The quartet of the STS-26 crew jogging on the side of a JSC roadway was a common site during the Shuttle period **(a)**. Pete Conrad has just finished a heavy session pre-flight on the first NASA cycle ergometer in space on SL-2 **(b)**. One of the ISS treadmills is also shown in use by European Space Agency (ESA) astronaut Frank De Winne **(c)**

was not identified as a problem. The first working extra vehicular activity (EVA) on Gemini showed that normal aerobic fitness could be defeated by inadequate suit and task design. Excess heart and respiratory rates, excess body temperatures and fatigue caused a premature termination of the EVA [3]. Pre- and postflight exercise testing of Gemini crews showed a decrement in aerobic fitness, and although it was not implemented, an in-flight cycle ergometer for exercise was scheduled for Apollo. Pre- and postflight exercise evaluations on Apollo concluded that these missions had been accomplished without significant loss of aerobic fitness [4].

A cycle ergometer for exercise and evaluation of exercise response plus elaborate respiratory gas analyses were part of a major study on all Skylab missions. It was concluded that the cycle ergometer exercise had preserved or in some cases actually increased aerobic fitness [5]. During this period a number of bed rest studies showed large losses in aerobic fitness to be only partially prevented by the aerobic exercise employed. A variety and number of studies of aerobic fitness loss and of countermeasures to it on Shuttle produced conflicting results.

During this period the Russians conducted similar studies, however the results of these studies are not readily available. Also during this time, a standard Bruce treadmill cardiac stress test was added to the annual astronaut physical. It was expanded to include a respiratory gas analysis and became a record of CR performance as well as a cardiac examination.

ISS includes bicycle ergometry for exercise and pre-, in- and postflight evaluations of aerobic fitness [6]. In-flight gas analysis is now in place, but few results were available until recently and these are from the first decade of ISS history. Three treadmills are on ISS, and these should be considered the primary means for aerobic exercise. However, effective coupling of the users to these machines has yet to be accomplished. Results of measurements from cycle and treadmill ergometry are almost certain to differ, especially under the circumstance.[2]

In summary, based on 1-g experience, absence of usual CR loads in weightlessness could result in rapid, significant decreases in aerobic fitness. Such large losses have not been previously documented in space either because of the lack of studies in the absence of exercise or because countermeasures were being used by the subjects.

8.2 Brief History

Stimulation of the CR system by exercise has been recognized for at least 2000 years. "If it doesn't increase breathing it is not exercise," said Claudius Galen, Greek physician, surgeon and philosopher. Objective investigation of the essential role of the CR system in exercise and physical work began around 1900 and rapidly accelerated during and after World War II in America, Europe, and especially in Scandinavia. In 1929 in New York Dr. Arthur Masters stressed the heart using aerobic exercise in his search for silent heart disease in his patients. This became the first cardiac exercise stress test.

In 1943 during World War II the Harvard Exercise Laboratory developed its exercise step test to evaluate the condition, or more correctly the capacity, of normal CR systems. Although both of these tests had several features in common they were designed for very different purposes; one was seeking clinical abnormalities

[2] This is discussed in the section on measurements, p. 266.

in the heart while the second was measuring functions of a normal system, which included the heart.

Unplanned cardiorespiratory stresses were sometime significant during Mercury and Gemini; e.g., Gus Grissom almost drowning on recovery, or the inability of others to complete an EVA. Aerobic fitness became desirable around the Astronaut Office, especially for the annual physical, which now combined a cardiac stress test and evaluation of the individual's aerobic fitness. Jogging and running became the most common exercises at the astronaut gym and some even ran a marathon.[3]

Combining exercise performance and cardiac exercise stress testing eventually produced problems at JSC with unforeseen consequences. Astronauts are by nature a competitive lot and some who were in the best condition pushed well beyond normal cardiac test limits into stress levels unseen by many cardiologists, into regions where these supra-normal hearts produced signs usually seen in diseased hearts. There were groundings and unnecessary catheterizations and unrest in the Astronaut Office. This was quelled by the promise of privacy agreements applicable to all data collected. After 4 years an unfortunate agreement was produced, which may be misapplied and prevent disclosure of harmless but much needed physiological data.

Gemini pre-post flight changes in heart rate during cycle ergometry were interpreted as a loss in maximum oxygen uptake. Various aspects of the flight were proposed as causes of this. It was during this time that Major Kenneth Cooper, cardiologist and Marathon runner at USAF's School of Aerospace Medicine, demonstrated the crucial role of cardiorespiratory capacity in cardiac rehabilitation and preservation—and altered forever its role in medicine and public health with his Aerobics.

Heart rate and oxygen uptake studies during pre and post flight ergometry on Apollo showed an immediate post flight loss followed by rapid recovery of work capacity. By the time Skylab flew reduced inflight physical activity, especially locomotor activity, was recognized as the major cause of this loss. Sky lab was launched with a cycle ergometer as the sole exercise and with a respiratory gas analyzer for studies of aerobic capacity. During Skylab one author (WT) demonstrated the shifts and losses and gains of fluids and blood with entry and exit to weightlessness and accompanying transient changes in exercise capacity.

It was claimed that Skylab demonstrated that aerobic fitness could be maintained or even increased inflight by cycle ergometry [4, 6]. On Shuttle flights most astronauts could get by without exercise, and in the absence of adequate hardware inflight, exercise was given little attention until the long duration flights of the Shuttle-Mir Program. Even then documentation of exercise remained a problem that continued into ISS where it was initially exacerbated by failures of JSC exercise hardware. The need for a controlled, more effective exercise program was recognized, and reorganization was affected with scheduled time and rules for pre- and in-flight exercise. Cycle ergometry now appears to be the primary Cardio Respiratory (CR) exercise inflight. All NASA astronauts have extensive aerobic training and testing preflight, in- and postflight. In most of the first decade of ISS only heart rate was available for inflight testing. Gas analysis equipment became available on ISS

[3]A notable exception to such activity was Neil Armstrong with his philosophy, "I have only so many heart beats, and I am not going to waste any running around a track."

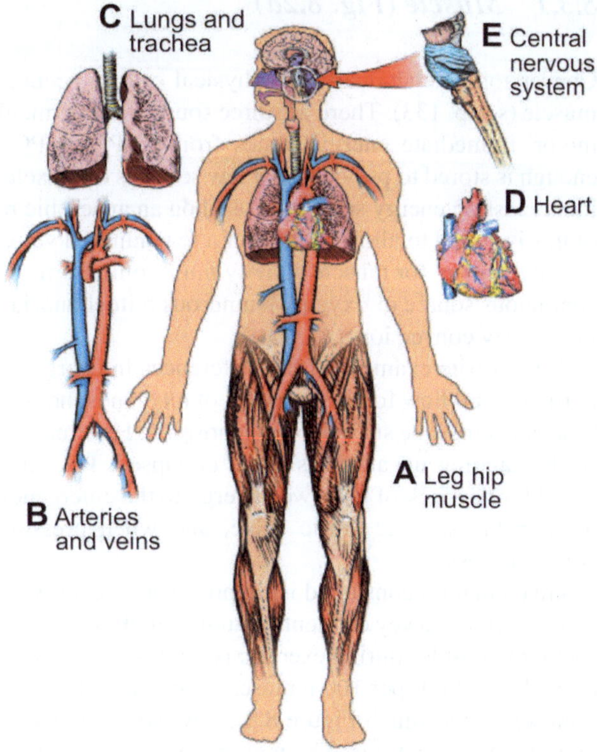

Fig. 8.2 Major components of cardiorespiratory system. Consumption of O_2 and generation of CO_2 is in the muscle mass shown here as (**a**). O_2 and CO_2 is transported byarteries and veins (**b**). Lungs exchange gases to and from the body with air (**c**). The heart propels the blood through lungs and then the body (**d**). Both respiration and blood flow are under control of the nervous system (**e**)

in 2009 and has been used to measure respiratory gas during calibrated cycle ergometer loads and measured results have been published.

8.3 Background

Energy is consumed for the body's various physical activities at a rate that primarily depends on the muscles producing the action, their mass and activity level. Increased cardiac and respiratory activity have been recognized from prehistoric times as a regular consequence of physical activity.[4] In aerobic activities that require large amounts of energy, maximum performance may be limited by the ability of the CR system to deliver oxygen to muscles for oxidative chemical energy reactions to support muscle activity. Major components in this energy transfer chain are shown schematically in Fig. 8.2. Function of these components will be examined in the order shown, then assembled and total performance and the means for testing their combined performance described.

[4] There is no longer any question that health, especially cardiac health, can be improved by regular physical activity.

8.3.1 *Muscle (Fig. 8.2a)*

Conversion of chemicals into physical energy occurs at the actin-myosin level in muscle (see p. 133). There are three sources of chemical energy for the actin-myosin motor. Immediate energy comes from ATP and PC[5] in the muscle fibre where enough is stored to provide for a few seconds of muscle activity. A less energetic but longer lasting energy source is based on an anaerobic reaction of glycogen. Enough of this is stored to allow a sprint of a few hundred yards. Yet within minutes of such activity, energy for muscle activity must come from aerobic reactions that require a continuous source of oxygen. Numerous mitochondria in the muscle are the sites of this energy conversion.

My favorite example of the differences in energy sources and the need for supporting blood flow for the transport of oxygen is the shrimp boat captain who played baseball across the street from the hospital. He stretched a single to second base, but while standing up after his slide, collapsed. His muscles were demanding more blood for payback of borrowed energy to the emergency stores than could be forced through his stenotic aortic valve, and when pressure to the brain fell, he lost consciousness.[6]

Given that O_2 consumed is proportional to energy produced, measurement of its consumption is a key element in studies of physical performance. Increased demand for O_2 by muscles during exercise is remarkable. Resting muscle O_2 consumption is normally 2–4 mL per 100 g muscle^{-1} per min^{-1} but can increase to 100 mL per 100 g muscle^{-1} per min^{-1}. Figure 8.3 shows some sports and exercises and their maximum O_2 consumptions. The maximum capacities shown are produced by years of aerobic training in champion athletes [7]. Such prolonged increased activity causes significant changes in muscle. Fiber types become predominantly red, slow twitch fatigue resistant, aerobic Type I. As noted, maximum performance in some of these activities are limited by ability of the CR system to provide O_2 to muscle. At the same time such trained muscle becomes both more efficient and able to extract more oxygen from blood.

8.3.2 *Vascular System (Fig. 8.2b)*

This system has two significant aerobic functions: (1) transport of energy substrates and (2) a compliant reservoir of blood. Components of energy production other than oxygen and their degraded products including carbon dioxide will receive little attention here since they are so effectively transported. Conversely O_2 is so poorly soluble in H_2O that only 3% is transported in solution by blood. Species adopted molecules of hemoglobin (Hb) contained in red blood cells carry 97%. At a normal partial pressure of about 135 mmHg, O_2 in the lung, Hb is fully saturated with oxygen and

[5] Adenosine Triphosphate and Phosphocreatine.

[6] One of his heart valves was later repaired in the hospital.

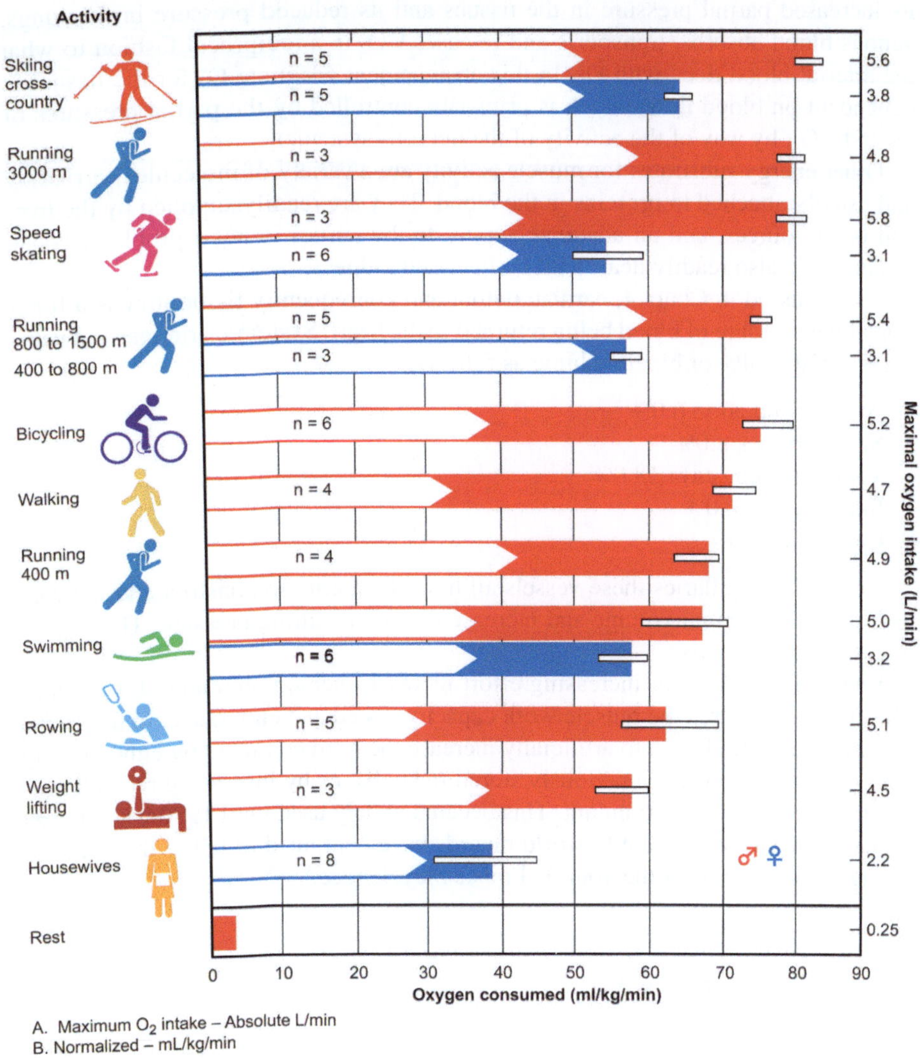

Fig. 8.3 Maximal oxygen intake for world class athletes, except housewives, on treadmill. These are near the upper limits of human capacity, achieved by a few after years of effort. The *red and blue bars* represent the range of measured values from male and female populations. This is also an indication of workloads imposed by a number of activities

carries 1.34 mL oxygen per g of Hb as HbO. Since blood normally has 16 g of Hb per 100 mL blood, the maximum capacity of blood is 20 mL O_2 per 100 mL of blood.

On reaching the tissues with their much lower partial pressures of 20 to 40+ mm of O_2, up to 80% of the O_2 is released. This is replaced by carbon dioxide.[7] Under

[7] There are complex interrelations of CO_2 and O_2 uptake and release that would add little to our concerns.

its increased partial pressure in the tissues and its reduced pressure in the lungs, venous blood absorbs, transports and releases CO_2 in a reciprocal fashion to what the arterial blood does with O_2. In this arrangement adequate O_2 for the tissues is dependent on blood flow, which is primarily controlled by the partial pressures of O_2 and CO_2, by way of the activity of the nervous system.

Other energy substrates for muscle activity are a variety of molecules carried by and usually attached to proteins in the blood. They are readily supplied by the liver and other sources, and an adequate supply to the tissues is not a problem. Waste products are also readily dealt with by liver and kidneys.

As discussed in Chap. 4, cardiac output and consequently tissue flow is a function of the pressure of blood being returned to the heart. Systemic arteries[8] and veins contain some 84% of blood volume as follows:

- Arteries and aorta 6.0%
- Small arteries 8.0%
- Veins and vena cava 18.0%
- Small veins 46.0%
- Capillaries 6.0%

Except for capillaries these vessels all have the ability to constrict and increase their pressure per unit volume and increase the heart's filling pressure. This occurs with work and exercise.

Endurance training at increasing effort increases hemoglobin and O_2 transport. This in turn increases the muscle work capacity, speed and endurance. In the 1970s endurance athletes began to artificially increase their red cell mass by either adding their own or compatible, previously stored red cells, or by increasing red cell production with a medical stimulant. This became widely used until 1980 when it was declared illegal. Subsequently world records have continued to fall to this practice including seven wins of the Tour de France bicycle race.[9]

8.3.3 *Lungs (Fig. 8.2c)*

Lungs provide the interface between blood and air, with a surface area of some 70 m^2, and miles of capillaries. Air is taken in and expelled through nose and mouth, conducted into the chest through a windpipe, the trachea, which divides into two bronchi, and continues to divide until it ends in the millions of microscopic air sacs or alveoli with their membranes that allow O_2 to pass out into a surrounding net of capillaries while CO_2 from blood passes into the alveolar air. Normally some 5 L of blood per minute are refreshed this way, but for a world class cross country skier or runner who has trained many years this can reach 30 L per minute and last for hours

[8] Vessels outside heart and lungs.

[9] It has been difficult to find valid studies of the increase in O_2 uptake or increase in performance with the increase in hemoglobin concentration.

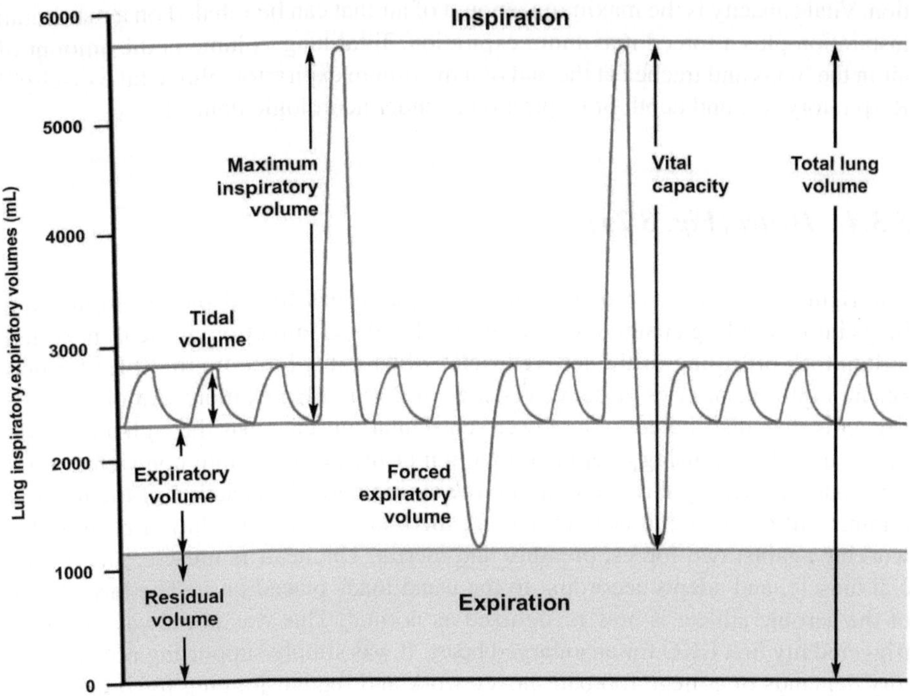

Fig. 8.4 Typical normal male inspiratory and expiratory volumes. This is a revision of the classic drawing of spirometer records with modified notations. Measurement of the residual volume (*yellow*), which cannot be expired, requires other measurement techniques (color figure online)

during a race or training. Figure 8.3 shows maximum oxygen consumption for individuals trained in various activities.

Air in the alveoli is normally exchanged for a fraction of a liter of fresh air some 16 times per minute, but this can reach 140 L or more with respiratory rates of 60 exchanges per min by some aerobic exercises. Air is pumped in and out of lungs by a musculoskeletal plus a muscle only arrangement [8]. At the end of expiration, ribs are pivoted up and out by action of the intercostal muscles. This increases thoracic volume[10] and creates a negative pressure in alveoli and lungs, which are rapidly filled by air under near atmospheric pressure. Volume and negative pressure is further increased by coordinated contraction of the domed diaphragm stretched across the bottom of the thoracic cage.

The amount of air that can be inhaled and exhaled by a normal average person is shown in the revised classic diagram of Fig. 8.4. Tidal volume is the volume normally inhaled and exhaled at rest. Maximum respiratory volume starts at the end of normal expiration and is a maximum intake while forced expiratory volume is measured from the end of a normal inspiration to the end of a forced maximum expira-

[10] A hallmark of aerobic athletes is their large chest (thorax).

tion. Vital capacity is the maximum amount of air that can be inhaled on a maximum inspiration plus a forced maximum expiration. Total lung volume is the amount of air in the lungs and trachea at the end of a maximum expiration plus vital capacity.[11] Respiratory rate and depth of respiration is under neurologic control.

8.3.4 *Heart (Fig. 8.2d)*

The right side (ventricle) of this muscular pump has forced blood through the lungs into the filling chamber (atria) of its left side. Contraction of the thin walled atrium tops off filling of the left ventricle, which must force up to 30 L of blood per min into the arteries at peak pressures of 200 mmHg or more. An electrical[12] trigger is then distributed to the lower ventricular muscle mass through a network of electric fibers causing a rapid contraction of the powerful interconnected muscle[13] mass to literally throw blood into the elastic aorta. This action of the heart is reminiscent of the leg muscle projecting the body into the air during running by working against two forces, pressure and inertia. The heart is muscle, albeit special muscle, and adapts according to the usual loads placed on it. The large heart of the aerobic athlete is now recognized as normal. This was not always so, and triggered my first EKG for an enlarged heart. It was simply supporting my circulatory demands of college football, heavy work and regular jogging for miles, also considered abnormal at the time.

8.3.5 *Controls (Fig. 8.2e)*

Chapter 4 describes some of the cardiovascular controls, but there is another level that controls response to aerobic demands. A nominal resting demand of 5 L blood flow per minute is met by a heart rate of approximately 70 beats per minute with approximately 70 mL of blood per beat. Increased O_2 demand produces an increase in rate that is almost directly proportional to demand. Normal heart rates range from 40+ bpm at rest in aerobic athletes to a maximum of approximately 205–215 bpm in the first year of life. Each successive year of life will reduce the individual's maximum rate by approximately 1 bpm irrespective of training.[14] Maximum cardiac

---

[11] Some designations have been changed from clinical usage and waveforms corrected.

[12] Lower chambers of the heart are electrically insulated from the upper to maintain the trigger system's control.

[13] Heart muscle is striated like the body's somatic muscle, but it is also interconnected such that contraction of one fascicle insures contraction of the adjacent connected fascicles.

[14] This rate reduction with age is considered obligatory in cardiology, but exceptions are seen in some athletes.

output is limited accordingly but somewhat compensated for by increased chamber volume with age.

Control of the blood levels of O_2 and CO_2 is critical. All tissues of the body are dependent upon adequate level of O_2 for energy while the body's chemical reactions are dependent on the number of hydrogen ions present, which is expressed as pH. This in turn is dependent upon CO_2 level in the blood.[15] A network of sensors plus the respiratory, cardiac and vasomotor centers in the medulla and brain stem control both O_2 and CO_2 level by varying rate and depth of respiration, cardiac stroke volume, heart rate, blood volume, flow and distribution [9].

Continued demands from regular aerobic activity or training produces increased in mass and size of the heart to increase stroke volume. The enlarged heart of the athlete is now usually recognized by physicians as the normal adaptation that it is. A number of aerobic fitness features, especially in exercise, are incompletely understood or appreciated. For example, when I applied telemetry to athletes at the University of North Carolina fifty plus years ago, one of the early findings was that the largest increase in heart rate in competitive mid-distance runners occurred in the blocks before starting. Obviously higher brain centers exert a significant influence.[16]

8.3.6 *Measurement of Aerobic Fitness*

Before looking at how well all these systems play together in 1 g and in space, a brief review of techniques for measurement of their function is included. As noted in the history of the problem of measuring aerobic fitness, there was some understanding of it when the first astronauts were selected. Physical stress tests were available and used for suspected cardiac malfunction and for evaluation of CR function. Applied stresses were similar and there was considerable overlap of protocols for application of the stresses. Such testing is now common in medicine and sports. There has been remarkable progress in the technology used.

To measure maximum aerobic fitness a large mass of muscle must be used in a way to consume the maximum O_2 and stress the cardiorespiratory systems. Only the masses of muscle in legs and low back are large enough for this, although other muscles may add to the total consumed. Measurement of aerobic fitness begins with a measured or known physical workload stress in a known schedule or protocol [10]. The two most common loads or stresses are walking on a treadmill and pedaling a cycle. These need to occur at known rates (speeds), inclines or pedal loads. Major errors are possible and all too common, for the equipment manufacturer's data is often used without calibration. A second common source of error is in the design and fabrication of the hardware. A common fault in both treadmills and cycles is lack of an adequate inertial component of the load. Running or pedaling

[15] CO_2 in water undergoes this reaction, producing hydrogen ions: $CO_2 + H_2O \rightarrow HCO^- + H^+$.

[16] However human runners' heart rate response is relatively crude to that of a trained horse whose total increase in rate occurs in one or two steps of starting and is appropriate to the gait selected.

against largely resistive components without an adequate inertial component load leads to muscle fatigue and premature termination of the test. The goal is to progressively increase muscle demand for O_2 until capacity is reached. This is designated as maximum oxygen consumption or Vee dot max in the usual jargon.

Measured input load quantities are:

- Treadmill: velocity in miles or kilometers per hour

  grade %: vertical versus horizontal distance traveled
  time (on treadmill): minutes and seconds

- Cycle ergometer: angular velocity in revolutions of pedals per minute

  load: watts/minute
  time (on cycle): minutes and seconds.

Load is usually increased in steps. Response is exponential and requires 3–4 min to reach a steady state. Three minutes per step is typical, but a few protocols that increase the load in smaller increments may use 1-min intervals. There are at least a dozen protocols for treadmill stress tests, but all use some combination of increasing speed and grade with fixed time increments or steps. Two that have stood the test of time are plotted in Fig. 8.5. The older Balke is a fitness test while the Bruce is a cardiac diagnostic test that has become the most currently used test in both fitness and cardiac testing.

Cycle ergometry protocols use a fixed pedal rate or preferably a cycle whose workload does not vary with pedal speed. Typical pedal rates are 60–80 revolutions per minute. Only the load is incrementally increased with time to the subject's capacity. Cycle protocols are more likely to be tailored to the situation and investigator. Ergometers vary widely in their design, especially in the physical characteristics of their load-and it must be said, in users understanding of the significance of these characteristics.

The primary measurement of the subject's response to the load is the rate of oxygen consumption. Once, large complex volumetric and gas analysis equipment was required to make these measurements, but now a flow meter with air sampling, spectroscopic gas analysis and a computer make even ambulatory monitoring possible. Oxygen consumption, often called oxygen or O_2 uptake, is calculated from the difference between oxygen content and volume of inspired versus expired air.

A variety of other measurements may be made including volume of air inspired or expired, CO_2 produced, heart rate and blood pressure. They are used to evaluate other aspects of the subject's performance. Oxygen consumption may be expressed as volume per unit time, liters or milliliters per minute (mL min^{-1}), $\dot{V}O_2$. Those in the trade often call this "vee dot O two". In evaluation of a subject's fitness this is the Maximum O_2 Intake (MOI) during the test. This may not coincide with the maximum work level for some trained individuals can briefly sustain a maximum work level by anaerobic means and actually reduce O2 intake at their maximum work load. See Fig. 8.8b. Taken without other individual data, or especially when combined with a group. Maximum oxygen intake is virtually meaningless and often

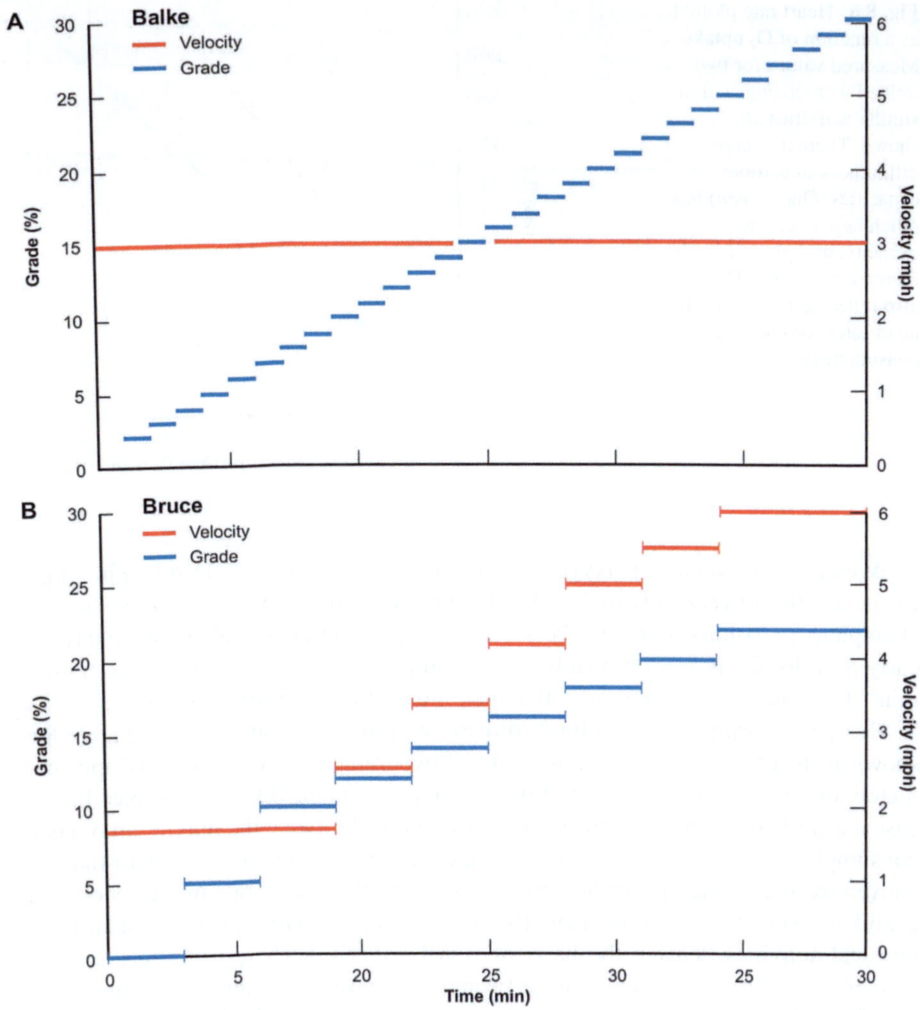

Fig. 8.5 Balke (**a**) and Bruce (**b**) Stress Tests Protocols for increases in speed and grade on a treadmill with time are diagrammed. There have been frequent changes made by others resulting in many variants of these basic tests. The Balke was originally a cardiorespiratory performance test, and the Bruce a cardiac stress test, but there are frequent crossovers in their usage. Note that the increase in load is linear and increases more slowly in the Balke test. Both grade and speed are increased in the Bruce test, which result in large exponential increases. While the Bruce test takes less time and is most popular, the Balke may produce more accurate measures of maximum capacity

misleading. For example, a massive body builder without aerobic exercise may have a MOI equal to a 135 lb female distance runner.

The first step in giving meaning to MOI is normalizing it by expressing it in terms of body mass; mL min^{-1} kg^{-1}. It may also be expressed in kilocalories. Calories may be approximately converted to liters O_2 by dividing by five.

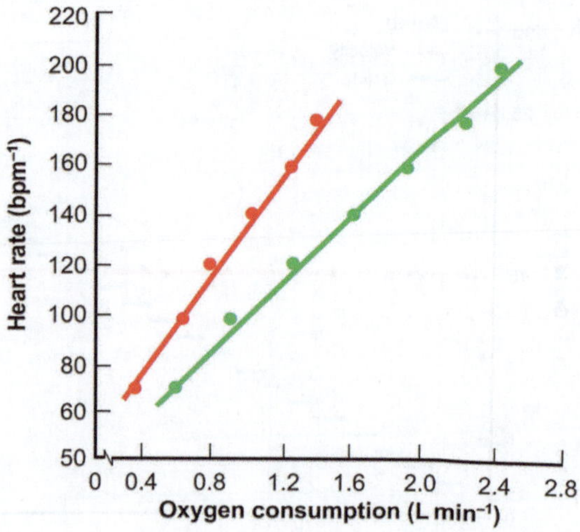

Fig. 8.6 Heart rate plotted as a function of O_2 uptake. Measured values for two trained women engaged in similar activities are shown. There are large differences in aerobic capacities. One (*green*) has much larger aerobic capacity, probably through increased training. This is also reflected in her lower heart rates for the same O_2 consumption

Average adult values of oxygen consumption max range 23–44 mL min^{-1} kg^{-1} for males 30–60 year olds to 21–39 mL min^{-1} kg^{-1} for females. Some values for champions in various sports can be seen in Fig. 8.3. Astronaut values in the days of early Shuttle flights were typically 40 mL min^{-1} kg^{-1} with a few reaching 60 mL min^{-1} kg^{-1} but at least one flew with a maximum of only 25 mL min^{-1} kg^{-1}.

Heart rate normally varies almost directly with oxygen consumption. Figure 8.6 shows the heart rate versus oxygen uptake of two trained female athletes of approximately the same size. A good bit of information is contained in such graphs. In this case one athlete has greater aerobic fitness as shown by twice the maximum oxygen consumption. Also her CR system is more efficient requiring a lower heart rate for a given oxygen consumption. When counted correctly,[17] heart rate in 1 g becomes a useful measure for the individual. Heart rate coupled with various standardized physical activities became the basis of Ken Cooper's Aerobics, one of the more valuable contributions to health in the twentieth century [11]. Heart rate is much too interdependent and variable to be used as some indication of MOI.

Benefits of aerobic fitness, such as escaping predators or chasing prey, have undoubtedly been recognized since prehistoric times. They directly increased chances of survival throughout the history of our species. Training to increase such capacity has also been part of human history, but only in the past hundred years have the complexities of this process been objectively understood with much of the practical understanding being discovered in the past 50 years.

Aerobic condition is determined by adaptation to the muscle activity of the individual, on Earth or in space. This adaptation is specific with two components: first the ability to extract oxygen from the air and transport it to muscle, which

[17] Accurate inexpensive EKG wristwatch monitors are readily available. Rate changes too rapidly with cessation of exercise for pulse rate to be accurate when counted after exercise has stopped.

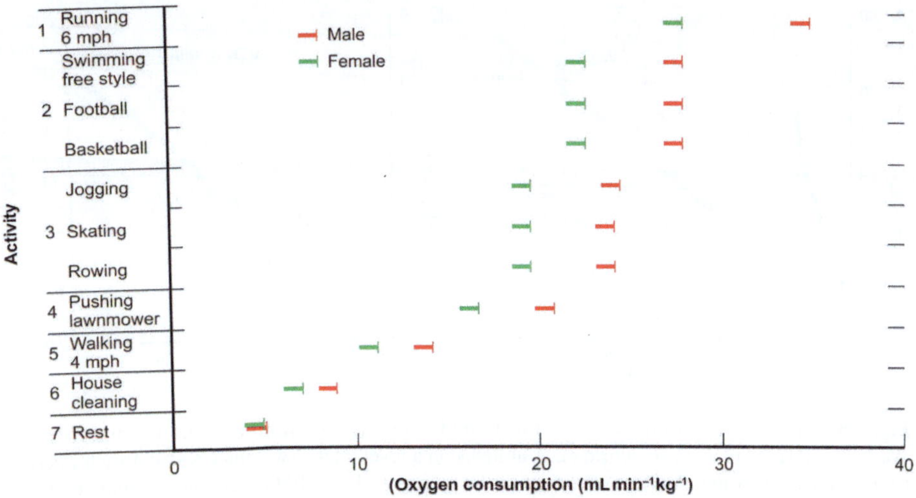

Fig. 8.7 Approximations of oxygen consumption for mature, untrained men and women for various activities. Actual intake will vary over a wide range depending on the intensity of effort. See Fig. 7.9 for more detail of locomotion exercises

converts its chemical bonds to mechanical motion, and the second capacity is endurance, the ability to continue the process for long periods, which may last for hours in long events such as marathons. Maximum capacity measured in terms of maximum oxygen intake normalized for subject mass is analogous to strength adaptation in muscle. Aerobic endurance is seldom measured since it could require significant time.

There is a chain of different tissues in aerobic fitness that includes muscle. In strengthening muscle, aerobic slow twitch fibers are converted into anaerobic fast twitch fibers with increased strength for adaptation to such activities as sprints or weight lifting. The opposite occurs in distance running, skiing, and cycling and other aerobic exercise that converts fast fibers to slow twitch fibers. Since aerobic fitness is dependent on a chain of tissues, each must increase its particular capacity, be it carrying oxygen or pumping blood. This chain is no stronger than its weakest link.

8.3.7 Aerobic Fitness in 1 g

Nature and duration of physical loads in work and exercise determine aerobic fitness. There is a vast literature available on the oxygen costs of various physical activities [12]. Figure 8.7 is a small example of such costs. While it is fascinating and tempting to explore the costs and benefits of such activities, only those of concern here will be examined.

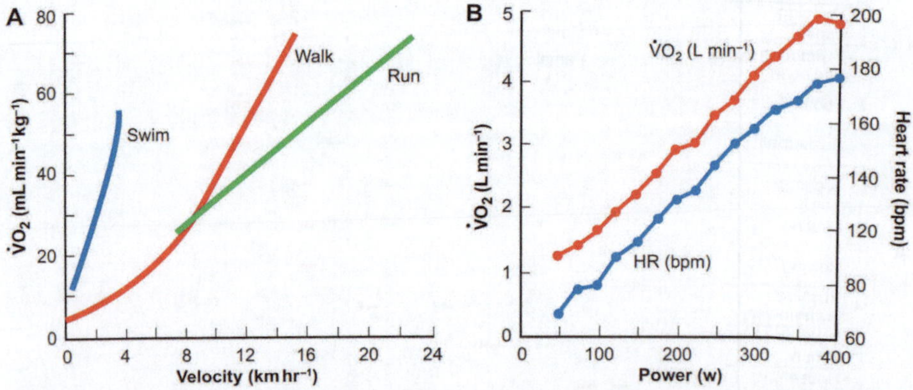

Fig. 8.8 (**a**) Oxygen consumption as a function of velocity in swimming, walking, and running. Walkers and runners were six world class athletes on a level track. Swimmer was highly trained in a tank with 18 °C water. Walkers were four race winning walkers. (**b**) Oxygen uptake and heart rate as a function of work load on a cycle ergometer. Pedal speed was constant. This is a classic record of a well trained cyclist who reached oxygen consumption max as shown by the plateaus in $\dot{V}O_2$ and HR. This same curve will be seen in a less trained individual who may reach 3–4 L min^{-1} $\dot{V}O_2$ with a more rapid rate of increase in HR versus load. Courtesy of Alan Moore, Ph.D.

Locomotor activity, walking, running, bicycling and swimming are common aerobic activities for astronauts in 1 g and two of these activities, running and bicycling, are carried out in space. Rapid, continued strength exercise can also produce large oxygen demands but will not be considered.[18] Conversely usual strength training is not aerobic (see Fig. 8.3).

Energy costs of bipedal locomotion and swimming are shown in Fig. 8.8a and for cycle ergometry in Fig. 8.8b. Basic locomotor training for increased aerobic fitness consists of regular sessions at greater than usual speeds for given distances or times, typically 2–3 miles or 30–45 min. This should be monitored by heart rate, which will decrease as capacity increases. Speeds are progressively increased until the desired level of fitness is reached. If increased duration is desired then training sessions of increasing distance are alternated with those of increased loads. As always, adaptation is specific.

If 90% of a subject's ultimate capacity is the goal then the training stimulus or countermeasures must provoke a 90% CR response. All too frequently duration is confused with maximum capacity. Eight hours per day at 50% will not produce 90% of capacity, while a few minutes a day at 90% can maintain that level. If endurance is desired then it too will have to be trained for by extending the time spent at 90% of maximum. This is why competitive distance runners spend years and countless miles running in practice.

[18] A total body exerciser in which the subject did rapid, full extension lifts was developed and tested including zero-g flights in 1966 for the USAF Manned Orbiting Laboratory (MOL) program. It could produce maximum oxygen intakes approximating those of running.

Such specificity is also present in measurement methodology. The treadmill is the all around best load for evaluation of O_2 uptake, but not the best for dedicated cyclists or swimmers, unless they cross train with running. On Earth such cross training is inevitable for the most dedicated cyclist or swimmer will spend a considerable amount of time walking. However, stress testing a runner who does not train on a cycle using a cycle ergometer will not measure his maximum capacity, nor will testing a dedicated cyclist on a treadmill.

The concept that cycling can maintain strength is a much more serious error. Pedal forces are too small to maintain leg strength adapted to walking or running. Apparently this was not recognized when Skylab was flown with only a bicycle ergometer, where duration was increased on each flight in an apparent attempt to increase leg strength. But locomotion combines strength and endurance exercise. In contrast to the low extended leg forces produced by cycling, walking requires that each leg alternately lift one body weight (BW) each step taken (see pp. 119 and 150). With running much larger forces are necessary to overcome the body's inertia and lift it clear of the ground. This can reach three or more BW.

8.3.8 *Effects of Bed Rest on Aerobic Fitness*

It had been long recognized that extended bed rest produces a number of deleterious effects. CR effects of bed rest were first measured in bed rest studies for space [13, 14]. But bed rest is not, by its nature, an analog of CR stress in spaceflight. Muscle activity, especially of the legs, is insofar as possible avoided. Activity, while even without locomotion, is if anything increased in weightlessness as shown by the increase in caloric demands inflight over those preflight. This was shown unequivocally by the Skylab mass measurements (see Chap. 10). While there is a plethora of bed rest data, with and without countermeasures, in this case it is meaningless as an analog to weightlessness and has led to errors such as the inadequate diet on early Skylab missions. Results of early bed rest studies with and without exercise are shown in Fig. 8.9.

One small bed rest study by Saltin in 1968 remains relevant as a definitive example of aerobic fitness loss with inactivity [15]. This remains a definitive study. Three sedentary and two athletic college males were measured pre-, in- and post-bed rest. After 19 days of bed rest they received intensive CR rehabilitation for 53–55 days. Effects of bed rest and exercise with the resulting loss and gain of aerobic fitness in this small study of five subjects remains the most rigorously conducted and documented to date.[19] Selected excerpts from the data illustrate the effects of physical activity on the major systems in the CR complex shown in Fig. 8.10a–d and described in the following paragraph. Unfortunately the effects on muscle were not

[19] "It is apparent that vast differences in scope and methodology, especially as regards the degree and duration of immobilization, make quantitative comparisons between [bed] rest studies difficult or impossible," Brent Saltin.

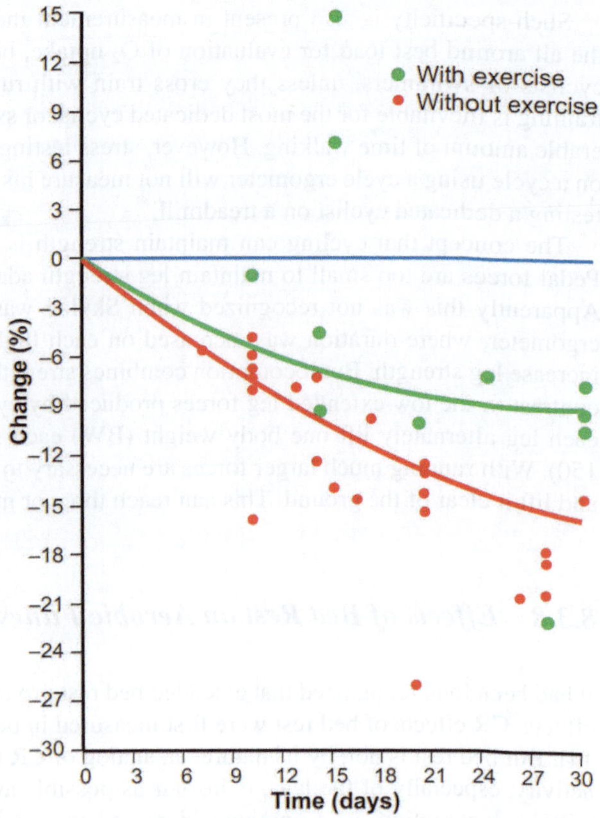

Fig. 8.9 Change in aerobic capacity in bed rest with and without exercise. Data are means from 33 early bed rest studies. Bed rest is inherently not an analog of loss of aerobic capacity in spaceflight

measured. An overview of the effects is seen in loss and gain of maximum O_2 uptake in Fig. 8.10a. In less than 3 weeks of bed rest there was a 30% reduction in the three nonathletic subjects and 21% reduction in the two athletic subjects. Note that the time for the non-athletes to regain their usual capacity was less than half the timerequired to lose it. Conversely it took much longer for the athletes to regain their greater losses. Nothing can be said about the final level that would have been lost with continued bed rest or the time it would have taken to fully adapt to it.

Looking through the systems involved in Fig. 8.2 during bed rest, the ability of muscles to use oxygen would have been reduced as would muscle strength and endurance. Moving to the oxygen transport system, little is known of changes in blood vessels except that change in size, compliance and distribution of small vessels in muscle do occur. Transport of oxygen depends on the number of red blood cells and total blood volume, both of which were slightly affected in this study (Fig. 8.10b). Loss of blood volume will be larger in weightlessness as a result of fluid shifts. (See Chap. 5.)

The most obvious changes occurred in heart size (Fig. 8.10c) and in heart rate (Fig. 8.10d). Clinical experience shows that the heart is capable of extraordinary

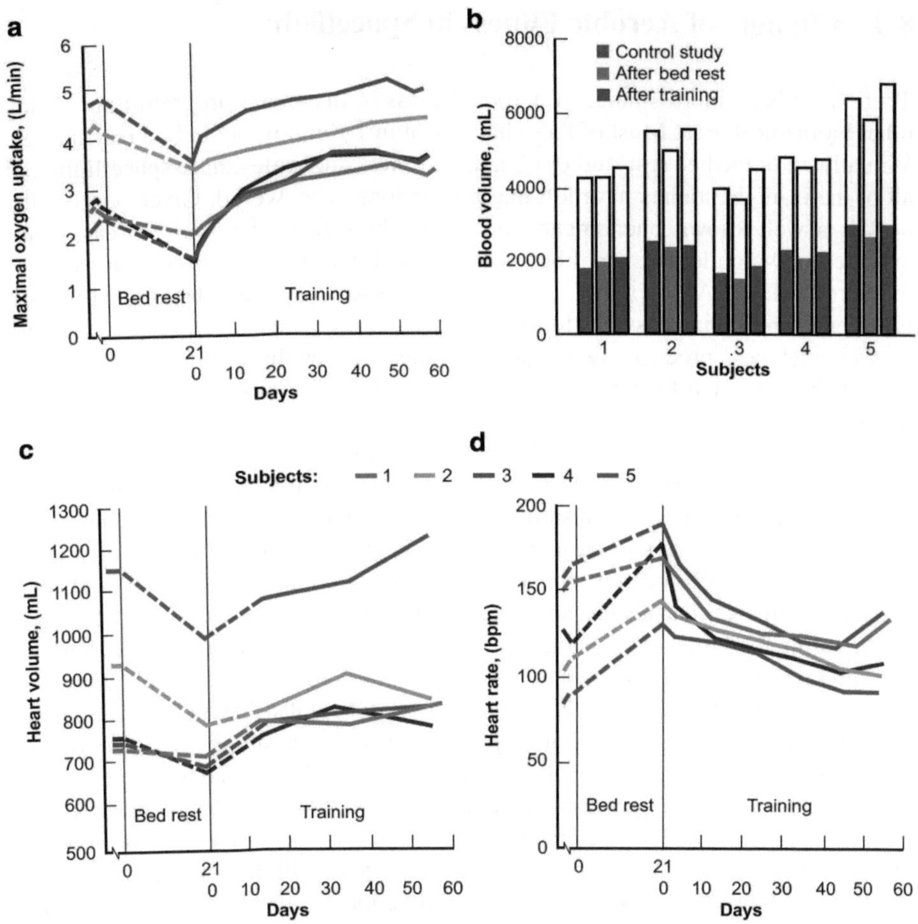

Fig. 8.10 Data plotted from Saltin study [15] of bed rest and rehabilitation. (**a**) Changes in maximal oxygen uptake with bed rest and training. Individual data before and after bed rest and at various intervals during training. (**b**) Blood volume (*upper panel*), red cell mass, and plasma volume (*upper panel*) in the control study (C), after bed rest (B), and after training (T). Individual data and mean values. (**c**) Changes in heart volume with bed rest and training. Individual data before and after bed rest and at different intervals during training. (**d**) Heart rate during exercise at a standard submaximal oxygen uptake and after bed rest and at various intervals during training. Heart rates at oxygen intake of 1.5 L/min calculated by linear interpolation from measurements at several submaximal oxygen uptake levels

changes in size, shape and pumping ability.[20] Size changes entailed changes in chamber and muscle volume and part in changes of muscle volume. Similar changes may be seen in adaptation to various conditions on Earth or in space. Again this represents a worst case for loss of aerobic fitness and best case for its gain. Unfortunately there is no such study in women.

[20] It is not unusual for a cardiac patient to be in shock from loss of infarcted muscle yet to walk out of the hospital a week later after the remaining muscle had increased in size and strength to carry the usual load.

8.4 Changes of Aerobic Fitness in Spaceflight

There have been more studies of aerobic fitness in the space program than in any other biomedical area. Most of these have been in 1 g and many in bed rest studies. Virtually all of the bed rest studies of aerobic fitness are irrelevant to spaceflight and all of the in-flight studies have left major questions unanswered. Given aerobic fitness is a well known phenomenon on Earth, the effects that can be reasonably expected in weightlessness will be described, and whenever possible, compared to measured effects. An appraisal of operational aspects of this adaptation for near Earth orbit and Lunar missions follow.

Weightlessness produces two significant effects on cardiorespiratory capacity in spaceflight. Legs and their mass of muscle are relatively unused. This removes the major stimulus to maintenance of aerobic condition. Some 2–4 L of fluid and 400–500 mL of blood are rapidly shifted out of the legs and then lost in the first 3–5 days of weightlessness. In 1 g maximum oxygen consumption is directly related to blood volume, and while this relation is presumed to hold in weightlessness, it remains unproven.

Exploration of these two basic effects will begin with energetics in weightlessness. Basal metabolism can be considered the lowest common denominator of normal oxygen uptake. The nearest approach to this inflight was measurement of uptake while resting for 5 min on the Skylab cycle ergometer prior to the start of the exercise measurement period. A mean value of 2.7 mL min^{-1} kg^{-1} was obtained. This is a normal value for 1 g. One might reasonably expect to also have O_2 consumption measurements for a range of activities given the availability, simplicity, and portability of commercial instruments, but such measurements still remain to be done.

Estimates of mean uptake can be made from food intakes. Using values for two crewmen in metabolic balance on the 84-day Skylab mission, one obtains mean O_2 intake of 6.5 mL min^{-1} kg^{-1} for a 16-h day[21] [6]. There will be some routine activities that require several times that, but none will approach the 30–40 mL kg^{-1} min^{-1} needed to maintain maximum oxygen intakes at 1-g prelaunch levels. On ISS missions of 80 days these numbers were smaller with mean uptake of 4.56 mL min^{-1} kg^{-1}, which is significantly less than that of the all male and very active SL-4 crewmembers.

Following Challenger's loss the crew was custom fitted with partial pressure suits for emergency egress during launch or landing. They were clumsy, heavy and initially, inadequately cooled. The energy required by these suits in simulated emergencies was larger than the capacity of some astronauts on return from a week's mission.

One thinks of EVA as placing very high energy demands on the user, and there were several Gemini EVAs where aerobic demand was greater than the astronaut's aerobic capacity or heat production was greater than suit dissipation. An inflight EVA typically requires the use of only the trunk and arm muscles with the legs providing stabilization, if used at all. While tasks may be strenuous and perceived as such, maximum O_2 consumption can be only a fraction of MOI measured by tread-

[21] O_2 consumption during 8 h sleep was assumed to be at BMR rate of 2.7 mL kg^{-1} min^{-1}.

Table 8.1 O_2 consumption of various EVA suits

	L min^{-1}	mL min^{-1} kg^{-1}
Apollo suit	1.0–2.5	–
Shuttle egress/entry suit	2.0–2.7	25.0–33.8
Shuttle/Mir EVA	0.7	8.8
U.S. suit (maximum activity)	1.7	21.3
Russian suit (maximum activity)	2.7	33.8

mill or cycle ergometry using the major mass of the body's muscle in the back and legs. This limited muscle involvement in EVA is reflected in the relatively modest O_2 consumptions shown in Table 8.1.

Routine EVAs have modest O_2 demands. When activity is at its maximum, the demands of NASA's suit is within the capacity of most astronauts. Russian suits are inflated to higher pressures and are more resistant to motion than NASA suits, as indicated by their O_2 cost at maximum activity as shown in the table.

Traditionally EVAs have been considered the upper limit of energy demands; however, unexpected demands do occur; e.g., Grissom in a water-filed suit after the Mercury hatch inadvertently blew off; fire in Mir; high CO_2 levels and landing errors without ground support: i.e., cosmonauts having to fight wolves, blizzards and snowshoeing out of a wilderness. Unplanned and unknown demands will recur in space operations and exploration. Such events [16] and the concomitant health benefits would seem to make it prudent to maintain the best possible aerobic fitness commensurate with other mission demands.

8.4.1 *Loss of Aerobic Capacity and Its Measurement*

The first NASA study of flight effects on the cardiorespiratory system was done on Mercury orbital flights. It consisted of 30 s of repetitive arm extensions of elastic tubing between hands and feet while heart rate was monitored. Flight produced no significant changes in response to this "stress test" but transient orthostatic heart rate increases were observed post flight.

Gemini repeated the Mercury inflight EKG stress tests several times on three flights and again no changes were noted. A pre/post flight ergometer cycle test was added. Subject load were increased to a cut off limit of 180 bpm heart rate. Post flight cutoff limits were produced by notably lower ergometer loading in all but one astronaut but quickly returned to preflight levels.

Twenty seven of the Apollo astronauts did submaximal exercise testing with a cycle ergometer and O_2 consumption, pre and post flight, at work levels which produced Heart Rates (HR) of 120, 140, and 160 beats/min. Six astronauts also went to 180 beats/min. Typically, O_2 consumption was reduced on landing but had recovered to preflight levels in 24 h. In summary, these flights produced no significant changes in the cardiorespiratory system except for post landing and this typically lasted 24 h. It was accompanied by lost weight and reduced plasma volumes.

Skylab used a cycle ergometer, a respiratory gas analyzer, and body mass measurement pre, and in and post flight. Diets were controlled and the exercise and testing done with a frequency and rigor not seen subsequently. The published result was that no changes were seen in exercise or cardiorespiratory capacity in flights of up to 86 days. One or two crew members may have gained capacity. This was based on published data and the primary investigators' conclusions from Skylab experiment M-171, which attempted to measure $\dot{V}O_2$ max on the SL-4 crew. There were limitations in the Skylab gas analysis equipment and in maximum cycle ergometer loads that combined to indicate 1-g maximum capacity had been reached inflight. In-flight fluid shifts and losses plus changes in capacity make interpretation of data complex. Closer scrutiny of this data plus application of fluid shift data from experiment M-174 and a DSO on fluid shifts do not support the principle investigator's conclusions that maximum oxygen intakes comsumption was measured. To add yet another complication, in weightlessness at low to mid levels of O_2 consumption, consumption versus indicated workload is lower inflight. How much of this is in the weight loads of the legs in 1 g versus its absence in weightlessness, and how much it had to do with cardiovascular differences is unknown. This is a significant detail that should have been measured before Skylab flew.

Shuttle had such a mélange of equipment, schedules, studies that chapters would be required to make a reasonable description and to disentangle and reconcile the results. This included exercise and cardiorespiratory experience and studies. These observations are experience based until 1994. After that information became increasingly difficult to obtain. The TW treadmill was on all flights until STS-63 after which various exercise modalities and studies were flown by the Extended Duration Orbiter Medical Project in preparation for ISS. Pre entry water and salt loading became standard practice. Attempts to modify transient post flight effects by inflight lower body negative pressure failed as could have been predicted from a series of in flight fluid shift studies. The usual results from exercise-O_2 consumption studies in this period was that they were little changed in 1–2 weeks time except for the immediate post flight transient losses.

ISS, with its capacity and potential resources promised to continue and dramatically extend Sklylab's legacy. In the cardio respiratory field the forty intervening years had seen remarkable advances in theory and practice, especially in measurement and data storage and processing. Ambulatory recorders now included respiratory gas analysis, activity, pulmonary, and the entire spectrum of electro physiology. Treadmills and cycle ergometer designs and prototypes for flight operation had been demonstrated as had accurate mass measurement devices.

During the first and second ISS Expeditions there were rumors and media reports of problems with exercise equipment. In Dec. 2015 the first official report on exercise in the first 10 years on ISS described some of the avalanche of problems, delays and program changes caused by the early failures and malfunctions of all three exercise devices on ISS and possibly why there was little reported work on loss of cardiorespiratory function in the first decade of ISS.

Now we turn to the major issue of this chapter, the real loss of cardiorespiratory capacity in weightlessness, its current measurement and prevention or modulation of this loss in space flight. Given the time and resources expended in preparation for such studies by JSC life sciences, the absence or failure of ISS facilities for the study and control of this problem in 10 years is remarkable. Preparation for ISS began in 1989 with the Extended Duration Orbiter Medical Project (EDOMP) [17].

In 2015 Moore [18] and colleagues, in "The First Ten Years of "Aerobic Exercise Responses to Long-Duration ISS Flights" sect 78 in the first decade supplement on exercise in ISS, reported on the first attempted aerobic study. Moore had previously published results of an earlier study. These studies are best treated in chronological order. Before reviewing their results, some scrutiny and comments on the methodologies and their analysis used seem appropriate. As noted in the History of functional physical examinations, physical stress tests became widely used as a diagnostic cardiac test with EKG, blood pressure, etc. providing cardiac signs. When respiratory gas analysis and body weight were added, much more information about the condition of other systems of the body became available. This can be seen in Fig. 8.2 that illustrates the major systems in the chain examined in an aerobic stress test. This test, usually taken to 100% of physical capacity, has long been used in the selection, annual retention and pre/post flight physical exams of astronauts. Note that the heart and its EKG is only about half way through the chain of events in the aerobic system and cannot provide a measure of aerobic function. To do this individually requires O_2 consumption and accurate body weight or mass. Validated body mass measurement not been available on ISS.

Cardiac and aerobic exercise stress tests were scheduled pre, in and post flight on ISS [18] but the inflight segment of aerobic testing was cancelled by delayed delivery of a gas analyzer. Pre and postflight testing began with a maximum work capacity determination on a cycle ergometer 9 months before flight. Individual maximum O_2s were used to derive work loads that would produce 25, 50 and 75% of MOI on submaximal testing 30–60 days pre flight and 5 and 30 days post flight. EKG monitoring was done on all tests. This protocol was used in flight every 30 days beginning with flight day 15 but without respiratory measurements. Thirty male and seven females were, tested, typically over 180 flight days. The data from each gender was separated and averaged for a number of statistical processes. While the results appear reasonable a more detailed set of data would prove valuable.

Moore and colleagues conducted another investigation of ISS inflight cardiorespiratory responses to cycle work loads to capacity during 2009–2012 with nine males and five female volunteers with most of the essentials for conduct of a valid maximum exercise stress test with measured O_2 uptake [19]. Validated inflight body mass was not available and the large intervals between measurements compromised assurance of some aspects. The study used an Ergometer with Vibration Isolation and Stabilization System cycle ergometer (CEVIS) and ISS Portable Pulmonary Function System (PPFS). Baseline 1-g measurements of oxygen intakes to individual maximum workloads were done approximately 90 days prior to launch, and repeated every 30 days inflight beginning 14 days after launch. Postflight measurements were conducted on and 10 and 30 days after recovery. Mean values are given in Table 8.2.

Table 8.2 Measured maximum O_2 uptake of 14 ISS astronauts (from Moore et al. [19])

Sample	$\dot{V}O_2$ max L min^{-1}		
Time	Mean	SD	%Δ
L − 90	3.14	0.22	–
L + 14	2.69	–	−17 ± 12
L + 60	Slight increase	–	–
R ± 2	2.66	0.17	−15.3
R + 30	3.14	–	–

L launch, *R* return

There was a very large variation in data, which is not surprising in a mixed gender population without documentation of countermeasures with baseline data gathered 3 months before flight. Most valuable of the results presented were the individual MOIs with time. While sample size was too small for statistical significance testing, several logically significant conclusions that are compatible with theory and prior experience can be drawn.

1. A rapid but limited loss of maximum capacity inflight occurs on exposure to weightlessness. This loss appears to remain more or less constant in approximately half of the subjects while two seemed to recover their lost capacity in 60 days or less. Three or four others with low capacity showed no loss. There are two phenomena that should be considered here. First is a loss of blood and fluid from the legs in response to the obligatory fluid shifts. Also loading by the ergometer may have been changed by biophysical interactions in weightlessness.
2. A second significant observation is that all subjects maintained a more or less stable capacity. The investigator's observation that four subjects with large capacity were able to do this is significant. Without higher time resolution and body mass information it is difficult to place import on the mean trends shown.
3. Post flight data shows the usual and now expected immediate loss and daily recovery of capacity. The stated mean loss on landing was 15%. Without hourly timing from landing and knowledge of post flight activities, especially rehydration status, one cannot separate effects of fluid volumes and shifts from inflight loss of capacity. Capacity increased by day 5 and probably represents other inflight effects. Its value was not given.

A second important component of this study [19] was the comparison of three commonly accepted extrapolations of measured heart rate to maximum $\dot{V}O_2$. These extrapolations are not valid in weightlessness, and may indeed have little validity in 1 g. After four plus decades this is a historic resumption of studies of an aspect of the human body in weightlessness that will be crucial to future space exploration. This long delayed confirmation of significant effects of weightlessness on one of the body's major functions should be refined, repeated and expanded to allow a validated understanding of the details of these phenomena including the following:

- Measurement of blood volume and hematocrit with $\dot{V}O_2$ max and other significant O_2 uptake studies in weightlessness.

- Valid histories of individuals usage of exercise countermeasures including type, level(s), and duration of use.
- A valid analysis of the total energy cost of cycle ergometry in 1 g versus weightlessness.

8.5 Countermeasures

Details and usage of countermeasures were not included in the above, but usage of the hardware is as critical to success as is the hardware. There are many or more or less effective programs to maintain or increase aerobic fitness on Earth. The principles involved and a proven program to increase aerobic fitness were demonstrated by Cooper and his aerobics by the late 1960s using a variety of devices and exercises [11]. Although there were no in flight means of preserving inflight aerobic capacity until Skylab, 1 g preflight aerobic fitness had become a way of life in the astronaut office from its beginning.

Currently two types of devices on ISS are theoretically capable of maintaining aerobic fitness, the NASA CEVIS and Russian cycle ergometers and both NASA and Russian treadmills. The treadmills are currently inadequate because of inability to provide 1-g weight equivalent loads on the majority of users. This leaves the cycle ergometers, which can generate adequate metabolic loads for both maintenance and testing.

Obligatory loss of blood and body fluids in weightlessness complicate maintenance of aerobic fitness [20]. Based on 1-g effects, an obligatory loss of 10% of blood volume in weightlessness should reduce maximum CR capacity by a similar amount. This in turn would initially reduce exercise level producing a loss in maximum capacity as illustrated in Fig. 8.11. If the subject were not already at absolute maximum capacity it should be possible but time consuming to "train up" to the desired capacity allowed by the reduced fluid volume volumes.

This assumption is based on loss of fluids having the same effects in weightlessness as on Earth. There are obviously large gaps in our basic knowledge of physiology in spaceflight that should be filled by measurements of CR capacities and fluid volumes as early as practical after launch.

On returning to 1 g fluid shifts and losses become even more complex. On landing the partially empty leg vessels will rapidly reclaim the 10% of lost blood volume. In addition some 2–4 L of body fluid must also be returned to the leg, and this is transported by blood. This fluid must be replaced from the upper body, but so long as replacement is in progress blood volume will be further reduced. Volume adjustments were still in progress when recovery $\dot{V}O_2$ max was measured during the ISS study. Of equal significance is loss of O_2 carrying capacity of the red blood cells (RBC). Gain of red cell mass is much slower than loss. Measurement of volumes and masses in this phase of recovery has not been done in sufficient detail to even speculate or, much less, to make quantitative estimates of their effect. Other aerobic adaptations to cannot be accounted for without records of such activity.

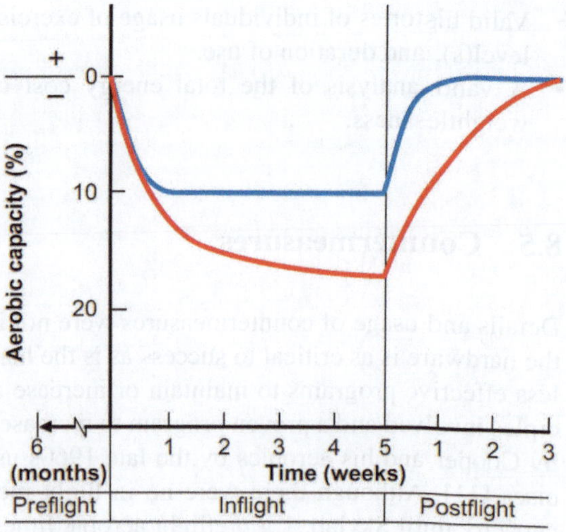

Fig. 8.11 A concept of aerobic losses in flight based on data available to date. The *blue curve* represents effects of fluid shifts based on 1-g studies. If this is valid, the subject may not be able to maintain his 1-g exercise level but will have to initially reduce the level and train up to it. The *red curve* is maximum capacity resulting from fluid losses and reduced exercise levels

In summary, based on theory and incomplete records we now have, there appears to be an obligatory loss of approximately 10% of aerobic capacity from changes in the blood and vascular system. Further losses on flights of 3 months appear to have been minimized by cycle and treadmill exercise. There will be a transient obligatory loss on recovery and this also needs further characterization.

8.6 Protocols for and Usage of Aerobic Countermeasures in NASA Space Programs

There were no inflight exercises prior to Skylab nor were there any protocols for preflight or exercise facilities until well into the Gemini program. Skylab had no preflight exercise protocols but monthly fitness tests made jogging a regular activity. In flight cycle exercise and testing was a high priority item on the crew schedule, and increased on each mission. Shuttle's second flight had the author's (WT) feasibility demonstration of aerobic exercise by jogging in place while axially loaded by bungee force equivalent to body weight. The third and subsequent flights had the TW treadmill until STS-63 when they were replaced by a series of JSC life sciences' devices. Except for various tests there was no recommended maintenance exercise and use of the treadmill was erratic.

For ISS, Russia had rather elaborate maintenance and test protocols with treadmills at the center [21, 22]. The initial plans were disrupted by failure of NASA hardware, work arounds, additions and replacements but after a decade, there are pre, in and post exercise and test protocols that are complex and requiring up to 2½ h per day [23].

8.7 Summary

Insofar as can be determined there has never been a statement of goals for countermeasures to effects of spaceflight in this or other systems of the body. For that matter the effects of weightlessness and cause of these effects have never been cleanly documented. Resutls have always been distorted by inadequate recording or some attempted countermeasure. It also appears that adaptation is usually considered pathological and something that must be repaired. Neither human life on Earth or in space exploration supports such thinking. It appears that the current goal is to maintain 1 g capacities, but the primary tool for doing this is still inadequate—the treadmills used cannot apply 1 g force loads to many of the subjects. The effects these items will be encountered in the informed speculation on future extended missions.

8.7.1 Aerobics on the Moon and Beyond

In spite of the large resources expended on advanced planning for a variety of exploratory missions, many aspects of the missions and their effects on the cardiorespiratory systems remain speculative. Aerobic effects of being on the Moon and Mars and on the flight to Mars will be briefly discussed to the extent allowed by theory and experience to date. There is little purpose in another listing of effects recorded to date unless they are generic. For example pressurized suits significantly increase O_2 consumption in usual activities and can in 1-g simulations of the 1/6-g lunar gravity adding a 1988 Mark III suit to treadmill walking increased O_2 uptake by 20%. In this simulation using astronauts experienced in EVA, level walking at their chosen speed in the MKIII suit at 1/6 g produced an average O_2 uptake of 2.0 L min^{-1}, or 50% of their $\dot{V}O_2$ max. There will be more generations of suits before the next lunar landing with different, presumably lower energy costs.

The primary differences between the Earth and the Moon are the absence of a lunar atmosphere and its reduced gravity. This dictates two modes of habitation, a pressurized indoor shirt sleeve environment and the pressure suit for outdoors. O_2 uptake indoors will depend on physical activity. Basal metabolic rate was not significantly changed in weightlessness and will probably remain the same here, or ~5 mL min^{-1} kg^{-1}. The energy cost of simulated lunar locomotion is reported to be substantially less than in 1 g. In contrast suited locomotion over difficult terrain and uphill may approach or even exceed an astronaut's 1-g maximum oxygen intake.

There are some potential changes in aerobic capacity and countermeasures that can be expected on the Moon.

One sixth of Earth's weight would be expected to very slightly reduce volumes of fluid shifted and lost in weightlessness. An obligatory loss of 10% of aerobic capacity is a reasonable estimate until more comprehensive studies are conducted. The reduced weight loads will rapidly reduce this capacity further in the absence of countermeasures.

Countermeasures in reduced gravity offer a variety of possibilities. Size and weight should preclude 1-g configured cycle ergometers, treadmills, and AREDs. By using the principles of mechanical impedance matching, it was possible to develop (see p. 196, Fig. 7.8) the Shuttle treadmill (1980). Relatively small, light unpowered countermeasure devices remain feasible as ever.

An interesting possibility is the use of Lunar mass to produce weights for strength exercise. Stone weights would have to be six times the 1-g mass to reproduce a 1-g load, but this increases the inertial force six times that in equivalent 1-g motion. This will provide some interesting possibilities for producing an optimum 1-g equivalent load.

A treadmill presents the same problem on the Moon because in 1/6-g body weight is also about one sixth of that on Earth but has 1-g equivalent inertia. Some acceptable simulation of 1-g equivalent loads on the subject's body will be necessary. In this case easily added mass would produce a large and cumbersome load. A constant-force hip and shoulder load such as is currently used, with a force of one half body weight plus and additional one third weight added by limb and trunk weight is one possible equivalent load.

8.7.2 *Earth to Mars Flight and Return*

There are precedents for this duration with four Russians and one American on spaceflights of a year or more and numerous cosmonauts and astronauts having flown more than 6 months. Also by the time a Martian mission is flown, one-way flight time is likely to be less than 6 months.

There is no precedent for the efficiencies demanded by a flight without resupply. Absence of resupply makes a great difference in allowable weight, volume, and power. Is the body going to be allowed to adapt to the local environment in an efficient fashion or will the mission pay the price of maintaining 1-g physical condition including aerobic capacity? This is a strategic question yet to be dealt with.

Consider the current countermeasure equipment that is large, heavy, inefficient and power hungry. The irony of this is that all of the current exercise gear also extracts significant power from its subject. A treadmill in 1 g requires power to drive the belt when it is level, but elevate it a few degrees and the belt is driven by the subject. Continue to elevate it and a braking force must be applied to keep the belt from "running away". Things are even simpler in weightlessness because the subject only had to lean forward to direct the load vector from the body harness against the belt to drive it. This is equivalent to elevating the front of the treadmill in 1 g.

An even more egregious waste of power occurs in the cycle ergometer, which uses a controlled friction to absorb up to 300 W of power. This aerobic exercise could generate kilowatt hours of power while smoothly and precisely providing any required resistive loads. Although not directly involved with aerobic maintenance, ARED is another example of using vehicle power to absorb muscle power. This source of power has been tapped in a few cases.

Looking at this ensemble of devices in further detail: A question that should be objectively answered is the need for cycle ergometry. A properly designed treadmill can provide a larger, more natural metabolic load to maintain aerobic capacity than the best of cycle ergometers. As for providing loads for metabolic testing, the treadmill is the gold standard on Earth for good reason. Of equal import is that it simultaneously provides the necessary loads to maintain strength and duration to maintain bone and muscle. With a properly designed and subject-loaded treadmill cycle ergometry may not be necessary.

Effects of gravity on aerobic fitness spans a current spectrum from 1 g to weightlessness. Lunar effects on maximum oxygen intakes are expected to be similar to but smaller in magnitude than those of weightlessness. Martian effects can be expected to be greater in magnitude than those of the Moon but less than on Earth. Based on the approximately direct relationship, the small but significant weight of blood and fluids should allow shift and lose two thirds of their leg volumes and reduce $\dot{V}O_2$ max by 5–6%. Space-suited physical activity should also have somewhat more O_2 uptake than in weightlessness. However, walking in a space suit in simulated 1/3 g produced an uptake equivalent to walking in 1 g. Also walking in a MKIII suit at 1/6 g increased uptake by 7 mL min^{-1} kg^{-1} or 20%, while Martian 1/3 g increased it to 20 mL min^{-1} kg^{-1} or 40% of subject's $\dot{V}O_2$ max. In current suits a high $\dot{V}O_2$ max may be required for suited locomotion. Countermeasures for use on the Martian surface should be easier to devise and use, but the same considerations discussed in Lunar countermeasures apply here but to a lesser degree.

In summary, investigation of the loss of aerobic capacity in space and control of such loss has been a slow and incomplete effort. Failures in exercise and test hardware have played a key role in this. It is known that aerobic capacity can be gained or lost in flight but are the mechanisms the same as on Earth? It is also known that changes in fluid volume and distribution affect aerobic capacity but is this the same as on Earth? These and other questions need answers before logical planning for exploratory missions can begin. Traditional methodologies for investigating these questions can be replaced by unobtrusive ambulatory devices that can sense and record such data under more realistic conditions. It is hoped that this may be in progress on ISS.

Glossary

Aerobic has to do with oxygen (in the air) consumption; e.g., aerobic exercise. Aerobic may now be used as a noun, such as in Ken Cooper's system. Kenneth H. Cooper, M.D. introduced the concept of aerobics. He authored the 1968 book Aerobics, which emphasized a point system for improving the cardiovascular system. His points system is also the basis of the 10,000 steps per day method of maintaining adequate fitness by walking.

Aerobic fitness is the capacity of the cardiorespiratory system's complex (Fig. 8.3) to use oxygen in converting energy into muscle activity.

Aerobic capacity is a measure of the cardiorespiratory system's ability to provide oxygen to muscles, usually measured in terms of oxygen consumption expressed as liters per minute, L min^{-1}, or milliliters per minute, mL min^{-1}. This may be normalized for differing subject masses by dividing oxygen consumption by subject mass or mL min^{-1} kg^{-1}, which is the usually preferred expression and used here whenever possible.

Cardiac Exercise Stress Test is one in which exercise is used to stress the heart, and is monitored by EKG, BP and other measurements. Respiratory measurements are not usually done.

Cardiorespiratory is more or less synonymous with aerobic.

Cardiorespiratory systems indicates the collection of systems that form the serial change that begins with intake of oxygen to forceful motion by muscle (Fig. 8.3). Note that the muscle mass may vary in location, but the other systems in the chain are fixed.

Maximum oxygen intake or MOI is the O_2 consumed during a maximum effort at some task that uses large masses of muscle such as treadmill locomotion or cycle ergometry.

O_2 may be used for oxygen.

Oxygen consumption may also be expressed in terms of energy transformed per unit of time, joules or kilojoules per hour or minute; i.e., J h^{-1}, kJ h^{-1} or kJ min^{-1} or J min^{-1}. This may also be expressed as kilo calories per unit of time; i.e., cal h^{-1} or cal min^{-1}. An approximate conversion factor to oxygen consumed is 5 cal = 1 L.

Oxygen uptake is a synonymous term for oxygen consumption.

pH is a figure expressing the acidity or alkalinity of a solution on a logarithmic scale on which 7 is neutral, lower values are more acid, and higher values more alkaline. The pH is equal to $-\log_{10}c$, where c is the hydrogen ion concentration in moles per liter.

Spirometer is a closed tank system that moves up or down according to the volume of gas in it.

References

1. Lovelace R, et al. Selection and maintenance program for astronauts for the National Aeronautics and Space Administration. Aerosp Med. 1962;33:667–84.
2. Wilson C. Physical fitness tests. WADC-TR 59-505, p 49–64.
3. Berry C. Man's response to long-duration spaceflight in the Gemini spacecraft. In: Gemini mid-program conference. MSC: Houston, TX. NASA SP–121, 25:135; 1966.
4. Rummel J, Savin C, Michel E. Exercise response. In: Biomedical results of Apollo. Washington, DC: NASA . NASA SP–368, 5:265; 1973.
5. Michel E, Rummel J, Sain C, Buderer C, Lem J. Metabolic activity. In: Results of Skylab medical experiment M171 1975. Washington, DC: NASA. NASA SP–377, 36:372–87; 1977.
6. Michel E, et al. Results of Skylab medical experiment 171-metabolic activity. In: Biomedical results from Skylab. Washington, DC: NASA. NASA SP–377; 1977.
7. Saltin B, et al. Maximal oxygen uptake in athletes. J Appl Physiol. 1967;23(63):353–8.

8. Guyton A, Hall J. Pulmonary ventilation. In: Textbook of medical physiology, 11th ed. Elsevier Inc., 37:471–80; 2006.
9. Guyton A, Hall J. Regulation of respiration. In: Textbook of medical physiology, 11th ed. Elsevier Inc, 41:514–22; 2006.
10. Wilson C. Treadmill stress test protocols. In: Social security advisory service physical fitness tests.
11. Cooper K. The aerobics way. Toronto: J. B. Lippincot; 1977.
12. Tikanen O. Muscle activity and inactivity periods during normal daily life. PLoS One. 2013;8(1):e52228.
13. Lee S, Moore AD, Everett ME, Stenger MB, Platts SH. Aerobic exercise deconditioning and countermeasures during bed rest. Aviat Space Environ Med. 2010;81(1):52–63.
14. Convertino V. Effects of deconditioning and reconditioning on aerobic power. In: Deconditioning and reconditioning. CRC Press LLC; 2004. p. 5.
15. Saltin B, Blomqvist G, Mitchell JH, Johnson RL Jr, Wildenthal K, Chapman CB. Response to exercise after bed rest and after training. Circulation. 1968;38(5 Suppl):VII1–78.
16. Shayler D. Disasters and accidents in manned spaceflight. Chichester: Praxis Publishing; 2000.
17. Sawin C. Extended duration orbiter medical project 1989–1995 NASA/SP-1999-534.
18. Moore AD, Lynn PA, Feiveson AH. The first 10 years of aerobic exercise responses to long-duration ISS flights. Aerosp Med Hum Perform. 2015;86(12):A78–86.
19. Moore AD, Downs ME, Lee SM, Feiveson AH, Knudsen P, Ploutz-Snyder L. Peak exercise oxygen uptake during and following long-duration spaceflight. J Appl Physiol. 2014;117(3):231–8.
20. Thornton W, Hoffler G, Rummel J. Anthropometric changes and fluid shifts. In: Biomedical results of Skylab, NASA Washington, DC, NASA SP–377, 32:330–338; 1977.
21. Kozlovskaya IB, Yarmanova EN, Yegorov AD, Stepantsov VI, Fomina EV, Tomilovaskaya ES. Russian countermeasure systems for adverse effects of microgravity on long-duration ISS flights. Aerosp Med Human Perform. 2015;86(12):A24–31.
22. Yarmanova EN, Kozlovskaya IB, Khimoroda NN, Fomina EV. Evolution of Russian microgravity countermeasures. Aerosp Med Hum Perform. 2015;86(12):A32–7.
23. Moore A, et al. Cardiovascular exercise. In: The US Space Program: past, present and future. Acta Astronaut. 2010;66:974–88.

Chapter 9
Neuromuscular Inhibition

9.1 Introduction

On Shuttle, after landing and roll out, one collected and rapidly stowed any remains of entry and flight in preparation for return to a familiar home. On attempting to stand it became a strange new world, for one couldn't stand. It was just as Ken[1] Mattingly had warned, "You may as well leave the straps on for you are not going anywhere the first time you try." After a minute or two of consideration and repositioning for maximum effort of both arms and legs, you made a shaky stand, feeling as of you were pulling two plus g's. After a few unsteady seconds you sat down with relief and a minute or so later made another shaky stand against a perceived decrease in load. After several repetitions the load became less, and you could walk and an hour later carefully descend the exit ramp stairs with a firm grip on the handrail. It still felt like you were pulling g's on the traditional postflight walk around. Next morning the next-door neighbor observed, "He was putting out the garbage can as usual." This was an unexpected experience that is still not in the space and life science books or reports[2] and was not talked about in the Astronaut Office (AO), even in postflight debriefing. When a written survey was made, this was found to be a virtually universal experience (Figs. 9.1 and 9.2).

A theoretical analysis led to the hypothesis that the cause of this postflight instability must be a neurological muscle control phenomenon. This led to a practical investigation. Unknown to the U.S space program at that time, those in the Russian program were conducting an investigation into the same effect but did not begin measurements until 24 h after landing by which time the effects had largely disappeared.

[1] Ken Mattingly, commander of the fourth Shuttle mission.

[2] A few of the later personal memoirs describe or mention it.

W. Thornton, F. Bonato, *The Human Body and Weightlessness*,
DOI 10.1007/978-3-319-32829-4_9

Fig. 9.1 Images from the three time sequences of a short-term Shuttle flight. The first panel is from the extended preflight preparation that included heavy weight training. The mid panel is typical of physical activity in weightlessness for a week in which rapid neuromuscular adaptation occurred. The third panel depicts what was typical of the first attempt to stand after landing because of neuromuscular inhibition from inflight adaptation. This was rapidly reversed in tens of minutes allowing the subject to stand and walk under body weight forces. Significant muscle strength was not lost in such a short exposure to weightlessness

This effect of weightlessness occurs in as little as 2 days time. It produces an estimated loss of 60% of maximum preflight strength, which is largely recovered in hours postflight. There are neurological mechanisms capable of inhibition, but none have documented losses and rapid recovery of such magnitude [1]. In the Shuttle program this was a potential threat to unaided emergency egress. One commander was unable to leave his seat for 30 min.

9.2 History

The first described experience of this phenomenon may have occurred on the first Shuttle flight when John Young warned his cohorts during the postflight debriefing, "The brakes on that thing sure are heavy. You better practice on them on the way down." Exercising on the brake pedals during entry became the standard but

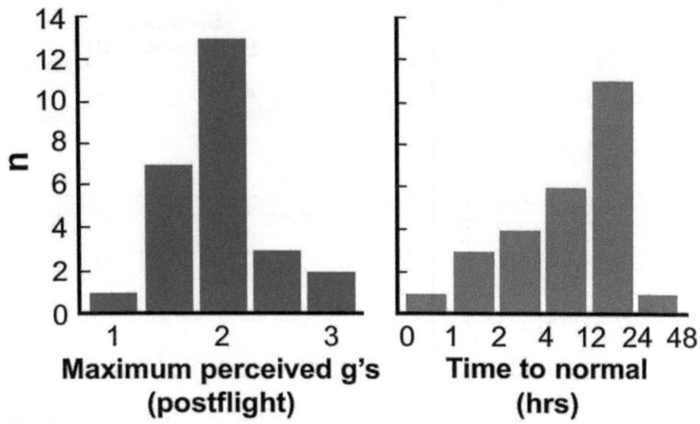

Fig. 9.2 Histograms of perceived g loads after landing (*red*) following 7 days of weightlessness and time to recover normal perception of weight (*green*) (Color figure online)

unpublished and undocumented practice until the end of the Shuttle program. It was the first effective countermeasure to neuromuscular inhibition.

An internal Astronaut Office program was generated to document these effects. Whitmore built simple hardware to document strength loss inflight and on entry and landing, plus an in-flight countermeasure. On a subsequent flight, after 5 days (in) orbit, one author, WT, who had experienced the phenomenon, was forced by muscle weakness to use a bungee force of only 40% of body weight (BW) on the first circumnavigation of the Earth on a treadmill, a slow jog of 90 min on STS-51B on the Shuttle treadmill which he and Whitmore had developed. Subsequent treadmill circumnavigations on Shuttle and ISS have received more or less attention. Other than times, none of the Boston Marathons in flight have reported details such as body weight equivalent which has potential value in several areas of weightlessness. For various reasons only a fraction of BW is still common practice in treadmill exercise bouts.

9.3 Background

This strength loss occurs after only a few days in weightlessness; therefore, it cannot be somatic. It must be neurological. There are both theoretical and practical examples of neurological strength inhibition in 1 g, although none of this magnitude. It is known that maximum strength of a muscle that can be elicited by electrical stimulation is normally limited by central neurological inhibition. Under great stress, such as in emergency situations like fires and accidents, there are sometimes well-documented examples of extraordinary strength. This limitation is a normal aspect of our musculoskeletal system. But it is postulated that the large loss of strength and the sensation of weight after weightlessness is the product of two well-known feedback circuits at the final level of muscle control.

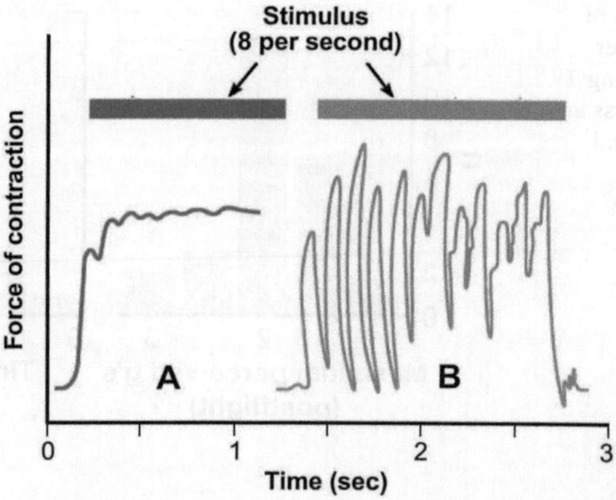

Fig. 9.3 Effects of negative feedback on muscle force. The importance of negative feedback in precise control of muscle action is shown by this demonstration of a muscle that is electrically stimulated to contract and does so in *B*. When feedback is added, peak force is reduced, but a useful controlled force is produced (*A*). (The order of events in the demonstration was reversed for clarity.)

The first of these circuits begins with a sensor (Golgi tendon organ) located in the tendons that transmit muscle force to bones. Moderate increases in force increase frequency of impulses in its neuron. These impulses are applied to the neuron controlling muscle force on the tendon, thus inhibiting it. It was first thought that this circuit was only a safety measure to prevent tendon damage, since it can only reduce muscle force. Currently there are some who think it may exert more or less continuous restraint on muscle throughout its force range. It also has connections to brain centers and affects their control and sensation of muscle activity [2].

More complex feedback circuit sensors are the muscle spindles. Action of these sensors is not fully understood, but they cause the muscle-control neurons to generate signals, which produce more or less force by negative feedback. Negative feedback circuits have been essential to life processes from life's beginning on Earth (see Chap. 2). They are present throughout our bodies and by the millions in our nervous system.

Feedback mechanisms have many benefits, one of which is graphically illustrated in Fig. 9.3 parts A and B, which show its smoothing function of a negative feedback. In B there is direct one-way control and in A the feedback circuit is present.

Muscle spindles are present in varied numbers depending on speed and precision of muscle action. For example, eye musçles have many more than an upper arm muscle. Feline species including domestic cats have many more than humans do. This species difference can be appreciated in the speed and coordination of a lion's motion compared to those of a grazing species (Fig. 9.4).

Fig. 9.4 Effects of spindle population in action. There are significant differences in the numbers of spindles in various species with felines, including domestic cats, having many more in their musculature than many other species. This can be seen in a cats' movements and allows the coordinated attack shown. Numbers of spindles in human muscles are less than the lion but more than the zebra

A simplified description of spindle action begins with Fig. 9.5, which shows the physical arrangement of tendon sensors and a simplified spindle. Figure 9.6 is a circuit to complete the connection shown [3]. A sensory neuron transmitting force from a tendon sensor is in blue. Spindles consist of modified muscle fibers physically connected in parallel to the mass of fibers producing force and motion. There are static and dynamic sensors attached to these spindles. Static sensors are sensitive to loads on a muscle such as holding a weight in position, while the dynamic sensors are sensitive to differences between length of muscle fibers to which the spindle is attached and the spindle fibers.

When a load is to be moved, a command from the brain with estimated force required is sent to the muscle, and a second command appropriate to the spindle is sent to it by a separate pathway such that the spindle action becomes a template for the desired motion. If now the load is lighter than expected and there is more rapid movement of muscle than spindle, an error signal is instantly transmitted to the muscle control neuron inhibiting the muscle force signals and reducing the difference between muscle and spindle fibers. Conversely if the load is heavier than was commanded for, signals are sent from the spindle sensor to increase muscle force; however, the tendon signals will be sending inhibitory signals if the muscle force approaches a level capable of damaging it.

Take the case of an astronaut who has just arrived on orbit and is 5 min out of his seat. He uses foot force to project himself across the cabin. If his brain sends a signal for 1-g force he would slam into the overhead. What happens is that a larger force than necessary is commanded producing a rapid muscle contraction, this leaves the spindle fibers slack, whereupon, they send an inhibitory signal to the motor neuron to reduce the force. In spite of this, over-controlling in weightlessness is common on first arrival. It is postulated that after hours of adaptation the sensitivity

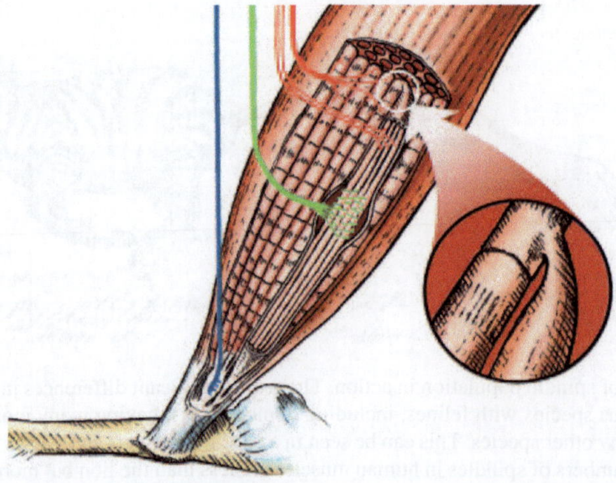

Fig. 9.5 Anatomical arrangement of tendon sensors and muscle spindles. Tendon sensors are in the tendon between muscle and bone and transmit their signals to the spinal cord via small nerve fibers shown in *blue*. Spindles are modified muscle fibers with two types of embedded sensors whose outputs are transmitted to the spinal cord by nerves shown in *green*. For simplicity the sensor outputs are combined here. Muscle spindles are attached in parallel to muscle fibers and are stretched or relaxed by action of the muscle. Contraction of the muscle is commanded through large (*red*) nerve fibers by a neuron in the spinal cord. A signal is sent by a separate small nerve (*red*) from a neuron in the cord and produces a force and motion template for the major muscle fibers to follow. Differences are detected and correction signals (*green*) are sent back to the muscle control neuron in the spinal cord

of the spindle sensors is such that contraction signals to the muscles are greatly reduced hence effectively reducing strength. Sensitivity of the tendon sensor may also be increased, thus decreasing and limiting maximum force.

While it takes months to years for the musculoskeletal system to adapt to weightlessness by losing unused strength and mass, the nervous system adapts much more quickly, increases the gain of the feedback loops, and effectively reduces muscle strength in much shorter but unknown time. Obvious effects of this have been noted.

9.4 One-g Effects

There are numerous reports of neurological inhibition of strength with forced disuse of muscle, but none approach the levels observed after spaceflight. However, the tendon organs and spindles play an essential role on muscle tone and all movement, constantly adjusting force to the situation.

Bed rest. There are no reported studies of this phenomenon in bed rest, but some level of inhibition can be expected.

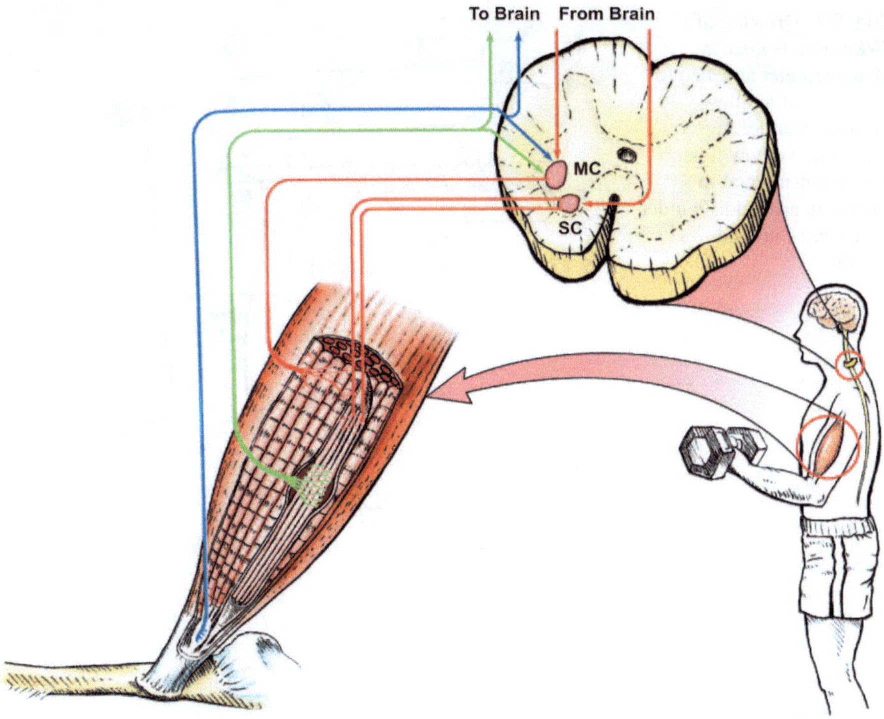

Fig. 9.6 Simplified diagram of feedback control of muscle force. Action, such as lifting a weight, begins with brain centers, which initiate the action triggering a learned pattern of muscle commands from a lower brain center. There are two commands sent, one to the involved muscle's control neuron (MC) and a second to an associated spindle's muscle fiber's command neuron (SC), which causes them to contract at the desired rate, becoming a template for the desired muscle contraction. Any difference the muscle action commanded by MC and the spindle muscle fibers connected to it (Fig. 9.5) will be sensed by the spindle and appropriate commands to increase or decrease the muscle's force will be sent to MC via the nerve shown in *green*, and to the brain for cognition. Should excess force be sensed by the tendon organ shown in *blue* and inhibitory signal will be sent to the muscle control neuron, MC. Since reduced forces are required in weightlessness, after a short time gain of the spindle and tendon sensors will be increased functionally reducing strength, an action which will persist for a time on returning to 1 g

In flight. This inhibition must be operative in flight, but the time course is unknown.

An investigation of this was initiated and Whitmore constructed and flight qualified the necessary flight hardware, which included an arm and leg isometric strength dynamometer. These were attachable to launch and entry seats, and could easily be set up and taken down (Figs. 9.7, 9.8, and 9.9).

The plan was to conduct a series of daily, or more frequently as necessary, strength measurements in order to assess the time course and magnitude of strength

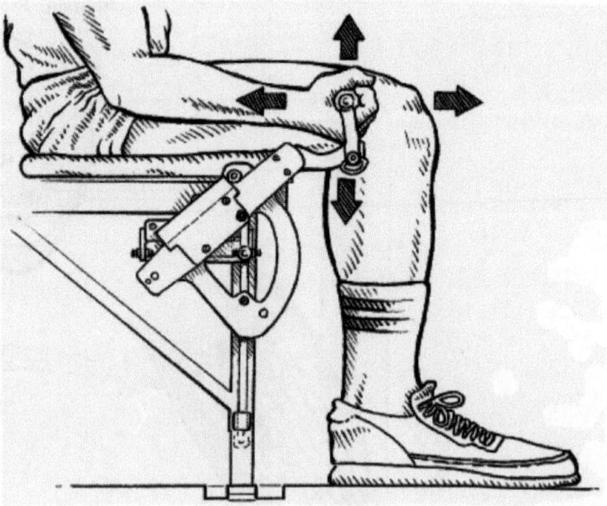

Fig. 9.7 Drawing of Whitmore isometric dynamometer attached to Shuttle seat to allow measurement of muscle strength in elbow flexion and extension and in eccentric arm and shoulder force, pre-, in- and postflight

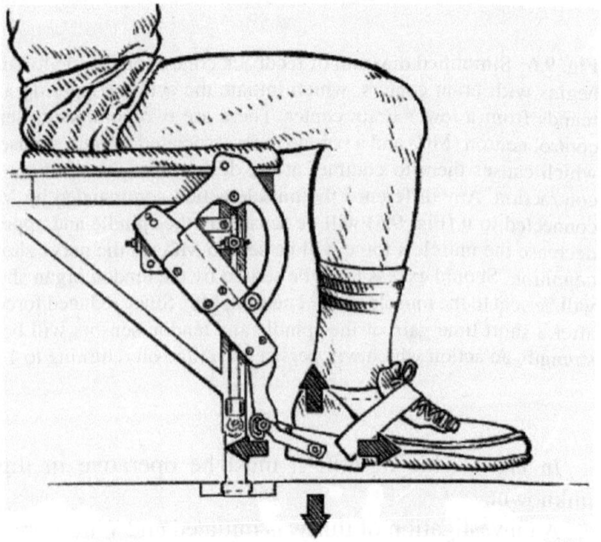

Fig. 9.8 Whitmore dynamometer on Shuttle seat for measurement of knee and hip flexion and extension

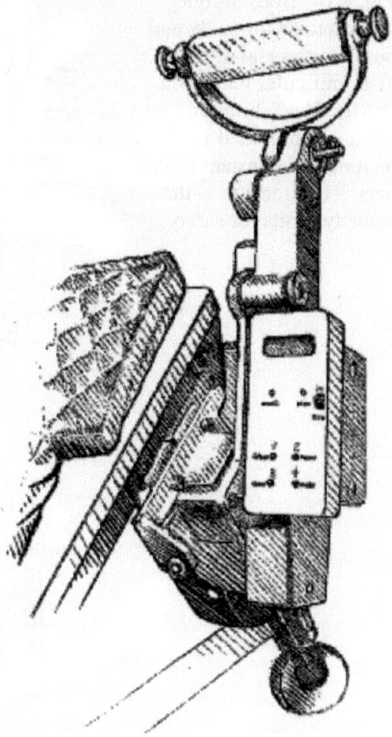

Fig. 9.9 Shuttle dynamometer control display panel and handle for strength measurement

loss inflight. A second phase of the investigation was to examine the effects of heavily loading the inhibited muscles. This was to be done with a constant force device with body mass providing inertia such that 1-g weight loads were simulated (Figs. 9.10 and 9.11).

While the inhibitions described must develop inflight, inhibition does not seem to be a problem on ISS. This may be hidden by the pre-exercise warm ups with progressive increase in loads, and since exercise is now a daily event, the inhibitory effects of an otherwise weightless environment may be greatly reduced.

The question of avoidance of 1-g loads on the treadmill remains. With the level of loading in squats, up to 400 lbs, there can be little inhibition so it must be a problem of harness load limitation. There are recent verbal reports that some ISS astronauts are exercising at 1 g equivalent BW loading with a new harness.

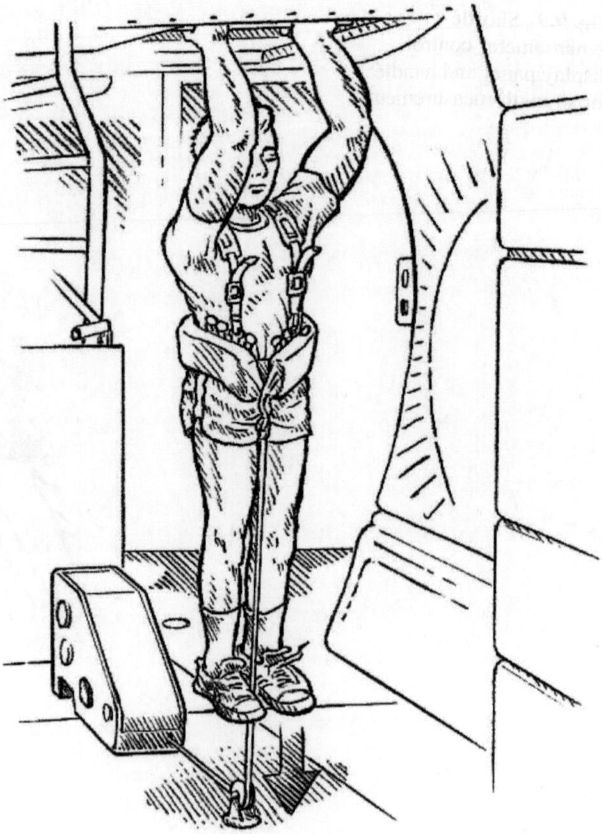

Fig. 9.10 Body weight arm exercise for study and possible reversal of neuromuscular inhibition (NMI). The Whitmore force unit provided incremental constant forces over 3-ft extension with velocity limits for safety

Postflight effects. It became obvious that a number of in-flight adaptive effects were sharply reducing physical performance in the immediate postflight period making unaided emergency impossible for a large number of the crew. A post-egress study confirmed this concern. Results are in the Extended Duration Orbiter report [4] of medical studies. Neuromuscular inhibition was not included in this study. The proposed study on Shuttle was opposed by JSC Life Sciences. The equipment produced for the study was transferred to Life Sciences but it was never used. Thus there is little more that can be added to this introductory description of the problem and the histograms in Fig. 9.2.

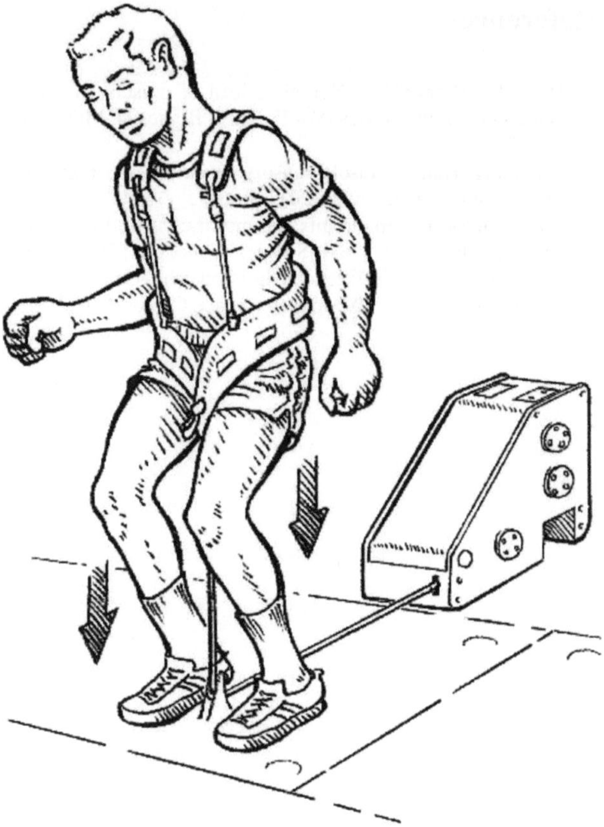

Fig. 9.11 Simulated body weight leg exercise for study of NMI on Shuttle or ISS. Inertia was provided by body weight. This unit had provision for addition of inertial load to simulate external weights

In casual conversation, returnees from ISS said that problems on first standing were still present, but without schedules of the exercise prior to landing little can be concluded.

9.5 Summary

Current space operations have apparently reduced the significance of this effect to a transient personal experience best forgotten. In contrast 1-g research in this incompletely understood area, especially spindle function, is very active. It might benefit from studies in weightlessness now instead awaiting its rediscovery at some future time such as when space tourism becomes a reality.

References

1. Houk JC, Rymer WZ. Neural control of muscle length and tension. In: Brooks VB, editor. Handbook of physiology. Vol II, Sec I. Bethesda, MD: American Physiological Society; 1981. vol 15, p. 257.
2. Guyton A, Hall J. Muscle sensory receptors. In: Textbook of medical physiology. 111th ed. Philadelphia, PA: Elsevier; 2011. p. 657.
3. Lower motor neuron circuits and motor control. In: D Purves et al., editor. Neuroscience. 3rd ed. Sunderland, MA, USA: Sinauer Assoc, Inc.; 2004. p. 371.
4. Brown JT, Sawin CF. Extended duration orbiter medical project: final report 1989-1995. NASA Lyndon B. Houston, TX: Johnson Space Center, NASA SP–1999–534; 1999.

Chapter 10
Loss of Body Mass During Weightlessness

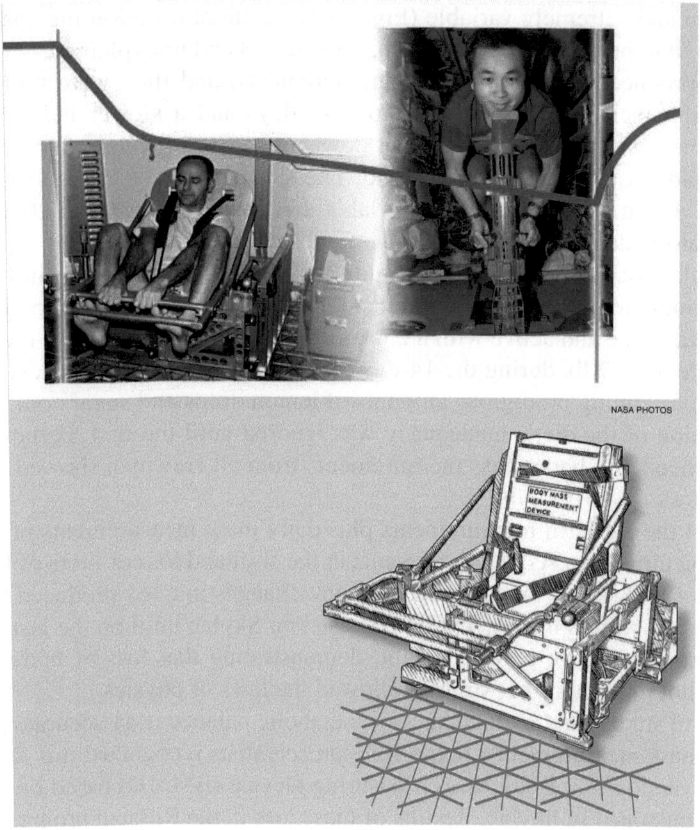

Fig. 10.1 Loss of body mass in weightlessness. The *red line* is a generic curve for the loss of body mass in weightlessness and its recovery on return to 1 g. Body weights were measured prior to launch (*green vertical line*), and daily body mass was measured on each crewman during Skylab. Postflight (*blue vertical line*), daily weights were resumed. This curve demonstrates the rapid obligatory loss and gain of fluids after entering weightlessness and on return to Earth. A continued slow loss from inadequate diet in space produces the negative slope shown however this slope depends only on metabolic balance and may be positive or negative. This curve was made possible by the Body Mass Measuring Device (BMMD) developed and designed by the USAF MOL Program in 1966 and is shown in use on SL-3 in the *left* panel. The Russian BMMD is on the *right*. The photographs shown grossly distort the relative size and apparent mass of the BMMDs which are approximately equal. Most of the Russian unit is hidden by the subject. The NASA SLAMMD is currently used on ISS. The drawing is of the BMMD on Skylab with the rack in place for independent calibration masses, an essential element of any means of mass measurement. Such in-flight calibrations have not been reported with SLAMMD or the Russian BMMD (Colour figure online)

W. Thornton, F. Bonato, *The Human Body and Weightlessness*,
DOI 10.1007/978-3-319-32829-4_10

10.1 Introduction

Postflight weight loss was the first physiological change noticed in the U.S. astronauts on return from their flights.[1] Such losses in the early programs were from dehydration and extremely variable (Fig. 10.2). As flight duration increased losses still appeared to be more or less random, unexpected and unexplained.

Skylab produced the first in-flight measurements, and they were biphasic. On entering orbit there was a rapid loss for 3–5 days and a slower individual loss. Postflight there was also a rapid and then slower regaining of weight. It was assumed that the rapid losses and gains were fluid. The slower more or less linear loss was familiar to anyone associated with Skylab's diet. Its basal diet was 2000 cal/day with any caloric deficits to be filled by lemon drops and sugar cookies.

The Skylab Medical Equipment Altitude Test (SMEAT), conducted in 1972, was a 56-day simulation of a Skylab flight including 5 psi O_2 atmosphere and diet. One crewman was large and active with a 2500–3000 plus calorie diet and a few percent body fat. He lost 17 lb. during the 14-day baseline and 45 days in the test chamber in spite of consuming prodigious amounts of lemon drops and sugar cookies.[2] This demonstration of the diet's inadequacy was ignored until the first Skylab manned mission, when daily body mass measurements from all crewmen showed the same pattern of loss.

On SL-4 the leg girth measurements plus daily mass measurements produced a complete picture of the sequence of events in the shift and loss of liters of leg fluids (see Chap. 4), which are superimposed on any changes in mass produced by metabolic losses. Food was incrementally increased on Skylab until on the last mission, one crewman may have gained weight, demonstrating that loss of body mass in space was not mandatory but simply followed the laws of physics.

Crucial to studies and maintenance of metabolic balance is an accurate practical means of mass measurements. Some Russian scientists recognized this and developed their version of a Body Mass Measuring Device (BMMD) based on the principles demonstrated in Skylab. Results of mass loss in the Russian program are not discussed here.

A revised food and drink system was developed for Shuttle with significantly increased caloric input. An unavoidable decrease and change in sense of taste and smell is produced in most individuals by the cephalic shift of fluid, which apparently produces edema around the nasal olfactory sensors. It can change the taste of any of the individually selected foods chosen in elaborate 1-g taste selections. This complicates any diet for space, especially for first time flyers (Box 10.1).

One of NASA's most knowledgeable and experienced astronauts made news with a large loss of mass on a Shuttle/Mir flight. However, it was taken for granted that all was well in this area with dietary experts providing all necessary nutrients, until

[1] When measured in weightlessness the terms mass and mass measurement (MM) will be used. In 1 g weighing, weight will be used. The unit of mass will be the pound in either case.

[2] Flight surgeons increased his food prior to the end of the 56-day chamber simulation.

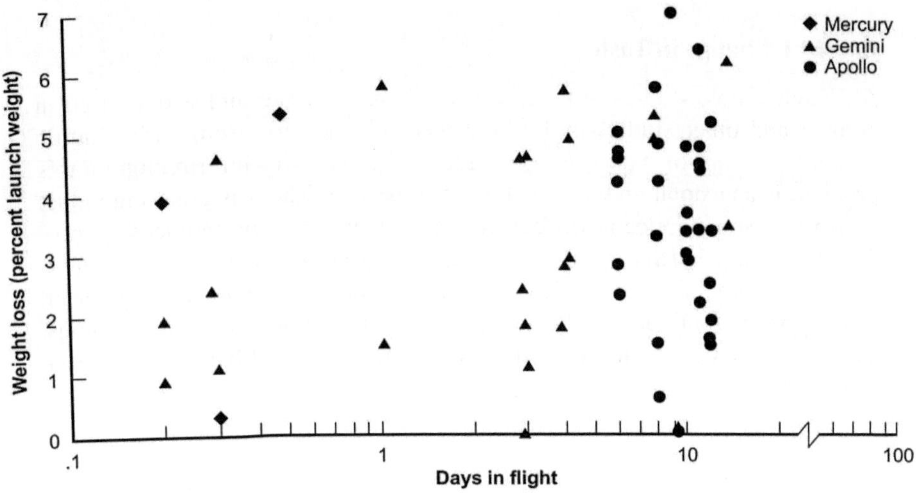

Fig. 10.2 Measured postflight losses of body mass in early space programs. This was the state of the art and understanding prior to Skylab. Measurement techniques may have contributed to the scatter

at least one study demonstrated serious deficiencies in both type and quantity of food in the NASA ISS diet. It was assumed that this would have been promptly corrected, but a current news report gives cause for concern. The first comprehensive report of human mass measurements on ISS gives even more cause for concern [1].

10.2 Brief History

Knowledge of effects and control of human body mass predates Human history by millennia. Measurement of mass is a much more recent accomplishment. More than 4000 years ago Egyptian tradesmen had devised means of measuring mass by weight, the Earth's attraction for objects on its surface. Scientific or even objective discussion of body mass is impossible without correct measurement. Measurement of mass without weight is a complex problem with a complex history and will be treated as briefly as possible.

The practice of weight control is now an international problem and there are countless experts and theories on this topic. When NASA and USAF consultants were confronted with the problem in an environment without weight 45 years ago they made a correct decision. In flight studies were needed and scheduled on USAF's MOL (Manned Orbiting Laboratory) and NASA's Skylab. By 1965 neither agency had been able obtain the instrument essential to such studies, a non-gravimetric Mass Measurement Device, MMD. My first project on MOL was development of workable MMDs. This is described briefly in the next section. The rigor and results obtained on the Skylab metabolic studies using the MOL MMDs remain

Box 10.1 Change in Taste

Although simple studies of taste and smell are available and widely used in clinical and other studies in 1 g and were planned for from early Shuttle flights, no record of their use can be found. Of necessity information on this problem is anecdotal and personal—but consistent. There is great individual variability. Some notice no difference while others' taste preferences are profoundly changed. This is not surprising during the first few days in weightlessness before fluid from the legs is lost, and runny, stuffed noses are present. Taste changes continue to indicate that the olfactory tissue may be edematous. This can be a significant factor in food consumption and body mass loss in some subjects.

unequalled; All intakes, outputs and body masses were individually measured and recorded daily. In flight two MOL small or Specimen Mass Measurement Devices, (SMMDs), and a Body Mass Measurement Device (BMMD) were employed. These instruments were regularly calibrated with masses traceable to the US NBS.

Other than the Russian BMMD and its studies, the next significant inflight body mass study was reported in 2014 [1]. A Russian BMMD and the JSC Space Linear Acceleration Mass Measuring Device (SLAMMD) were used to gather 55 data sets on 50 subjects on ISS expeditions 1–37. Results of this study are discussed in the section on loss and gain of mass in space flight.

10.3 Background

There are two obvious causes of loss of mass in weightlessness [2]. One is the adaptation of fluids to weightlessness resulting in a shift of fluids to the upper body. These fluids are then lost over the first 3–5 days in flight and regained in approximately the same time on return to 1 g. Russia uses thigh cuffs, Brazlets in an attempt to control this. The second cause is the metabolic difference between food consumed and energy expended, just as in 1 g, and is reflected in changes in body mass. This process is controlled by the subject and the given situation. Measurement of mass is a part of the process, and although not immediately obvious, the waste collection system can play a significant role in short-term flights.

After a brief review of basics, mass measurement and food systems are described, with the results they produce. It was felt necessary to include waste collection systems and their effect. We will be dealing primarily with metabolic mass losses for the remainder of the chapter. Fluid losses were treated in Chap. 4.

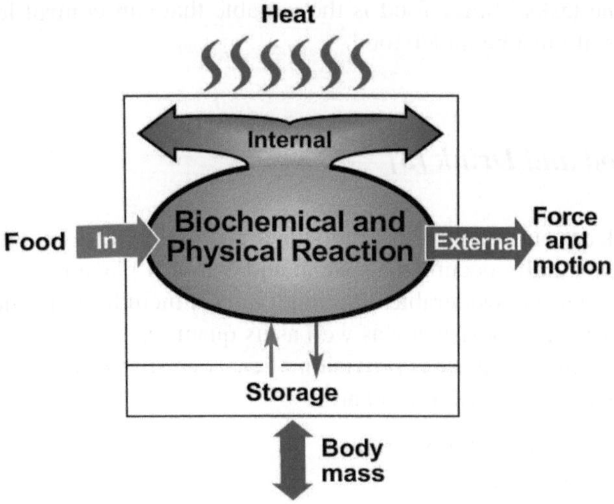

Fig. 10.3 Energetics of all body mass loss or gain. This gross simplification of energy transfers is often lost sight of, but it defines the biophysics of energy exchange on Earth or in space. Energy intake is in the chemical bonding of food. Incredibly complex chemical reactions transfer this binding energy into bonds in new chemical complexes most of which transport energy to sites and uses in the body including transformation of bond into mechanical energy by muscles for both internal and external motions. If intake exceeds energy consumed, it is stored in the chemical bonds of fat, and body mass will be changed accordingly

10.3.1 Energetics

In the human body, on Earth or in space, metabolic gain or loss of mass is the practical result of a fundamental concept of physics; conservation of energy, or to put it crudely, what goes in must come out. Figure 10.3 is a simplistic diagram of energy balance often lost sight of, but the sum of the intake and output plus or minus storage will always be zero. Energy input is in the chemical bonds of food that when broken and reformed by the body's organic chemistry liberates chemical energy for various functions of the body. Conversion to mechanical energy is not a very efficient part of this process, about the same as the approximate 20+ percent efficiency of an old internal combustion engine.

Control of body mass begins with knowledge of intake. Depending on their physical activity level and muscle mass, lumberjacks or some athletes in training can utilize 6000/day. Yet even a well conditioned astronaut who may expend 500 calories running for an hour still expends much more on the great variety of other activities during a day on Earth. When the astronaut is in flight with 8–10 h /day of activities time lined to the minute, energy consumption increases sharply such that the energy saved by the absence of an hour per day of jogging is consumed by the stress of routine tasks.[3] Since food is the variable that can control loss of mass in weightlessness, it must be understood.

10.3.2 Food and Drink [3]

Food and drink are our second most immediate requirement for maintenance of life itself and have naturally occupied a special and complex place in space operations. In addition they have considerable emotional import including gratification or lack of it from its taste and consistency as well as its quantity.

Food's most immediate use is provision of energy [4], and the energy equivalent of the three major categories of food are:

Carbohydrates [5]—4.1 calories/gram
Fats [6]—9.3 calories/gram
Protein—4.4 calories/gram

A second function is provision of building and rebuilding material for the body, most of which is protein [7].

Human bodies have a fantastic capacity for disassembling and reassembling the molecules in food into those needed by the body. Much of the substrate needed for synthesis of muscle protein is derived from byproducts of the conversion of carbohydrates and fat into high-energy molecules used in muscle action. Many of the body's complex reactions are reversible, such that when protein in food is inadequate, a body's muscles are literally consumed.

This happened in SMEAT and Skylab and still happens on ISS. If there is a difference between energy intake and output the body has a buffer system in its fat storage (Fig. 10.3). Thus body mass is the primary indicator of energy balance. Any extended energy demand, such as jogging, is initially met by easily available yet limited energy stores. As exercise continues, fatty acids become the energy source and the stored fat is utilized. Conversely if food is greater than required it will be converted to fat and stored with a corresponding increase in body mass. There are also much smaller protein energy buffers; however, their relatively minuscule short-term exchanges are not significant.

Protein synthesis requires up to 20 amino acids, 10 of which are easily fabricated by the body. There are ten essential amino acids whose quantity must be supplemented by food; e.g., meats and/or some fruits, nuts and vegetables. Muscle will be lost without them. On SMEAT and Skylab, 70 grams/day was considered an adequate amount of protein, but that is inadequate by today's standards. In addition the level of protein also apparently affects the density of some bones and is discussed in the section on flight effects.

[3] World renowned metabolic experts who designed the Skylab diet did not appreciate this hence the inadequate caloric allowance. Apparently their diet was based on bed rest.

Tiny quantities of complex molecules are required for the efficient conversions and syntheses just discussed, the vitamins. Some can be produced internally, but others are totally dependent on external sources. Scurvy was the first such recognized consequence of vitamin deficiency. It was once the scourge of sailors on long sea voyages whose diets consisted basically of salt meat and hard biscuits. Scurvy is easily cured by restoring the missing vitamin C with citrus fruit or juice. Another example, relevant to current spaceflight, is vitamin D, which is required for adequate bone mineralization.

Another class of required food molecules are minerals. The two most prominent minerals are calcium for bone structure and iron to carry O_2 in the blood. There are a number of others that are more or less critical.

Knowledge and control of diet are the beginning of health maintenance and was demonstrated on Skylab where errors were made and corrected, but some of its lessons appear to have been forgotten.

Results of the diet are measured in various ways. The effectiveness of vitamins and minerals in a diet is normally determined from blood samples. Measurement of energy balance by mass measurement inflight has been, except for Skylab, a continuing problem.

10.3.3 Body Mass Measurement

Measuring human mass on Earth is a simple weight dependent procedure with a 4000 year history of development (see Chap. 1), but one that often contains procedural errors. Current medical scales are typically accurate and reproducible to ±0.1 lb. The best home scales usually approach this accuracy; however, typical weights are usually 2–10 lb. greater than basal body weight, caused by errors from two sources, clothing and variable diurnal effects. Basal weight is obtained by the subject arising from sleep, voiding and if necessary defecating and then weighing immediately without clothes. If this is not possible and the weight is measured during the day, the subject should void, defecate if necessary, remove as many clothes as possible, and record the time of day and the clothes worn.

A typical day's record in 1 g is shown in Fig. 10.4. In 1 g there are also large day-to-day variations in basal weight (Fig. 10.5). Reliability of the scale is another potential source of error. The scale must be calibrated with known masses over the range of the subject's weight. If only one subject is being measured, a single calibration point near the weight measurements being taken is adequate. If multiple subjects are being measured the scale must be calibrated over the range of measurement. Variation with time and temperature must also be known. These procedures and precautions are especially applicable to in-flight mass measurements.

Pre- and postflight conditions impose many compromises, and practical considerations must be accommodated. Preflight measures are usually made over an extended period of weeks or months before a flight to accommodate the conditions

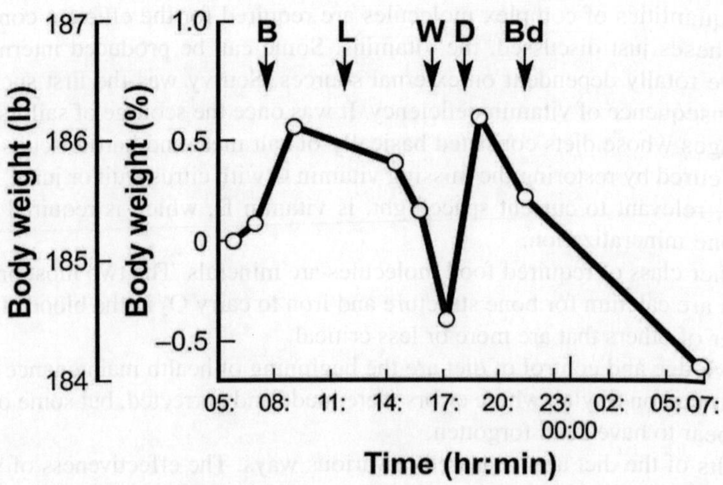

Fig. 10.4 Normal body weight changes throughout the day in 1 g. A typical record from a normal moderately active male is shown here. There is a continuous mandatory loss of mass with periods of increased loss by physical activity and of gain through intake of food and drink. *B* breakfast, *L* lunch, *W* walk, *D* dinner, *Bd* bed. Start and end of measurements are on arising from bed

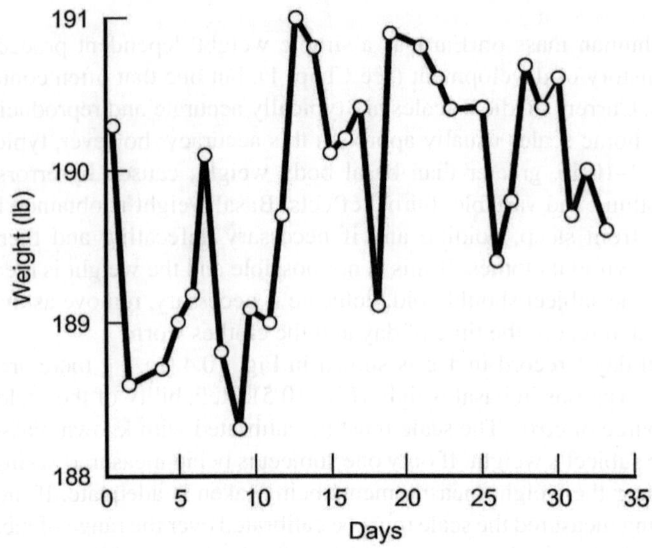

Fig. 10.5 Younger, more active men have larger variations. The larger sudden increase could have been from large evening meals; e.g., a Chinese or other caloric dinner, and the large sudden losses are probably longer than usual jogs or runs

usually crowded with activity near the time of launch. To obtain the desired real preflight weights, measurements should be taken under basal conditions during the week before launch. Astronauts can easily do this in their quarters using small portable calibrated scales.

Postflight weight measurements are more difficult because conditions are again busy, and the astronauts should be rehydrating and their weight rapidly increasing. Repeated weight measurements should be taken until they are stable, which may require a day or more.

10.4 Mass Measurement in Weightlessness

As noted in the introduction, loss of body mass in flight has been and remains complicated by the necessity for a nongravimetric mass measuring device. Currently there are two potential means. Dual Energy X-ray Absorption (DEXA) provides an approximation of ±2 lb. of body mass and also of fat and lean body mass. It is used to make measurements on the astronauts before and after their ISS mission, but the radiation exposure limits preclude repetitive measurements. Inertial methodologies have demonstrated accuracies of ±100 grams for body mass measurement. Details of the methodologies would occupy a large chapter or a book and are not included here.

In 1966 both NASA's Skylab[4] and USAF's Manned Orbiting Laboratory (MOL) had programs underway to study changes of mass inflight using the mass measurement of crewmembers correlated to food intake. Multiple contracts for specimen and body mass measurement devices had been unproductive, and the MOL medical support group at Brooks Air Force Base began an in-house development with support of the School of Aerospace Medicine (SAM) machine shop and its other facilities. Two methods of mass determinations were theoretically possible, high-energy photon absorption or some arrangement for inertial measurements, but photon scanning equipment was not practical at that time. The inertial method chosen was a mass-controlled, linear spring mass oscillator. With support of the SAM shop this method was validated, and a demonstration of the first mass measurement in weightlessness was made at Wright Patterson Air Force Base in zero-g aircraft in 1966. Measurement of the nonrigid human body was more complex, and a prototype with the requisite accuracy required another 6 months to produce.

In a joint USAF-NASA agreement, two Small Mass Measurement Devices (SMMD) for food residue and fecal mass and a BMMD were incorporated in Skylab as experiments M-074 and M-171 with their designer as the Principle Investigator (PI). In 1967 the PI became an astronaut and was followed by some of the MOL astronauts when that program was cancelled in 1969. A SMMD was included in 1972 Skylab Medical Experiments Altitude Test (SMEAT), and its operational performance was demonstrated with calibration masses, food residues and fecal bags [7, 8].

[4] Then known as Apollo Applications.

The three mass measuring devices functioned as designed for all three Skylab missions with three crewmen making daily basal body mass measurements plus mass measurement of all feces and residual food. Frequent in-flight calibrations verified their accuracies with masses traceable to the National Bureau of Standards. This allowed the first characterization of mass losses in space, and BMMD measurements plus leg volumes on SL-4 demonstrated the fluid shifts for the first time and accounted for the rapid initial mass loss and rapid postflight recovery of it. A backup Skylab BMMD was flown twice on Shuttle by NASA Life Sciences without calibration or the knowledge of operation. Although erroneous, the data obtained was published, and the instrument apparently was discarded. Studies and contracts were let for alternative devices. Russia built and flew a spring- mass oscillator BMMD on Salyut and MIR missions [9].

In anticipation of ISS, one author (WT) demonstrated a Linear Accelerating Mass Measuring Device (LAMMD) with performance equal to or better than that of Skylab instruments. NASA gave this work to a contractor who changed some key features and the name to Space Linear Acceleration Mass Measurement Device (SLAMMD) and is used on ISS [10]. The Russian BMMD was in place for the first ISS expedition and SLAMMD has been in place since Expedition 11. Both instruments appear to be in more or less regular use.

10.4.1 Space Food Systems

There are three often conflicting requirements for space food:

1. Accommodation for it on the vehicle is more complex than one might expect. Size and weight plus long-term storage logically lead to freeze drying, since there was a surplus of onboard water in Shuttle. Other problems are packaging and constraint of food and liquids during consumption. Heating and cooling facilities for it are desirable.
2. Provision of essential nutrients employs a study and analysis team to meet the myriad requirements of energy, substrate for body maintenance, vitamins and minerals. Tracking individual intake was an apparently unsolved problem on ISS for some time in spite of the labeling and recording systems available. In 2014 it was reported that crewmembers make a weekly dietary intake report using a Food Frequency Questionnaire which provides an estimate of caloric and other intakes.
3. Meeting the taste and other psychosensory requirements of the crew requires balancing the individual needs and taste requirements. This requirement consumed untold amounts of time and effort by both crew and nutritionists. Unfortunately their efforts are frequently frustrated by losses in the senses of smell and taste caused by fluid shifts and the alteration of nasal smell sensors when the food is consumed inflight.

The first foods on Mercury were pastes in squeeze tubes [11]. Packaged cubes and small plastic bags of powdered, freeze dried juices, soups and meats that had to be rehydrated were added. Gemini continued this system with a planned 2500 cal/day intake, but that level was not reached in practice. As usual the astronauts found a more palatable solution—across the street in Cape Canaveral in Wolfie's Yiddish Delicatessen—until the food scientists overheard John Young on the intercom in Gemini III asking Wally Schirra if he wanted a share of his corned beef sandwich. The food scientists went into such a public tirade over their science being threatened that Deke Slayton put a letter of reprimand in John's record folder. Some generations of food scientists later, non-crumbling bread and other ingredients for a sandwich to be made in space were flown with great fanfare. This was a poor substitute for a good Yiddish Deli sandwich whose flavors are robust enough to overcome the reduced sense of smell and taste in space in addition to providing enough protein for several days. While the Gemini foods were unpalatable, they kept men alive long enough for the planned Lunar mission.

Initially the Gemini foods were used on Apollo, which had hot water to rehydrate and warm the powdered food as desired. Irradiated food in flexible packets were added to later flights plus a "spoon-bowl" plastic package. The bowl contained freeze dried food and was covered with a plastic membrane. Water for rehydration was added through a sealed port. Its plastic membrane cover was then slit. Surface tension and adhesion held the food in the bowl or in a carefully handled spoon. However, at the end of Apollo, astronauts were losing mass through inadequate intake of unappetizing food.

A major effort [12] to correct this on Skylab resulted in the epitome of food in space to date and became the basis for much of the current food systems. Packaging, preparation and serving of food was revised beginning with a fixed table to which individual food trays with timed heaters could be attached. There were fitted depressions in the trays for rectangular plastic cups or round aluminum pop-top cans. Freeze-dried beverages, soups and foods were in the plastic cups and rehydrated by hot or cold water through a sealed port. Food could then be carefully consumed with ordinary utensils through a slit membrane. With intact membranes the cup became a beverage container with a "straw" inserted through the liquid port for consumption. Irradiated foods such as filets and shrimp were in cans and could be consumed directly from the can. Frozen desserts including ice cream were also stored and served in the cans. Menus that involved elaborate taste testing by the crews were produced. Other than occasional bits of gravel in the spinach and green beans, the food was quite acceptable. However, there was consternation in the SMEAT and Skylab crews when NASA HQ ordered the wine removed.[5] The irony of all this preparation was the inertia of the metabolic investigators who did not allow an adequate caloric intake until the last Skylab mission.

---

[5] A French reporter was interviewing a SMEAT crewmember prior to entry. "You are going to be locked in that tank 24 h a day?" "Yes." "That will be difficult for 56 days. How many days conjugal leave?" "None." "Incredible!" "What wines will you have?" "None, the director is a Mormon." The reporter slapped the desk and all but shouted, "But that's barbaric!"

Shuttle initially had the same plastic cup arrangement for freeze-dried food. A forced air oven replaced the hot water. Frozen foods were replaced by a number of dried and thermostabilized items in flexible sealed packages. A personal taste preference program was maintained. Spices and condiments were made available to compensate for the reduced senses of taste and smell. There were individual food trays with velcro and elastic constraints and more or less regular conventional utensils. After Skylab, food and caloric allotments were increased.

One lesson learned on Shuttle involved carbonated beverages. Early in the Shuttle Program Coca Cola had made an elaborate drinking device and an arrangement to document a Coke in space. Two days before our launch Coke was removed by a White House order. Eventually a Coke and Pepsi were flown, although without consideration of what CO_2 in a warm beverage agitated in weightlessness would do. The whole liquid mass becomes a low order explosion with a rapidly expanding cloud of foam. Karl Henize took a hefty mouthful through the drinking apparatus and his cheeks became those of an overstuffed chipmunk with geysers of foam from each nostril. The Pepsi remained unopened.

Russia tended to fly more traditional food stuffs and with its vehicles' increased mass capacity did not seem to be as concerned with dehydrated items. On the Shuttle/Mir missions food for the participating astronauts and cosmonauts was 50% NASA and 50% Russian.[6] Food on ISS is a composite of Russian and NASA food, with additional European and Asian inputs. Since water is now limited, there has been a shift to thermostabilized food with a combined selection from 185 NASA items and 100 Russian items that are initially available for crew preference selection. Also there are spices and sauces with continued additions to the lists including those for other ISS participants. Frozen items are not available. One initial plan was for individual menus on a 8-day cycle, but this was later changed at the crews' request to a pantry system with on-orbit selection. This generated additional problems involving record keeping as noted in a news release.[7] It also resulted in continued crew body mass loss, which was exacerbated by the absence of knowledge of body mass measurement. The ISS food system is still evolving so rapidly that any detailed description is likely to soon be outmoded. An example of a typical 6-day menu is shown by Table 10.1. Russian and NASA dining tables are shown in Fig. 10.6. Large waste bags are clearly visible. Handling of waste packaging adds to the complexity of any food system. The large plastic cups on Skylab and early Shuttle were replaced by smaller flexible membrane packages. Metal cans add a considerable weight load, and individual packaging increases that load. On ISS

[6] This may account for the large body mass loss experienced by the first astronaut to fly a Shuttle/Mir mission.

[7] According to recent news releases this is yet to be achieved on ISS. For example, Adrienne Burke in Forbes writes, "Imagine having to recount everything you ate in a week while orbiting the Earth. That's what astronauts do in a weekly 'food frequency questionnaire'. But diet logging isn't rocket science, so NASA is turning to 'the crowd' for help. The NASA International Space Station Food Intake tracker or ISS Fit-Challenge, launched 2013, is the latest open-innovation contest sponsored by the NASA Tournament Lab, a partnership between NASA, Harvard's Institute of Qualitative Social Service, and competitive software development TopCoder."

expeditions this waste was put in a Russian vehicle, which was incinerated in the atmosphere on an Earth return trajectory.

10.5 Loss and Gain of Mass

10.5.1 Loss and Gain in 1 g

This is one of the most studied and demonstrated areas of human activity. The information about this activity is also one of the most flagrantly confused and abused in human history, by the food industry, by consumers, and by an army of experts including many scientists who should know better.[8] The problem of body weight or mass loss, or gain, on Earth or anywhere else in the universe is as simple as the diagram in Fig. 10.3. Energy intake that exceeds energy used will produce an increase in body weight or mass, and vice versa. This balance can be controlled to a fraction of a pound if the amount of intake is known, and if there is a reliable means available for measuring basal weight or mass accurately to about ±0.2 lb. Weight or mass can be controlled by qualitative means to a pound or so by an observant, knowledgeable person, although less precisely.

10.5.2 In Bed Rest

Bed rest is not a valid analog of spaceflight for changes in body mass. Physical activity is deliberately minimized in bed rest, in contrast to activity in space. In addition, the shift and loss of fluids is a fraction of that in weightlessness (see Chap. 4).[9]

10.5.3 In Space Flight

There was confusion about causes of loss of mass in weightlessness at the start of Skylab. The most prominent feature of in-flight mass measurements on SL-2 (Fig. 10.7) was a more or less linear decrease in mass from the beginning of the diet until recovery, which clearly demonstrates the diet's caloric inadequacy. Salvage

---

[8] It was "common knowledge" among many professionals that weight loss in space could not be avoided.

[9] The principal investigator (PI) who set inadequate limits on SMEAT/SL diets apparently based them on the numerous BR studies he had done.

Table 10.1 A 6-day cycle astronaut menu for ISS

Meal	Day 1	Day 2	Day 3
Breakfast (T)	Blueberry/raspberry yogurt (T)	Dried peaches (IM)	Blueberry-raspberry yogurt (T)
	Granola with blueberries (R)	Oatmeal with raisins (R)	Granola with raisins (R)
	Orange drink (B)	Orange drink (B)	Orange drink (B)
	Kona Coffee with cream, sugar (B)	Kona Coffee with cream, sugar (B)	Kona Coffee with cream, sugar (B)
Lunch	Beef fajitas (I)	Smoked Turkey (I)	Chicken strips in salsa (T)
	Tortilla (FF) × 2	Tortillas (FF) × 2	Tortillas (FF) × 2
	Applesauce (I)	Dried pears (IM)	Applesauce (T)
	Almonds (NF)	Almonds (NF)	Cashews (NF)
	Lemonade (B) × 2	Orange-grapefruit drink (B) × 2	Lemonade (B) × 2
Dinner	Shrimp cocktail (R)	Vegetarian vegetable soup (T)	Chicken noodle soup (T)
	Grilled chicken (T)	Chicken fajitas (T)	Beef stroganoff w/ noodles (R)
	Macaroni and cheese (R)	Tortilla (FF) × 2	Broccoli au gratin (R)
	Green beans with mushrooms (R)	Cherry-Blueberry cobbler (T)	Dried peaches (IM)
	Candy coated chocolates (NF)	Orange drink (B)	Apple cider (B)
	Lemonade (B)	Tea with lemon and sugar (B)	Tea with lemon and sugar (B)
	Tea with lemon and sugar (B)		
	Day 4	Day 5	Day 6
Breakfast	Dried pears (IM)	Dried peaches (IM)	Blueberry-raspberry yogurt (R)
	Oatmeal with Brown sugar (R)	Oatmeal with raisins (R)	Granola bar (NF) × 2
	Orange drink (B)	Orange drink (B)	Dried peaches (IM)
	Kona Coffee with cream, sugar (B)	Kona Coffee with cream, sugar (B)	Orange drink (B)
			Kona Coffee with cream, sugar (B)
Lunch	Peanut butter (T)	Beef fajitas (I)	Chocolate brownie cliff bar (FF)
	Grape jelly (T)	Tortilla (FF) × 2	Vanilla breakfast drink (B)
	Tortilla (FF) × 2	Dried pears (IM)	Almonds (NF)
	Trail mix (IM)	Almonds (NF)	Orange-mango drink (B)
	Grape drink (B) × 2	Orange-grapefruit drink (B) × 2	

(continued)

Table 10.1 (continued)

Meal	Day 4	Day 5	Day 6
Dinner	Shrimp cocktail (R)	Grilled pork chop (T)	Split pea soup (T)
	Crawfish étouffée (T)	Mashed potatoes (R) T	Teriyaki chicken (R)
	Vegetable risotto (R)	Broccoli au gratin (R)	Rice pilaf (R)
	Creamed spinach (R)	Peach ambrosia (R)	Broccoli au gratin (R)
	Dried peaches (IM)	Apple cider (B)	Peach ambrosia (R)
	Apple cider (B)	Tea with lemon and sugar (B)	Apple cider (B)
	Tea with lemon and sugar (B)		Tea with lemon and sugar (B)

Definition of fresh food is not obvious. Current policy allows meals to be selected from pantry stock, a process not without potential problems including accurate intake records, not required except for research

B indicates beverage; *FF* fresh food, *I* irradiated, *IM* intermediate moisture, *NF* natural form, *R* rehydratable, *T* thermostabilized

Fig. 10.6 NASA and Russian dining areas. Typical packages of food from each are shown in foreground. A variety of condiments including commercial hot sauces are shown, popular items to override reduced sense of smell and taste

operations on the spacecraft precluded measurements for the first 3 days inflight. Other in-flight features show the increases in loss just prior to return. Probably the most significant aspect of these curves is the greater rate of loss inflight than on Earth. Body mass records from the first Skylab crew (labeled SL-2) show some of these features with greater clarity. Those losses forced an increase in the diet from 18 to 19.4 cal/lb. of BW per day for the next crew (labeled SL-3). In two of the three crewmembers the curves became flat except that, as in SL-2, there were significant losses in the period of reentry preparation. This was a time of increased activity. Pre- versus in-flight data again showed that energy requirements were higher inflight. Rapid losses of fluid mass in the first days of flight were matched by equivalent postflight gains. On SL-4 these data were coupled with leg volume measurements, and the rapid mass losses over the first days were consistent with volumes of blood and water lost from the legs in the first 24 h of fight.

On the SL-4 mission, mean caloric intake was increased to 21.0 cal/lb. of BW per day. Figure 10.8 is the commander's loss curve, and if the fluid losses are accounted for, he actually gained approximately 1 lb. The numerical losses quoted are based on the 1-g measurements made on recovery day before fluids were fully replaced. Using these numbers, Fig. 10.9 is a three point "curve" showing an expected and almost linear relationship between caloric intake and mass loss in space and on Earth.

Fundamental facts are demonstrated by these still unique Skylab data. There are two components to mass lost in space. One is a rapid obligatory loss of fluid shifted from the legs. This normally amounts to 4–6 lb. of individual loss, and should be returned to the body before and during entry. Secondly, if the diet is controlled, a more accurate measure of in-flight mass can be obtained 2–5 days postflight. Once this is accounted for gains and losses are dependent on food consumed versus energy expended.

As noted, results of mass measurements on Russian programs are not reported here. One secondary report claims losses of 15% body weight on 3 month missions. The only studies of mass loss on post Skylab missions are pre- and postflight measurements (Fig. 10.10). Two significant reports describing changes in body mass on ISS are available. The first by Smith, et al. [13] was an investigation of the effects of exercise and nutrition on loss of bone in 13 astronauts on ISS expeditions 14 (2006) through 20 (2009) with 48–215 days duration. They were divided into groups of five based on usage of the IRED and eight who used the ARED strength exercise devices [14].

In addition to using different strength exercise devices (IRED and ARED) (see Chap. 7) both treadmill (strength) and cycle ergometers (aerobic) were also used, but time and load levels are not reported. Food intake was based on weekly recall and completion of a Food Frequency Questionnaire (FFD). Body mass was calculated from density determinations, and from pre-and postflight weight measurements done at unspecified times using unreported procedures. The disparity between body mass and weights shown in Table 10.2 can be explained by the delay in mass measurements after landing. Attribution of differences in mass lost inflight to dietary differences is consistent with the more rigorously done intake and body mass changes documented in the nine crewmen on Skylab (see p 218 and Fig. 10.9).

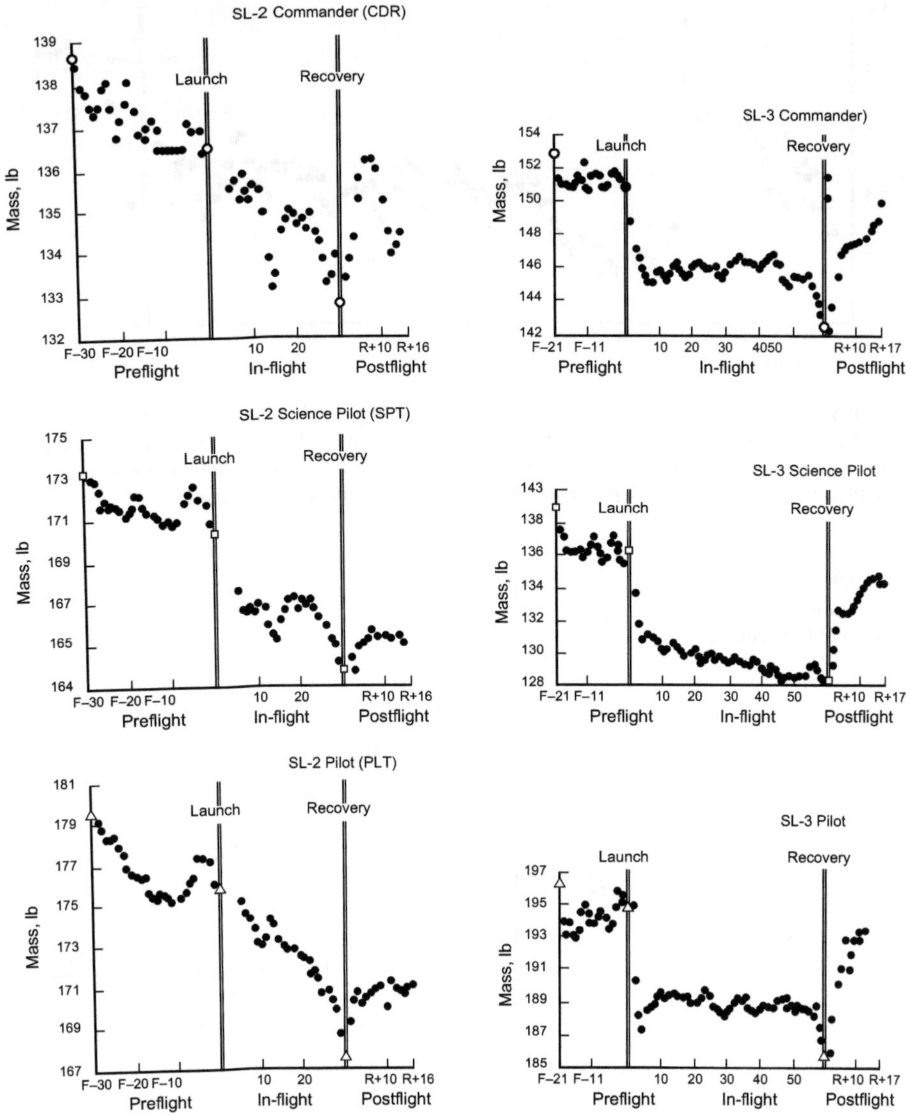

Changes in body mass of SL-2 crewmen, where F – 10 is 10 days before lift-off, R + 10 is 10 days after recovery, etc.

Fig. 10.7 Records of body weight and mass pre-, in-, and postflight from SL-2 and SL-3. Shown are key features of such records. Obligatory fluid losses are difficult to distinguish from very rapid metabolic losses. On SL-3 with increased caloric input the rapid initial losses and early postflight gains from fluid may be clearly distinguished from the new much slower metabolic loss. Two other features are (1) rapid losses during the pre-entry period in two members of SL-3 and all crewmembers of SL-2 and, (2) the increased rate of metabolic loss inflight versus 1 g in four of six crewmembers

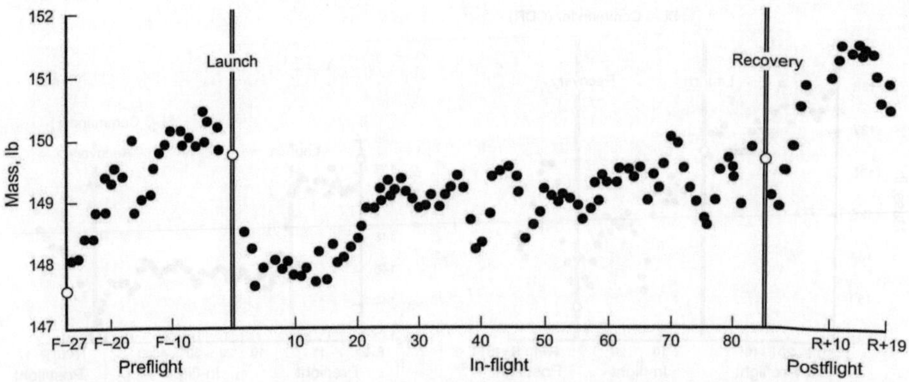

Fig. 10.8 Shown is a record of the SL-4 commander who gained tissue mass in weightlessness. Note the positive slope of the in-flight metabolic curve. When fluid losses were restored by day 6, there was a total gain of approximately 1 lb. These records illustrate the need to allow several days for replacement of fluids before pre- and postflight body weights are compared

Fig. 10.9 Mean body mass loss versus energy intake on SL missions.

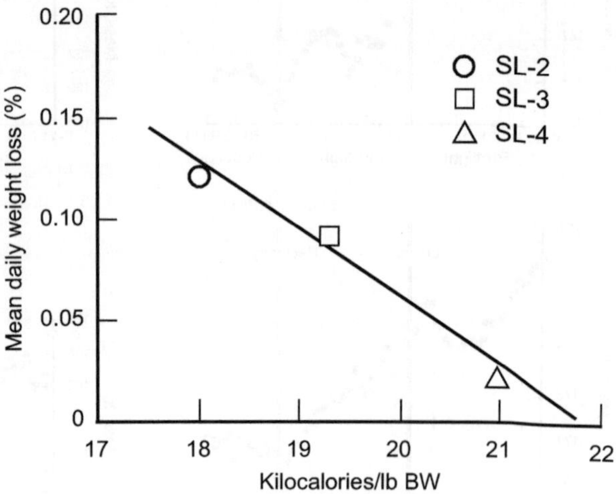

The author of this paper correctly recognizes another problem in this paper with the statement; "Its one important limitation is the inability to separate the influences of resistance exercise and adequate nutrition."

The definitive publication on Body Mass Measurement on ISS to date [1] is unusually well produced but its content and conclusions raise questions [15, 16].

It is a study of 55 subjects, 44 males and 11 females, with ISS flights of 93–199 days, body mass of all subjects was measured inflight by the Russian BMMD, and on 25 subjects, 19 males and 6 females, by the JSC SLAMMD. One g reference or base-line weights were made 74 ± 21 days preflight and 3–24 h and 30 days postflight. Scales used for subjects' 1 g measurements were allowed calibration errors of ±0.23 kg (0.5 lb) Precision was not measured. Other potential sources of base line or

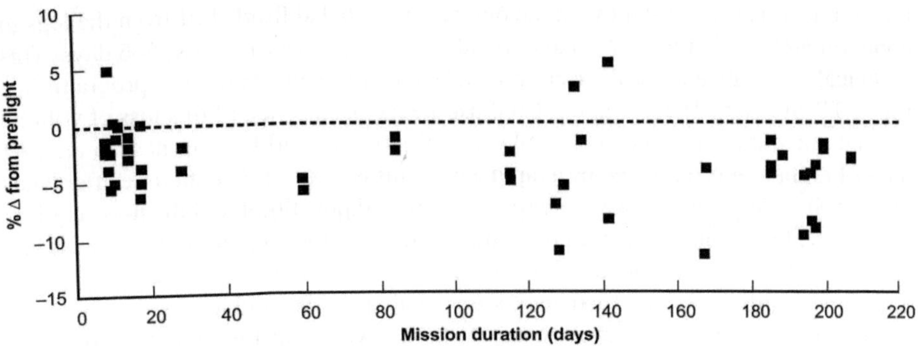

Fig. 10.10 Postflight weight loss. Assuming time had been allowed for postflight fluid equilibration, inadequate food intake is the rule

Table 10.2 Reported intake and body mass of two groups on ISS

Group (n)	Calories (kcal/day)	Protein (gm/day)	Weight (kg)	Δ (%)	
				Pre	Post
8	2.23 ± 0.36	85 ± 13	77	79	−2.5
5	2.43 ± 0.87	102 ± 57	74	73	1.5

reference errors are the procedures used, including time of day and time before launch, which in this case was 3–6 months. The differences from the actual body masses at launch were not addressed but could reasonably amount to more than a kilogram.

Validity of the inflight measurements becomes even more problematic. A long paragraph describes some characteristics of the inflight instruments but fails to mention that neither can function normally in 1 g. nor have been calibrated with known masses in flight, although the need for such calibration was noted in the papers conclusions. Instead various statistical methods were employed in an attempt to derive instrument validation from the operational results they produced. Statistics of precision or reproducibility of repeated measurements of the same subject produced coefficients of variation of 2.1% for the BMMD and 1.6% for the SLAMMD. Such variation becomes more meaningful when the data points are plotted from both instruments. While the mean difference in indicated masses is plus 1.1% for the SLAMMD, the 95% confidence intervals of the points are 3.5 and −5.8 kg. The result of such potential errors may be seen in the 25 thumb nail plots of the data points from both instruments plotted as a function of mission duration.

While the various data points were scheduled, many appear to have been random occurrences. SLAMMD data appeared too erratic to consider. We made an attempt to correlate these records with the validated Skylab records and other known aspects of loss of BM in space flight to illustrate several problems in this study. Eighteen records had a data point in the first 10 days after launch. Seven of these showed no significant change from baseline If this were true it would be a new phenomenon. It is far more likely explained by baseline procedural error or BMMD instrument error. Fifteen showed losses of 4–6 kg with a median of 5 kg (11 lb) in this period. Except for magnitude, this is consistent with the obligatory cephalad fluid shift from the legs in approximately 12 h after launch and its subsequent loss over the next 5–6 days. This is typically 4–6 lb. and easily seen in an adequate mass measurement program. See Figs. 10.7 and 10.8. These figures also illustrate the persistence of this loss of volume through the duration of the 86-day Sky lab mission. A rapid loss of mass in the first days of flight is approximated by a rapid gain of mass in an equivalent time after landing. On SL-4 leg volumes were measured pre in and post flight and the mass equivalents of the fluid volume changes individually matched the changes in body mass just discussed. These are cardiovascular fluid phenomena that are discussed in Chap. 4, and are very different from food and tissue metabolic gains and losses.

One major point is that more needs to be known about the nature of losses of mass in space. The obligatory loss of several pounds of fluid is an appropriate adaptation of the CV and produces a new normal baseline for the duration of the flight. At the end the flight in 1 g, fluid will be rapidly restored and the subject's mass compared to the preflight baseline. The reference for inflight control of mass should be that measured at the end of the transient fluid loss, not the preflight baseline. To replace normally lost fluid with fat is a mistake. Figure 10.8 is an example of this. Examples of loss of tissue mass in flight may be seen in Fig. 10.1 and the SPT in SL-3 in Fig. 10.7.

A second major point is that information not looked for will not be found. Single measurements 74 days preflight, sometime in the first 2 weeks and then every month with single measurements cannot give rise to meaningful mission records and makes misinterpretation probable. Timing and frequency in relation to relevant events is essential to production of useful information.

A third major point is that the data can be no better than the instruments that produce it. The current SLAMMD design does not take into account key principles and after almost two decades and untold amounts of money have been unable to produce usable results. While calibration of the Russian BMMD would produce more meaningful data, its records will always be too noisy for statistical correction. It remains in use only because there was nothing else after MOL Skylab instruments were discarded.

The first two points raised can be easily remedied if so desired. The third is more problematic. It has been a half a century since the USAF MOL BMMD was demonstrated and none of the newer instruments have come close to its performance on Skylab. Moreover, until such performance again becomes available, the only progress in understanding and properly controlling loss of body mass in space will be empirical or let's try it and see. This definitive paper describes a regression in both understanding and practice.[10]

[10] Some 2 years ago while helping to close Brooks A.F.B. the author found the prototype of the MOL BMMD on a heap of metal to be discarded. Permission to salvage it was obtained. The electronics were missing but it was mechanically intact and sent to Dr. Yusaku Fugii, Gunma Univ., Japan who used laser interferometry to verify its function with solid and human masses to verify its performance, which exceeded the flown Skylab BMMD.

10.5.4 Waste Collection Systems

It is tempting to avoid this subject. The relation of waste collection in space to weight loss is psychological not physical. Eating and elimination are simple physical imperatives for all on long duration flights but not for all on shorter flights. Experiences on a 7-day, six man Shuttle flight illustrated this. A military astronaut developed low abdominal pain after not using the Waste Collection System (WCS) for 4 days. After repeated failures, the WCS's fecal processing mechanism had been removed and there remained a big pot (Fig. 10.11c), which soon filled with a floating mass of feces and wipes. A long handled kitchen spatula had been added to punch down a space large enough for whatever contribution one expected to make. It was messy. The astronaut was given a laxative and instructed to use the WCS. Next morning he was outraged, "Your pill made me use that thing two times!"

During reentry preparation a civilian mission specialist was well behind the time line. He was found eating rabidly with food in both hands—after eating nothing for 2 days to avoid using the WCS. This could again become a problem with the advent of tourist flights. An improved urine collection system had been added to the same flight for a functional test. It promptly choked and regurgitated urine over the user's flight suit and into floating globules.

These incidents illustrate both the problem and resulting difficulties. Contractor engineers have not done well in this area. They have depended on computer design and simulation rather than taking their pants down and realistically trying their designs. An example of this was the original Skylab urine collection and measuring system. It was functionally tested in SMEAT and both the urine separation and measuring system failed completely when urine rather than water was used. A simple in-house system was thrown together at great additional cost just in time for Skylab's launch.

America's first astronaut launched with a wet back after hours of unplanned delays. The use of a condom, tube and bag became and remained standard practice for males on launch through Shuttle. This was also the urine collection system during the early missions and through Apollo. Fecal bags with adhesive edges for temporary sealing to one's buttocks completed the early waste collection system. Such bags have been used for several periods during failures of ISS waste collecting systems. Skylab's urinal was a funnel and tube with a rapid airflow that carried the urine into a centrifugal air/fluid separator where it was processed and stored in individual collection bags. Skylab's fecal collection was a conical porous fiber bag beneath a seat with rapid airflow through it. On SMEAT and Skylab every fecal bag was sealed after use, its mass measured, vacuum dried and returned for study.

Real problems with waste collection systems began with General Electric's big spherical pot for Shuttle. It was so complex that a mock up with under-seat TV camera was considered necessary to train users to properly position themselves for the system's use. Its failure inflight triggered a series of fiascos.

These failures triggered the collaborative design of a simple workable system by Henry Whitmore of Whitmore Enterprises, Inc., which was flown on two missions [17], to acclaim from the crew (Box 10.2).

There were competing big contractor designs, one of which was a large complex disposable "Dixie cup" arrangement that may have been flown on late Shuttle and

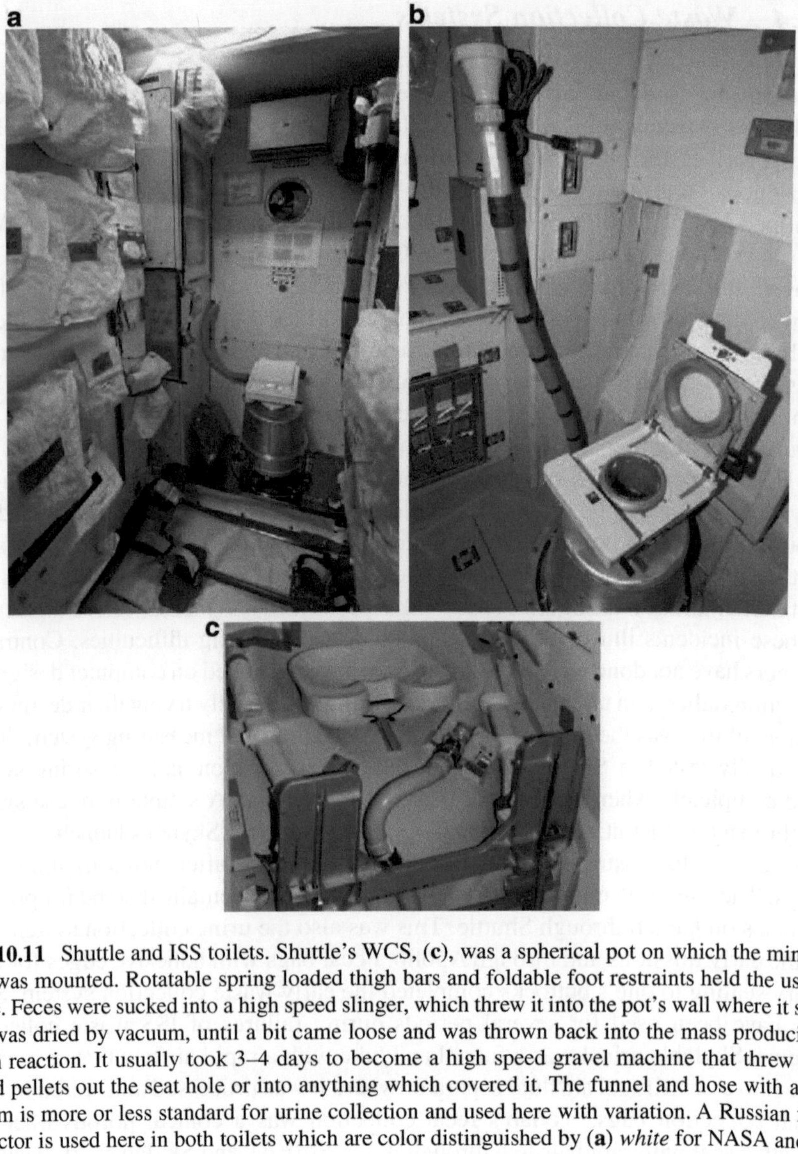

Fig. 10.11 Shuttle and ISS toilets. Shuttle's WCS, (**c**), was a spherical pot on which the minimal seat was mounted. Rotatable spring loaded thigh bars and foldable foot restraints held the user in place. Feces were sucked into a high speed slinger, which threw it into the pot's wall where it stuck and was dried by vacuum, until a bit came loose and was thrown back into the mass producing a chain reaction. It usually took 3–4 days to become a high speed gravel machine that threw high speed pellets out the seat hole or into anything which covered it. The funnel and hose with an air stream is more or less standard for urine collection and used here with variation. A Russian fecal collector is used here in both toilets which are color distinguished by (**a**) *white* for NASA and (**b**) *blue* for Russia (Colour figure online)

early ISS. "May have been flown" because the story of WCS is especially opaque, and the few comments that follow are gleaned from press reports. It is known that the $100 million costs of WCS development produced two Congressional investigations in the 1990s.

It appears that whatever was initially flown on ISS was replaced by a $19 million unit, which failed causing a major problem when the Soviet WCS also failed after 7 years use. It was back to stick-on fecal bags, condoms and urine bags or diapers. NASA finally balked at the $216 million cost of an "upgraded" WCS and instead

Box 10.2 The Whitmore Improved Waste Collection System (IWCS)

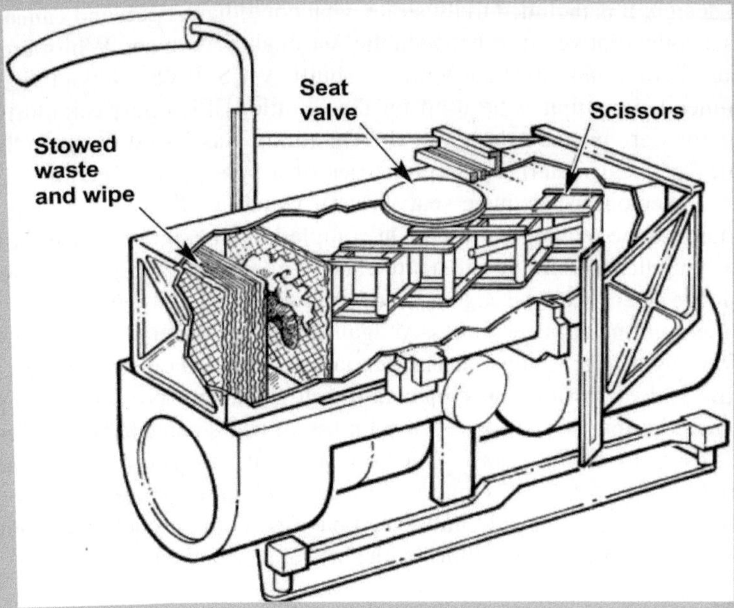

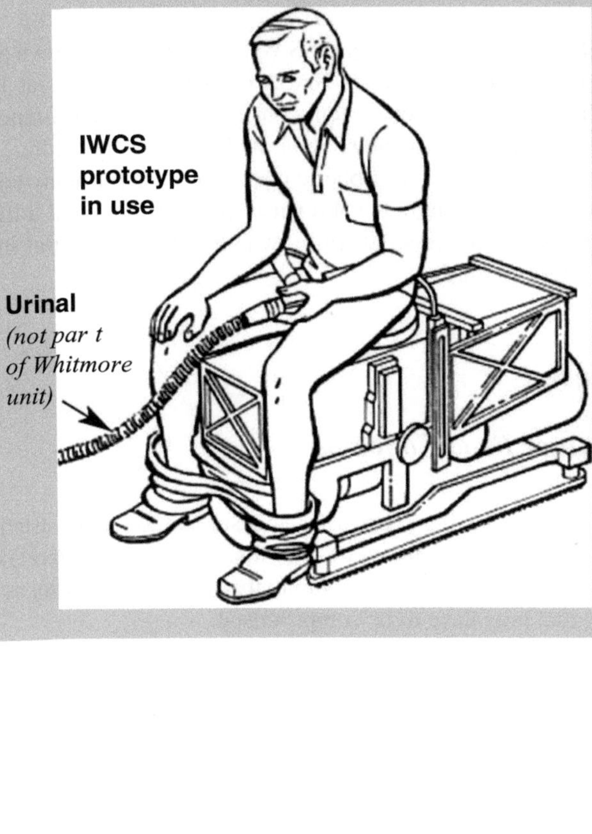

The Whitmore IWCS had features that remain in advance of any subsequent design. It is included to illustrate what could have been and can still be. It was a collaborative effort between the Astronaut Office and Whitmore after a series of continuing mishaps with the Shuttle WCS. It was flown twice with recommendations that it be used on the Shuttle. Efficiency, reliability and simplicity were primary design goals. Operation was based on the adhesive qualities of fecal material plus properties of a paper pad. Operation began with the user on an adequate seat through which air flowed carrying fecal material into the rectangular tube. On completion an extension arm moved a thin rectangular piston through a stretched fiber seal and wrapped it over the piston face and around the edges. As it continued forward it wiped all surfaces until the fecal material was pressed against another sheet from its previous use, forming a large sandwich. The piston was retracted closing all ports and the now sealed interior was subjected to a low vacuum pressure drying the fecal material into a more or less rigid mass. It had a capacity for a nominal week's usage by six subjects at which time the end plate was removed and the material pressed out of its temporary storage en bloc into a rectangular sealable bag. Single use specially fabricated bags cannot approach the efficiency of this design nor its inherent mechanical reliability and ease of use.

purchased a Russian unit. The current ISS WCS facilities are shown in Fig. 10.11 and consist of identical WCS units in the NASA (A white) and Russian (B blue) toilets.

The Russian urine collection system consists of a hose and funnel with air transport of urine to a liquid separator. Separated urine is then filtered into reusable water. Russia has reclaimed atmospheric water for reuse for many years. Their WCS has an airflow system through its solid waste compartment and a fluid separator. A bag beneath the seat catches solid waste, which must then be put in an impervious over bag for return to Earth.

10.6 Future

10.6.1 Future Food Concerns

Exploratory missions lasting years will increase the food and waste problems by an order of magnitude. Packaging waste [18] that occupies an appreciable fraction of the food mass cannot be used. The complexity of gratifying taste, as important as it is, is a luxury that may have to be compromised.

Food may have to be concentrated to its essence so that the indigestible fibers and other components that form much of fecal mass is eliminated (see Table 10.3). Recycling and preservation of every possible drop of water will be mandatory.

Table 10.3 Human waste (amount and composition)

Fluid 9 (extremely variable)	Urine, volume 1.5 L/day (minimum)	Sweat and respiratory 1.0 L/day (typical)
Fecal mass	City population, processed food	Third world urban population, unprocessed food
	Median	Mean
Male	104 gm	72–470 g/day
Female	99 gm	80–120 g/day
Typical composition		
Water	75%	
Solids	25%	
Solid composition		
Insoluble fiber	30%	
Dead bacteria	20%	
Fat	10–20%	
Inorganic material	10–20%	
Undigested protein	10%	

The only ace in this hand of cards is the incredible adaptability of the human body and mind. History is filled with stories of survival and function in the face of privation. Long sea voyages with diets of hard tack and salt beef come to mind. There are records of even greater enforced privation. In the same way explorers have always been known for their irrational willingness and their determination and ability to reach extreme goals, a combination of these factors plus intelligent usage of experience in space makes a Martian voyage inevitable.

Significant advances will have to be made in this area before exploratory missions can be considered. But based compared to Skylab decades ago the state of the art may have regressed in this area for lessons already learned have been forgotten or ignored. It is hard to imagine a inaccessible without accurate mass measurement. This depends on the ability of engineers and craftsmen.

10.6.2 Summary of Weight Loss in Space

These losses have two components, which are (1) a mandatory loss of several liters of fluid from the legs, and (2) a metabolic loss or gain dependent on food and activity just as on Earth. Waste collection can play a role in metabolic loss. Arguably the metabolic problems are hardware problems; i.e., the need for a means of monitoring losses, adequate food and drink, and a waste collection system, because in this the theory is simple and has been demonstrated. But just as in exercise counter measures, solution of this problem is now dependent on provision of adequate support systems in weightless and not on the human body's ability to accommodate to weightlessness.

References

1. Zwart SR, Launius RD, Coen GK, Morgan JL, Charles JB, Smith SM. Body mass changes during long-duration spaceflight. Aviat Space Environ Med. 2014;85(9):897–904.
2. Thornton W, Ord J. Physiological mass measurements in weightlessness. In:Biomedicalresults from Skylab NASA SP–377. Houston, TX: NASA Johnson Space Center; 1977. p. 19.
3. Guyton A, Hall J. Energetics. In:Textbook of medical physiology. 11th ed. Philadelphia: Elsevier Saunders; 2006. p. 72.
4. Guyton A, Hall J. Dietary balances. In:Textbook of medical physiology. 11th ed. Philadelphia: Elsevier Saunders; 2006. p. 71.
5. Guyton A, Hall J. Metabolism of carbohydrates. In:Textbook of medical physiology. 11th ed. Philadelphia: Elsevier Saunders; 2006. p. 67.
6. Guyton A, Hall J. Lipid metabolism. In:Textbook of medical physiology. 11th ed. Philadelphia: Elsevier Saunders; 2006. p. 68.
7. Guyton A, Hall J. Protein metabolism. In:Textbook of medical physiology. 11th ed. Philadelphia: Elsevier Saunders; 2006. p. 69.
8. Johnson RS. Skylab Medical Equipment Altitude Test (SMEAT). Houston, TX: LBJ Space Center; 1973.
9. Sarychev VA, Sazonov VV, Zlatorunsky AS, Khlopina SF, Egorov AD, Somov VI. Device for mass measurement under zero-gravity conditions. Acta Astronaut. 1980;7(6):719–30.
10. Smith DC, Kaufman KA. Space linear acceleration mass measurement device (SLAMMD) for the human research facility (HRF). SAE Technical Paper; 1998 Jul 13.
11. Perchonok M, Bourland C. NASA food systems: past, present, and future. In:Nutrition, vol. 11: Elsevier Sciences, Inc; 2002. p. 913–20.
12. Smith SM, Zwart SR, Kloeris V, Heer M. Nutrition of spaceflight. In:Space science, exploration and policies series. New York: Nova Science Publishers, Inc.; 2009.
13. Smith S, et al. Benefits from resistance exercise and nutrition in long duration spaceflight. J Bone Miner Res. 2012;27(9):1896–905.
14. McCrory JL, Baron HA, Balkin S, Cavanagh PR. Locomotion in simulated microgravity: gravity replacement loads. Aviat Space Environ Med. 2002;73:625–31.
15. Campbell M. Body mass changes during long-duration spaceflight: letter to the editor. Aerosp Med Hum Perform. 2014;85(12):1229.
16. Thornton W. Comments on body mass changes during long-duration spaceflight. Aerosp Med Hum Perform. 2015;86(12):1070–1.
17. Improved waste collection system DTO (IWCS). In: Project summary, performance evaluation, and close-out report. Houston, TX: NASA Johnson Space Center, CTSD–SH–522; 1991.
18. Strayer RF, Hummerick ME, Richards JT, McCoy LSE, Roberts MS, Wheeler RM. Characterization of volume of trash from the three FY11 STS missions: trash weights and categorization and microbial characterization. Cocoa Beach, FL: NASA Kennedy Space Center, AIAA; 2001.

Chapter 11
Accommodations to Weightlessness

11.1 Introduction

Unlike the anatomical and physiological changes necessary to adapt to weightlessness covered in the previous ten chapters, there are other significant changes that the body makes that are within its existing 1-g capacity to accommodate the new force environment [1]. Our bodies are elastic and plastic, and although seldom noted, they are constantly changing size, shape and posture under the influence of their varying orientation to the 1-g weight vector. These accommodations are typically rapid, seconds to hours, and except for posture are small (Fig. 11.1).

In weightlessness these accommodations are larger and may become operationally significant as indicated by John Young's comments on preparing for Charlie Duke's Apollo 16's first Lunar EVA, "We didn't think we were going to get Charlie into that thing (a precisely cut EVA suit fitted in 1 g)." Charlie Duke was tall and had probably "grown" two plus inches taller in weightlessness before his lunar landing and the lunar 1/6 g did not reduce the increased height by more than 1/3 in.

These accommodative changes are all plastic and elastic except for posture, which is functional. They were all initially documented on Skylab by Astronaut Office Detailed Supplementary Objectives (DSOs) [2], and are periodically rediscovered by new astronauts and investigators.

Table 11.1 lists these accommodative changes.

W. Thornton, F. Bonato, *The Human Body and Weightlessness*,
DOI 10.1007/978-3-319-32829-4_11

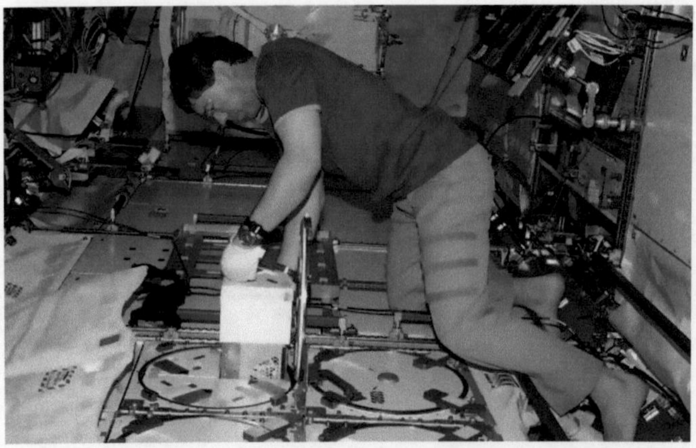

Fig. 11.1 In 1 g there is a variety of postures that under the weight of gravity are comfortable and minimize muscle counterforces. In weightlessness attempts to use straps and other devices to replace weight forces to make the body conform to 1-g chairs etc. are typically uncomfortable. Without weight the body segments move to positions established by relaxed muscle tone as shown here by Michael Barratt aboard ISS. The result is a single body posture to which "furniture" should be fitted

Table 11.1 Accommodative changes in the human body in weightlessness

Accommodation	Time constant	Δ %
Height (stature)	Immediate + slow component (hours)	3–4
Girths	Immediate	<1 to 10
Center of Mass	Biphasic components	
	Rapid (hours)	2
	Slow (days)	4
Posture	Immediate	–

11.2 Changes in Height

Changes in height are a well-known accommodation in 1 g, and the mechanics were known before the larger changes were seen in weightlessness [3]. They are the product of changes in spinal length and curvature. Their mechanisms are complex and require care in measurement.

Figure 11.2A illustrates the two sources of change in height on Earth and in space. The flexible spinal column, B, has three curvatures, cervical, thoracic and lumbar, which vary with bodyweight and load. Lordosis that occurs in response to corpulent bellies or pregnancy is easily seen, but the changes in curvature are complex and not easy to measure. Changes in spinal curvature can result in detectable

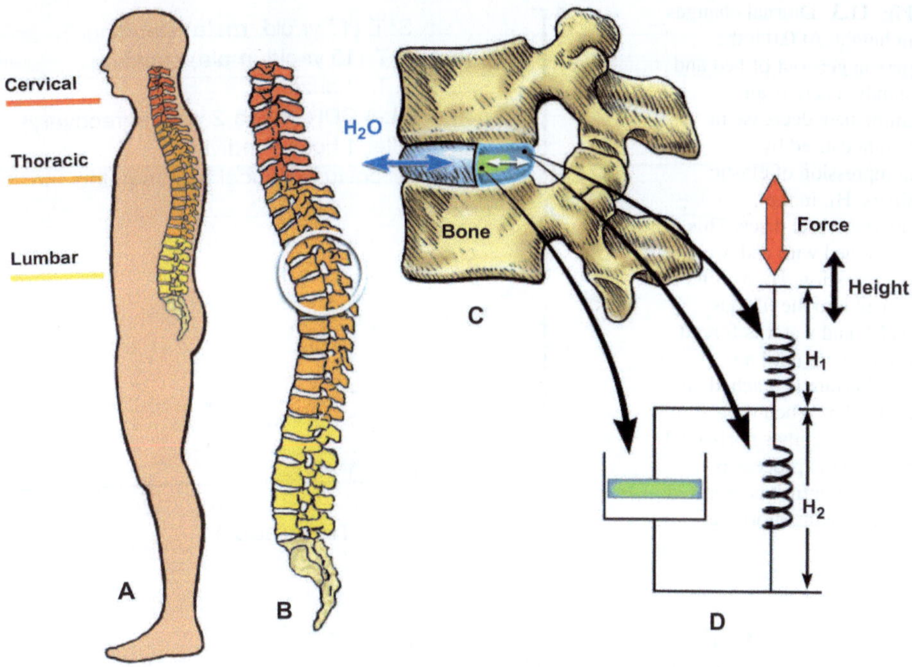

Fig. 11.2 Schema of height changes. Total change in height (**a**), occurs in the spinal column (**b**), whose length varies with its curvature and height of its intervertebral discs [5] (**c**), which have elastic elements that respond immediately to changes in force loads and a viscously damped elastic component, which responds more slowly. An analog model of this is shown in (**d**)

changes in height, but only the larger changes in the spine's 23 intervertebral discs will be examined here. Rigid vertebral bodies are separated by elastoplastic discs, C, which consist of a thick annular wall (annulus fibrosis) formed of dense intermeshed elastic collagen fibers. It encloses a space capped by thick tough membranes forming a reservoir that contains a viscous hydrophilic semi fluid. D is a mechanical analog of the disc with this reservoir. When vertical forces are reduced as in horizontal posture in sleep in Fig. 11.3, the elastic walls immediately expand axially as the H_1 spring in the analog does, causing an immediate increase in length of all the discs and a cumulative increase in height. Simultaneously, dense viscous fluid that had been forced into the surrounding fibers is slowly squeezed out by the elasticity of the fibers as represented by the piston/cylinder (dashpot) and spring, H_2. Water is now absorbed by the disc to fill its expanded volume [4]. The result is a slow additional exponential increase in height over a matter of hours. H_2 is the mechanical analog for this process. Diameter of the disc also increases as water is absorbed.

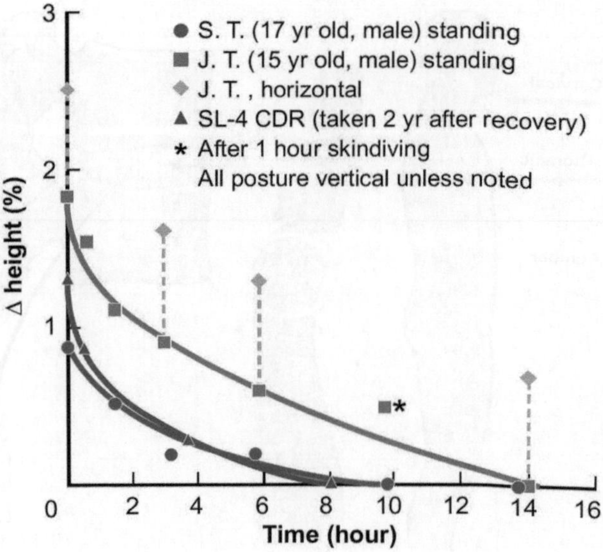

Fig. 11.3 Diurnal changes in height. At 0.0 h the person gets out of bed and stands. There is an immediate decrease in height caused by compression of elastic fibers, H_1, in the intervertebral discs. This is continued with a slower compression, H_2, as fluid is forced into the fibrous walls, and water is forced out of them until an equilibrium is reached, at about 7 h. When the subject becomes horizontal to sleep the process is reversed until it is repeated the following morning

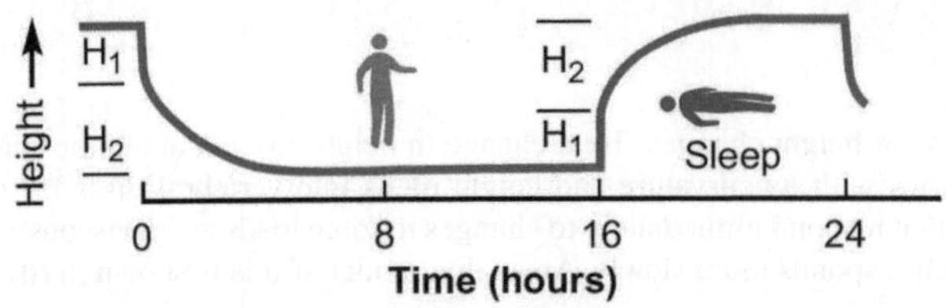

Fig. 11.4 Height changes in subjects in 1 g. Height changes throughout a day were measured with a precision jig, accuracy ±1 mm, throughout 1 day using three subjects. They began on first arising from sleep. In the 15-year old heights were measured in horizontal and vertical position at four points through the day. The aberrant point was after an hour of scuba diving

After a period of horizontal posture, such as in sleep, the process is reversed when the person arises and the discs are reloaded with body weight (Fig. 11.3). There is an immediate decrease in height as the pure elastic component ($H_{1)}$ is compressed. This is followed over a normal day by the continued exponential loss in disc height and body stature until a force equilibrium between load and disc is reached. Figure 11.4 shows measured changes in three subjects during a normal day. A 15-year-old male showed the largest change and required the longest time to reach equilibrium. Upright and supine measurements of his height were made at three times during the day, allowing measurement of the rapid elastic component, H_1. The "damped" spring's (H_2 in Fig. 11.2) response was much larger. There was a slight increase in the H_2 response following an hour's skin diving (asterisk). Changes

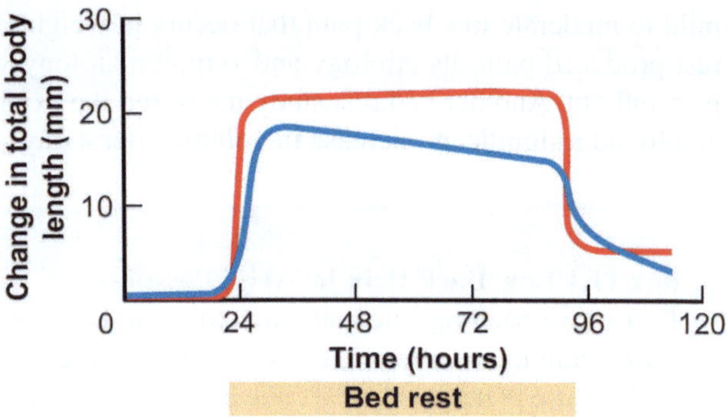

Fig. 11.5 Change in height with horizontal (*red*) and 6° head-down tilt (*blue*). During tilt subjects had a counterweight attached to legs to counterbalance the vertical tilt component

in the 17-year-old male and an older astronaut were smaller and required less time. This is a dynamic process, which responds biphasically to any change in load; e.g., to weight, work or to muscle loads in exercise.

Although not measured, the response to horizontal posture in sleep or bed rest can be reasonably assumed to follow a reversed exponential course (see Fig. 11.3). In the normal course of a 24-h day on Earth there is a gain of approximately an inch in height when horizontal during sleep, which is lost in upright posture in a matter of hours. When carefully measured the rapid H_1 phase and slow H_2 phase of gain loss are easily seen. There can be additional smaller losses with weight and muscle loads in work and exercise during the day. This varies significantly with individuals and with age, when changes are decreased, and the H_2 phase is more rapid. There are also inverse changes with changes in body weight. Normally these actions are not perceived, but various failure modes of intervertebral discs are a major cause of discomfort and[1] more or less disability.

11.2.1 Bed Rest Studies

When properly measured, stature in 1 g is always less than in weightlessness. In an effort to better simulate weightlessness a study at NASA ARC [6] compared the effects of 3 days of horizontal bed rest with 6° head down tilt in which the subjects were "counter balanced" by 10% of body weight traction on feet and legs. Results are shown in Fig. 11.5. There was little difference between head down tilting and horizontal bed rest, and it produced less than 50% of typical changes produced by weightlessness. Another objective of this study was attempted reproduction of a

[1] One of the most common failure modes is a ruptured disc in which a sidewall fails and no longer contains its viscous fluid. It may then collapse and put pressure on the spinal cord and nerves.

mild to moderate low back pain that occurs in weightlessness. While the tilted bed rest produced pain, its etiology and symptomatology were not consistent with the pain inflight. Another bed rest study measured the expansion of intervertebral discs and found a significant increase in volume after 4 days bed rest (Box 11.1).

Box 11.1 Low Back Pain in Weightlessness

There have been repeated attempts to relate low back pain that affects a number of individuals after several days in weightlessness to stretched nerves by the increase in height. At least one large series of in-flight studies was initiated by a neurosurgeon. None of the studies have been able to demonstrate objectively any anomaly in the functions or sensations of nerves that were supposed to be stretched by increase in spinal length. The concept is not consistent with normal anatomy. Symptoms from a nerve under stress are not consistent with those of the discomfort in weightlessness, as personally experienced or described in real time discussion with others. The discomfort is that of an over used muscle—a dull ache relieved by forceful extension of the legs. Weightless posture seems a more likely cause and this pain is discussed in that section.

11.2.2 Inflight

Unknown to us at the time, Russian investigators did a few postflight girth measurements on cosmonauts as early as Soyuz 3 (1968) and also measured pre- and postflight height changes. The latter were apparently conducted on R + 1 by which time such changes would have disappeared (Box 11.2).

Box 11.2 Measurement of In-Flight Changes

Accurate, reproducible measurement of crew stature on Earth or in space is a challenge. The earliest measurements were taken with the back against a flat surface and feet secured to the deck and a book on the vertex parallel to the deck with the back surface marked at the book edge. A steel tape was used to measure the marked distance above the deck. Results were surprisingly reliable.

A variety of jigs have been used to improve stature measurement, but the best foot to vertex scale was developed by Whitmore and extensively used by the Astronaut Office during Shuttle until 1994. It consisted of a thin metal footplate which contained a height scale graduated and accurate to 1 mm, a precision reel and two thin cables to a head bar. Operation on Earth or in space consisted of standing on the footplate, placing the bar on the vertex, standing erect on the plate and actuating the measurement by a slight toe motion. It was accurate and repeatable to ±1 mm.

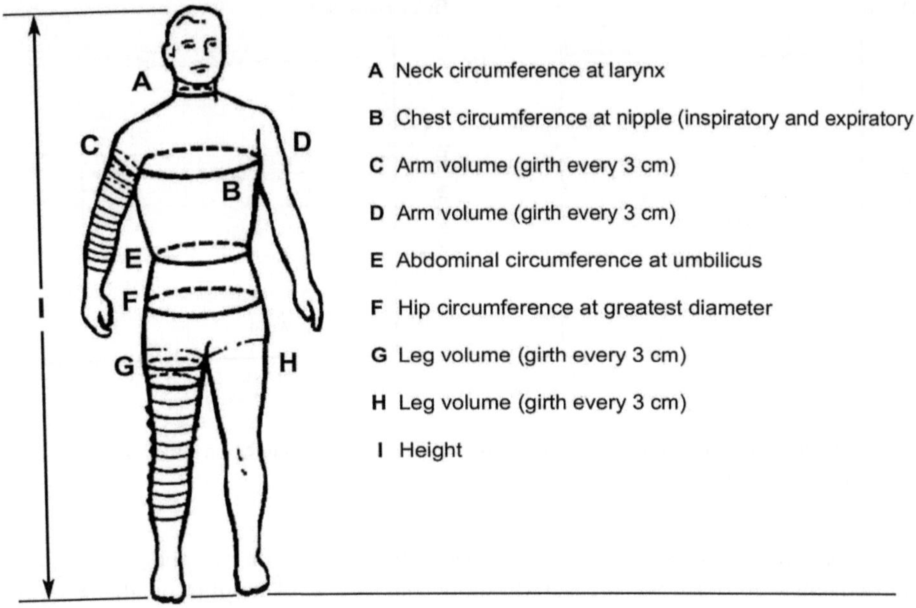

Fig. 11.6 Anthropometric measurements made on the Skylab crewmen. Circumferential measurements were made with a flexible scale to nearest mm. For height a block was placed on the vertex of head and normal to a vertical wall. This level was marked on the wall, and the distance to the deck was measured and calibrated. Legs were measured by a series of circumferential tapes attached to a longitudinal tape

After difficulties in various tasks and complaints from the crew, the Astronaut Office began a series of DSOs to document changes in crew size and posture [7]. A first attempt at posture was made on SL-3, but the result was a picture of a taller upright 1-g posture with flatter back and stomach. Jerri Brown of JSC also needed some of the measurements including stature, and their requirements were combined with those of the Astronaut Office DSO. A more extensive set of instructions and fabric measurement jigs with a miniature steel tape[2] were squeezed aboard SL-4, and the crew did remarkable work with them producing the first proof of fluid shifts, showing increased height in weightlessness, changes in body size and shape and the posture of weightlessness. Figure 11.6 shows the measurement sites, and Fig. 11.7 show results from two of the crew. Average gain in height was 2 in. (50 mm) or 3% of 1-g height. There was no diurnal variation, only small, random AM to PM fluctuations in the range of a millimeter or two, apparently caused by muscle loads from various activities. The height gain was reversed in a matter of hours on return to 1 g.

An immediate result of the Skylab findings was an increase in the truncal length of "space suits" by 2 in. length over their 1-g fittings. This remains standard practice to date. Designers of new vehicles are taking such changes into their design. At least

[2] In spite of being the world's most advanced laboratory there were basic elementary tools missing, such as a simple scale. Unfortunately the tradition continued on ISS.

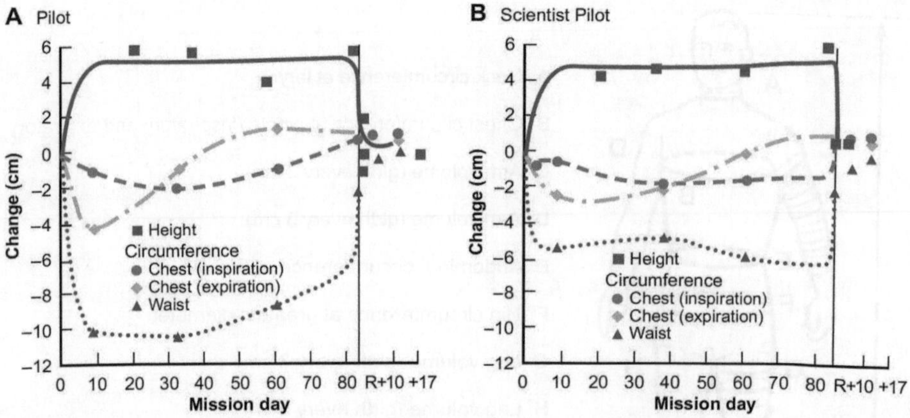

Fig. 11.7 Plots showing in-flight and postflight anthropometric changes on SL-4. The scientist pilot was very athletic

one patent has been issued on a 1-g chair design based on weightless posture. Subsequently Jerri Brown measured sitting and eye level heights on the Apollo Soyuz Test Project [8] and a number of other studies have been conducted.

In 1992 on the STS-42-IML-1 mission changes in height, spinal curvature and presence of back pain were evaluated in three astronauts by Canadian researchers who postulated that back pain in orbit was produced by the stretching of the associated nerves. Stereo-photogrammetry recorded height increases of 48 to 74 mm. This study was continued with two astronauts on IML-2 using direct manual measurement as described earlier. Leg to brain nerve transmission was measured by somatosensory evoked potential. Stable changes in height of 40 mm were measured in both subjects. Their nerve transmission time was normal indicating no change in length, and the project was terminated.

NASA obtained seated and stature heights [9] from crewmembers on Shuttle and ISS. Only maximum changes were reported: 3% in stature and 6% in seated height. Another study of measurements of Shuttle crewmembers also led to reported standing height increases of 3% and seated height increases of 6%. Details and techniques were not given. The standing heights are consistent with the Skylab findings. An ISS ultrasonic study of the spinal column is in progress "to discover the mechanism of height increases".

Change in size and shape. Figure 11.5 shows significant changes in two crewmen on SL-4. There were differences in the magnitude of changes among the three but much in common. Overall crew size and weight were very similar. Changes in stature were virtually identical. Also, changes in chest diameter during expiration followed a similar time course. The initial decrease could reflect the increase in length of the spinal column, but reasons for the return to preflight level are not obvious.

Decrease in diameter of the abdomen may also be in part attributed to a lengthening of the spinal column. A larger component is reduction of counter forces to the abdominal muscles' normal tension. In 1 g abdominal cavity contents behave like a

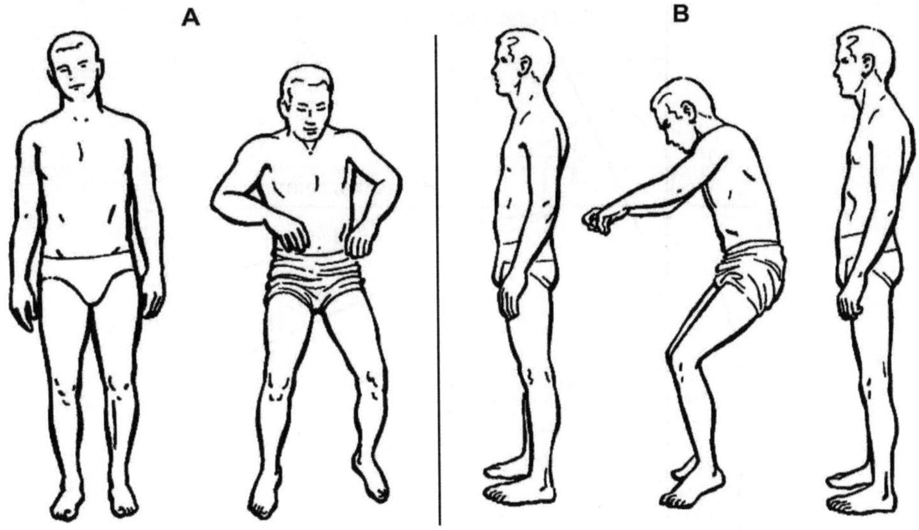

Fig. 11.8 (**a**) Illustrates a front-view comparison of 1 g and weightlessness posture in the Skylab SPT (tracings from photographs). (**b**) Illustrates a side-view comparison of 1 g and weightlessness posture in the Skylab SPT. Illustrations of in-flight posture show subject "floating," relaxed and eyes closed

liquid and exert considerable hydrostatic pressure against the side and front walls of abdominal muscle, which are under constant tension in 1 g. On removal of weight the abdominal hydrostatic pressures are removed, and these muscles act as a girdle. There is a remarkable difference between the two crewmen. One crewman was a very active athlete, which may account for differences.

Although not yet documented, any tissue that is pendulous in 1 g will move upward in weightlessness. All of these changes may be approximated underwater or in a free fall.

11.2.3 Posture in Weightlessness

Probably most unexpected of the accommodations was the single preferred posture in weightlessness. There had been numerous crew comments on the effects of posture and work positions from first and second manned Skylab missions. A photo of Al Bean at the telescope control console showed that they had discarded the carefully designed 1-g seat, which was not being used, and instead work took place from a modified upright posture. Attempts to get a postural picture on the second manned SL-3 produced only a posed 1-g stance.

Details were added to the instructions for postural photos on SL-4 including closed eyes, complete relaxation and floating freely. Once seen (Fig. 11.8), many of

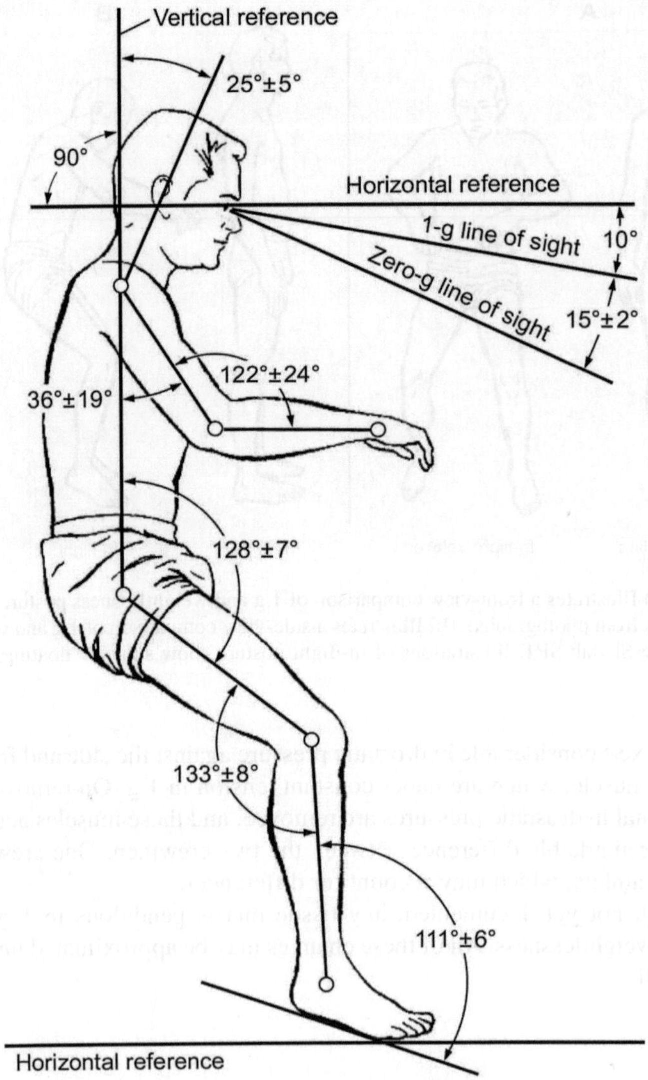

Fig. 11.9 The segment angles of the weightless neutral body position

the postural and work position questions were answered. Previous casual photographs also showed the weightlessness posture once it had been recognized. Photos underwater showed similar, although smaller, changes from 1-g posture.

Figure 11.9 was a measured summary of the postures of the three SL-4 crewmen. As a result of these findings on Skylab, there were significant changes in specifications for man-machine hardware in NASA's Man-System Integration Standards. At least one commercial patent was based on this finding. In 2002 a NASA[3] study of

[3] The National Space Biomedical Research Institute (NSBRI) and Johnson Engineering.

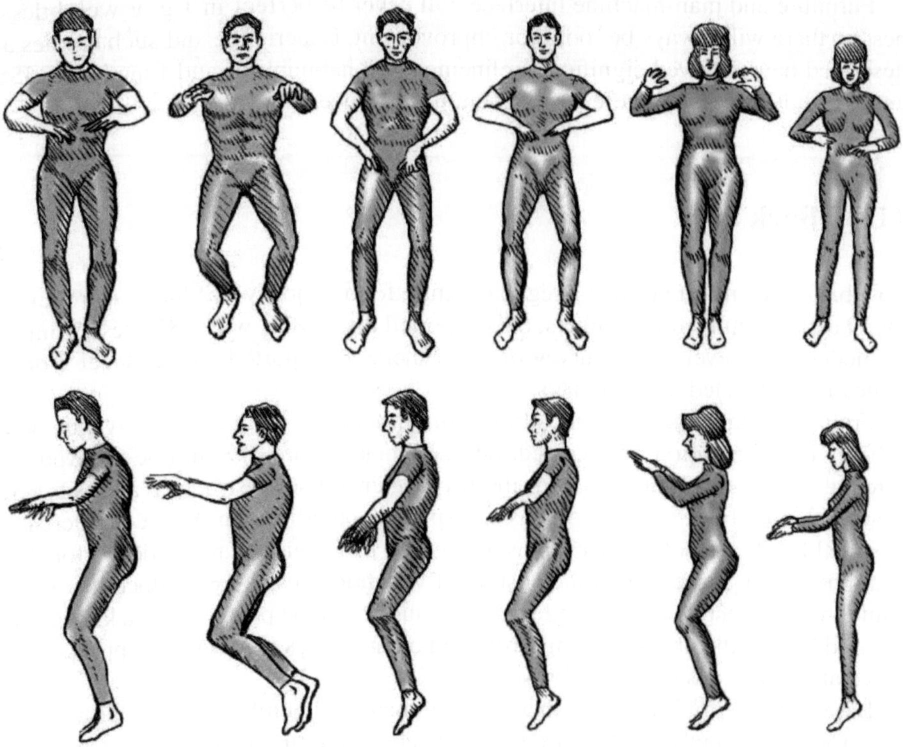

Fig. 11.10 This study of six astronauts demonstrated a range of variations in the weightless posture

the crew on STS-57, four men and three women, was conducted [10]. A wider range of joint angles in the same basic posture was shown (Fig. 11.10). Japan Aerospace Exploration Agency (JAXA) astronauts have also taken photos on ISS revealing the same posture.

In 1 g there are several postures, all designed to minimize muscle effort, whether standing, sitting or lying. Apparently this economy of effort is continued in weightlessness. It appears that relaxed muscle tone determines limb position. In a number of astronauts the head is tilted down, probably through straightening of the cervical spine. Feet are plantar flexed, possibly by increased force from the resting tone of the larger mass of the flexor calf muscle; however, applying this argument to flexion of the thigh and calf is problematic. This is an area that needs further study.

There have been significant complaints against stoop (bending forward) labor in weightlessness. In 1 g as the trunk is flexed forward and down by abdominal muscles the motion is aided by trunsal weight, but in weightlessness the abdominal muscles must assume all the bending loads and soon tire, resulting in complaints made of that position.

At rest in weightlessness the resting tone of larger muscles may stretch smaller opposing muscle masses possibly producing fatigue and the low back pain experienced by some.

Furniture and man-machine interface will never be perfect in 1 g or weightlessness so there will always be room for improvement. Experience and such studies as described have allowed significant refinements in habitablilty and function in ISS and weightlessness, but there is still room for improvement.

11.3 Back Pain

Low back discomfort became a regular feature for a majority of Shuttle astronauts. Of 59 early Shuttle crewmembers, 68% reported discomfort with 28% describing it as moderate to severe. In a survey of 19 of them, 14 reported pain at level 2 on a scale of 5 that lasted one to 6 days.

From personal experience and clinical discussions with others, it is a vague dull ache giving a sensation of overused, fatigued musculature. It could be relieved by forceful extension of the legs with the back against a solid structure and was not exacerbated by full extension of legs and spine or other motion. While an increase in spinal length was often invoked as a cause by investigators, no evidence for this could be found. It seems that the posture of weightlessness may produce low back pain through constant unbalanced stress on muscles in the pelvis and back. A more focused look at the problem using EMG and other studies would seem appropriate, especially if space becomes a tourist site.

Based on informal sampling there was an increased number of acute disc problems following Shuttle flights of one to 2 weeks duration. There are demonstrated biophysical changes in the discs with continued expansion and thinning of the intervertebral disc walls and increased hydration, Since there is a demonstrated rapid decrease in height postflight, there may be bulging of thinned walls at a time when the body is still dehydrated. If the ultrasonic studies of the discs now in progress on ISS are continued through the recovery period, some insight into this may be gained.

11.4 Summary

Again the human body displays remarkable abilities in accommodation to the new environment of space, but operations in space will benefit from a better understanding of them than are provided by the glimpse in studies to date. In contrast to the inflight pencil marks on structure and clandestine minature steel tape used to gather a good bit of the data here JSC now has a large fully staffed anthropometric facility with state of the art equipment. For both static and dynamic inflight anthropometry a few images and available photo grammetric programs can replace the time and tedium of manual measurement. Also anthropometry in Lunar habitation may produce some interesting findings.

References

1. Thornton W. Anthropometric changes in weightlessness. In: Anthropometric source book. Vol 1, Anthropometry for designers. Washington, DC: NASA Scientific and Technical Information Office. NASA RP-1024; 1978. p. 1.
2. Thornton W, Moore T. Results of the life sciences DSOs conducted aboard the space shuttle 1981–1986; 1987. pp. 55–7.
3. Kishnan KA, Kishnan V. Diurnal variations in stature. Anthropologist. 2007;9(2):113–7.
4. Dangerfield PH, et al. Diurnal variation in water content of the intervertebral discs. In: Studies in health technology and informatics. Vol. 15. Landsdale, PA: IOS Press Books, West Pt. Commons, S-201.
5. LeBlanc AD, et al. Changes in intervertebral disc cross-section area with bed rest and space-flight. Spine. 1944;19(7):812–7.
6. Hutchinson KJ, Watenpaugh DE, Murthy G, Convertino VA, Hargens AR. Back pain during 6° head-down tilt approximates that during actual microgravity. Aviat Space Environ Med. 1995;66(3):256–9.
7. Thornton W, Hoffler W, Rummel J. Anthropometric changes in fluid shifts. In: Biomedical results from Skylab. Washington, DC: US GPO. NASA SP–377; 1977. p. 32.
8. Brown JW. Crew height measurement. In: Nicogossian A, editor. The Apollo-Soyuz test project medical report, vol. 16. Washington, DC: NASA; 1977. p. 119–21.
9. Young KS, Rajulu S. The effects of microgravity on seated height (spinal elongation), In: NASA Human research program investigators' workshop. Houston, TX: Lockheed Martin and NASA Johnson Space Center; 2012.
10. Mount FE, Whitmore M, Stealey SL. Evaluation of neutral body posture on shuttle mission STS-57 (SPACEHAB-1). Houston, TX: NASA Johnson Space Center. NASA TM-2003-104805; 2003.

Index

W. Thornton, F. Bonato, *The Human Body and Weightlessness*,
DOI 10.1007/978-3-319-32829-4

MIX
Papier aus verantwortungsvollen Quellen
Paper from responsible sources
FSC® C105338

If you have any concerns about our products,
you can contact us on
ProductSafety@springernature.com

In case Publisher is established outside the EU,
the EU authorized representative is:
Springer Nature Customer Service Center GmbH
Europaplatz 3, 69115 Heidelberg, Germany

Printed by Libri Plureos GmbH
in Hamburg, Germany